Hidden Markov Processes

PRINCETON SERIES IN APPLIED MATHEMATICS

The Princeton Series in Applied Mathematics publishes high quality advanced texts and monographs in all areas of applied mathematics. Books include those of a theoretical and general nature as well as those dealing with the mathematics of specific applications areas and real-world situations.

A list of titles in this series appears at the back of the book.

Hidden Markov Processes

Theory and Applications to Biology

M. Vidyasagar

PRINCETON UNIVERSITY PRESS
PRINCETON AND OXFORD

Published by Princeton University Press, 41 William Street,
Princeton, New Jersey 08540

In the United Kingdom: Princeton University Press, 6 Oxford Street,
Woodstock, Oxfordshire OX20 1TW

press.princeton.edu

Library of Congress Cataloging-in-Publication Data

Vidyasagar, M. (Mathukumalli), 1947–
Hidden Markov processes : theory and applications to biology / M. Vidyasagar.
p. cm. – (Princeton series in applied mathematics)
Includes bibliographical references and index.
ISBN 978-0-691-13315-7 (hardcover : alk. paper) 1. Computational biology. 2. Markov processes. I. Title.
QH324.2.V54 2014
570.285–dc23
2014009277

British Library Cataloging-in-Publication Data is available

This book has been composed in LaTeX.

The publisher would like to acknowledge the author of this volume for providing the camera-ready copy from which this book was printed.

Printed on acid-free paper ∞

Printed in the United States of America

10 9 8 7 6 5 4 3 2 1

To the memory of
Jan Willems (1939–2013)
Scholar, friend, and role model
He will be missed sorely

यद्यदाचरति श्रेष्टस्तत्तदेवेतरो जनः ।
स यत्प्रमाणम् कुरुते लोकस्तदनुवर्तते ॥

Whatever a great man does, that alone the other men do.
Whatever he sets up as the standard, that the world follows.

Bhagavadgita, Chapter 3, Verse 21.
English translation by Alladi Mahadeva Sastry in 1897.

Contents

Preface

Every author aspiring to write a book should address two fundamental questions: (i) Who is the targeted audience? (ii) What does he wish to say to them? In the world of literature, it is often said that every novel is autobiographical to some extent or another. Adapting this maxim to the current situation, I would say that every book I have ever written has been aimed at readers who are in the situation in which I found myself at the time that I embarked on the book-writing project. To put it another way, every book I have written has been an attempt to make it possible for my readers to circumvent some of the difficulties that I myself faced when learning a subject that was new to me.

In the present instance, for the past few years I have been interested in the broad area of computational biology. With the explosion in the sheer quantity of biological data, and an enhanced understanding of the fundamental mechanisms of genomics and proteomics, there is now greater interest than ever in this topic. I got very interested in hidden Markov processes (HMPs) when I realized that several researchers were applying HMPs to problems in computational biology. Thus, after spending virtually my entire research career in blissful ignorance of all matters stochastic, I got down to try to learn something about Markov processes and HMPs. At the same time, I was trying to learn enough basic biology, and to read the literature on the applications of Markov and hidden Markov methods in computational biology. The Markov process literature and the computational biology literature each presented its own sets of problems. The choices regarding the level and contents of the book have been dictated primarily by a desire to enable my readers to learn these topics more easily than I could.

Hidden Markov processes (HMPs) were introduced into the statistics literature as far back as 1966 [12]. Starting in the mid 1970s [9, 10], HMPs have been used in speech recognition, which is perhaps the earliest *application* of HMPs in a nonmathematical context. The paper [50] contains a wonderful survey of most of the relevant theory of HMPs. In recent years, Markov processes and HMPs have also been used in computational biology. Popular algorithms for finding genes from a genome are based either on Markov models, such as GLIMMER and its extensions [113, 36], or on hidden Markov models, such as GENSCAN [25, 26]. Methods for classifying proteins into one of several families make use of a very special type of hidden Markov model known as a "profile" hidden Markov model. See [142] for a recent survey. Finally, the BLAST algorithm, which is universally used to carry out sequence alignment in an efficient though probabilistic manner, makes use of some elements of large deviation theory for i.i.d. processes. Thus any text aspiring to provide the mathematical foundations of these methods would need to cover all of these topics in mathematics.

Most existing books on Markov processes invariably focus on processes with *infinite* state spaces. Books such as [78, 121] restrict themselves to Markov processes with *countable* state spaces, since in this case many of the technicalities associated with uncountable state spaces disappear. From a mathematician's standpoint, the case of a finite state space is not worth expounding separately, since the extension from a finite state space to a countably infinite state space usually "comes for free." However, even the "simplified" theory as in [121] is inaccessible to many if not most engineers, and certainly to most biologists. A typical engineer or mathematically inclined biologist can cope with Markov processes with *finite* state spaces because they can be analyzed using only matrices, eigenvalues, eigenvectors, and the like. At the same time, books on Markov processes with *finite* state spaces seldom go beyond computing stationary distributions, and generally ignore advanced topics such as ergodicity, parameter estimation, and large deviation theory. Yet, as I have stated above, these "advanced" ideas are used in computational biology, even if this fact is not always highlighted. There are some notable exceptions such as [116] that discusses ergodicity of Markov processes, and [100] that discusses parameter estimation; but neither of these books discusses large deviation theory, even for i.i.d. processes, let alone for Markov processes.

Thus the current situation with respect to books on Markov processes can be summarized as follows: There is no treatment of "advanced" notions using only "elementary" techniques, and in an "elementary" setting. In contrast, in the present book the focus is exclusively on stochastic processes assuming values in a finite set, so that technicalities are kept to an absolute minimum. By restricting attention to Markov processes with finite state spaces, I try to capture most of the interesting phenomena such as ergodicity and large deviation theory, while giving elementary proofs that are accessible to anyone who knows undergraduate-level linear algebra.

In the area of HMPs, most of the existing texts discuss only the computation of various likelihoods, such as the most likely state trajectory corresponding to a given observation sequence (also known as the Viterb algorithm), or to the determination of the most likely parameter set for a hidden Markov model *of a given, prespecified order* (also known as the Baum-Welch algorithm). In contrast, very little attention is paid to realization theory, that is, constructing a hidden Markov model on the basis of its statistics. Perhaps the reason is that, until recently, realization theory for hidden Markov processes was not in very good shape, despite the publication of the monumental paper [6]. In the present book, I have attempted to remedy the situation by including a thorough discussion of realization theory for hidden Markov processes, based primarily on the paper [133].

Finally, I decided to include a discussion of large deviation theory for both i.i.d. processes as well as Markov processes. The material on large deviation theory for i.i.d. processes, especially the so-called method of types, is used in the proofs of the BLAST algorithm, which is among the most widely used algorithms in computational biology, used for sequence alignment. I decided also to include a discussion of large deviation theory for Markov processes, even though there aren't any applications to computational biology, at least as of now. This discussion is a compendium of various results that are scattered throughout the literature, and is based on the paper [135].

At present there are several engineers and mathematicians who would like to contribute to computational biology and suggest suitable algorithms. Such persons face some obvious difficulties, such as the need to learn (I am tempted to say "memorize") a great deal of unfamiliar terminology. Mathematicians are accustomed to a "reductionist" approach to their subject whereby everything follows from a few simply stated axioms. Such persons are handicapped by the huge differences in the styles of exposition between the engineering/mathematics community on the one hand and the biology community on the other hand. Perhaps nothing illustrates the terminology gap better than the fact that the very phrase "reductionist approach" means entirely different things to mathematicians and to biologists. To mathematicians reductionism means reducing a subject to its core principles. For instance, if a ring is defined as a set with two binary associative operations etc., then certain universal conclusions can be drawn that apply to *all* rings. In contrast, to biologists reductionism means constructing the simplest possible exemplar of a biological system, and studying it in great detail, in the hope that conclusions derived from the simplified system can be extrapolated to more complex exemplars. Or to put it another way, biological reductionism is based on the premise that a biological system is merely an aggregation of its parts, each of which can be studied in isolation. The difficulties with this premise are obvious to anyone familiar with "emergent behavior," whereby complex systems exhibit behavior that has not been explicitly programmed into them; but despite that, reductionism (in the biological sense) is widely employed, perhaps due to the inherent complexity of biological systems.

Computational biology is a vast subject, and is constantly evolving. In choosing topics from computational biology for inclusion in the book, I restricted myself to genomics and proteomics, as these are perhaps the two aspects of biology that are the most "reductionist" in the sense described above. Even within genomics and proteomics, I have restricted myself to those algorithms that have a close connection with the Markov and HMP theory described here. Thus I have omitted any discussion of, to cite just one example, neural network-based methods. Readers wishing to find an encyclopedic treatment of many aspects of computational biology are referred to [11, 53]. But despite laying out these boundary conditions, I still decided not to attempt a thorough and up-to-date treatment of all available algorithms in genomics and proteomics that are based on hidden Markov processes. The main reason is that the actual details of the algorithms keep changing very rapidly, whereas the underlying theory does not change very much over time.

For this reason, the material in Part 3 of the book on computational biology only *presents the flavor* of various algorithms, with up-to-date references.

I hope that the book would not only assist biologists and other users of the theory to gain a better understanding of the methods they use, but also spur the engineering and statistics research community to study some new and interesting research problems.

I would like to conclude by dedicating this book to the memory of Jan Willems, who passed away just a few months before it was finished. Jan was a true scholar, a personal friend, and a role model for aspiring researchers everywhere. With his passing, the world of control theory is infinitely poorer. Dedicating this book to him is but a small recompense for all that I have learned from him over the years. May his soul rest in peace!

M. Vidyasagar
Dallas and Hyderabad
December 2013

PART 1
Preliminaries

Chapter One

Introduction to Probability and Random Variables

1.1 INTRODUCTION TO RANDOM VARIABLES

1.1.1 Motivation

Probability theory is an attempt to formalize the notion of uncertainty in the outcome of an experiment. For instance, suppose an urn contains four balls, colored red, blue, white, and green respectively. Suppose we dip our hand in the urn and pull out one of the balls "at random." What is the likelihood that the ball we pull out will be red? If we make multiple draws, replacing the drawn ball each time and shaking the urn thoroughly before the next draw, what is the likelihood that we have to make at least ten draws before we draw a red ball for the first time? Probability theory provides a mathematical abstraction and a framework where such issues can be addressed.

When there are only finitely many possible outcomes, probability theory becomes relatively simple. For instance, in the above example, when we draw a ball there are only four possible outcomes, namely: $\{R, B, W, G\}$ with the obvious notation. If we draw two balls, after replacing the first ball drawn, then there $4^2 = 16$ possible outcomes, represented as $\{RR, \ldots, GG\}$. In such situations, one can get by with simple "counting" arguments. The counting approach can also be made to work when the set of possible outcomes is *countably* infinite.[1] This situation is studied in Section 1.3. However, in probability theory infinity is never very far away, and counting arguments can lead to serious logical inconsistencies if applied to situations where the set of possible outcomes is *uncountably* infinite. The great Russian mathematician A. N. Kolmogorov invented axiomatic probability theory in the 1930s precisely to address the issues thrown up by having uncountably many possible outcomes. Subsequent developments in probability theory have been based on the axiomatic foundation laid out in [81].

Example 1.1 Let us return to the example above. Suppose that all the four balls are identical in size and shape, and differ only in their color. Then it is reasonable to suppose that drawing any one color is as likely as drawing any other color, neither more nor less. This leads to the observation that the likelihood of drawing a red ball (or any other ball) is $1/4 = 0.25$.

Example 1.2 Now suppose that the four balls are all spherical, and that

---

[1] Recall that a set S is said to be **countable** if it can be place in one-to-one correspondence with the set of natural numbers $\mathbb{N} = \{1, 2, \ldots\}$.

their diameters are in the ratio 4 : 3 : 2 : 1 in the order red, blue, white, and green. We can suppose that the likelihood of our fingers touching and drawing a particular ball is proportional to its surface area. In this case, it follows that the likelihoods of drawing the four balls are in the proportion $4^2 : 3^2 : 2^2 : 1^2$ or 16 : 9 : 4 : 1 in the order red, blue, white, and green. This leads to the conclusion that

$$P(R) = 16/30, P(B) = 9/30, P(W) = 4/30, P(G) = 1/30.$$

Example 1.3 There can be instances where such analytical reasoning can fail. Suppose that all balls have the same diameter, but the red ball is coated with an adhesive resin that makes it more likely to stick to our fingers when we touch it. The complicated interaction between the surface adhesion of our fingers and the surface of the ball may be too difficult to analyze, so we have no recourse other than to draw balls repeatedly and see how many times the red ball comes out. Suppose we make 1,000 draws, and the outcomes are: 451 red, 187 blue, 174 white, and 188 green. Then we can write

$$\hat{P}(R) = 0.451, \hat{P}(B) = 0.187, \hat{P}(W) = 0.174, \hat{P}(G) = 0.188.$$

The symbol $\hat{P}$ is used instead of P to highlight the fact that these are simply *observed frequencies*, and not the *true but unknown probabilities*. Often the observed frequency of an outcome is referred to as its **empirical probability**, or the empirical estimate of the true but unknown probability based on a particular set of experiments. It is tempting to treat the observed frequencies as true probabilities, but that would not be correct. The reason is that if the experiment is repeated, the outcomes would in general be quite different. The reader can convince himself/herself of the difference between frequencies and probabilities by tossing a coin ten times, and another ten times. It is extremely unlikely that the same set of results will turn up both times. One of the important questions addressed in this book is: Just how close are the observed *frequencies* to the true but unknown *probabilities*, and just how quickly do these observed frequencies converge to the true probabilities? Such questions are addressed in Section 1.3.3.

1.1.2 Definition of a Random Variable and Probability

Suppose we wish to study the behavior of a "random" variable X that can assume one of only a finite set of values belonging to a set $\mathbb{A} = \{a_1, \ldots, a_n\}$. The set $\mathbb{A}$ of possible values is often referred to as the "alphabet" of the random variable. For example, in the ball-drawing experiment discussed in the preceding subsection, X can be thought of as the color of the ball drawn, and assumes values in the set $\{R, B, W, G\}$. This example, incidentally, serves to highlight the fact that the set of outcomes can consist of abstract *symbols*, and need not consist of *numbers*. This usage, adopted in this book, is at variance from the convention in many mathematics texts, where it is assumed that $\mathbb{A}$ is a subset of the real numbers $\mathbb{R}$. However, since biological applications are a prime motivator for this book, it makes no sense to restrict $\mathbb{A}$ in this way. In genomics, for example, $\mathbb{A}$ consists of the four symbol set of nucleic acids, or nucleotides, usually denoted by $\{A, C, G, T\}$. Moreover, by allowing $\mathbb{A}$ to consist of arbitrary symbols, we also allow explicitly the possibility that *there is no natural ordering of these symbols.* For instance, in this book the nucleotides are written in the order A, C, G, T purely to follow the English alphabetical ordering. But there is no consensus on the ordering in biology texts. Thus any method of analysis that is developed here must be *permutation independent.* In other words, if we choose to order the symbols in the set $\mathbb{A}$ in some other fashion, the methods of analysis must give the same answers as before.

Now we give a general definition of the notion of probability, and introduce the notation that is used throughout the book.

Definition 1.1 *Given an integer n, the* **n-dimensional simplex** $\mathbb{S}_n$ *is defined as*

$$\mathbb{S}_n = \{\mathbf{v} \in \mathbb{R}^n : v_i \geq 0 \,\forall i, \sum_{i=1}^{n} v_i = 1\}. \tag{1.1}$$

Thus $\mathbb{S}_n$ consists of all nonnegative vectors whose components add up to one.

Definition 1.2 *Suppose $\mathbb{A} = \{a_1, \ldots, a_n\}$ is a finite set. Then a* **probability distribution** *on the set $\mathbb{A}$ is any vector $\boldsymbol{\mu} \in \mathbb{S}_n$.*

The interpretation of a probability distribution $\boldsymbol{\mu}$ on the set $\mathbb{A}$ is that we say

$$\Pr\{X = a_i\} = \mu_i$$

to be read as "the probability that the random variable X equals x_i is μ_i." Thus, if $\mathbb{A} = \{R, B, W, G\}$ and $\boldsymbol{\mu} = [0.25 \quad 0.25 \quad 0.25 \quad 0.25]$, then all the four outcomes of drawing the various colored balls are equally likely. This is the case in Example 1.1. If the situation is as in Example 1.2, where the balls have different diameters in the proportion $4 : 3 : 2 : 1$, the probability distribution is

$$\boldsymbol{\mu} = [16/30 \;\; 9/30 \;\; 4/30 \;\; 1/30].$$

If we now choose to reorder the elements of the set $\mathbb{A}$ in the form $\{R, W, G, B\}$, then the probability distribution gets reordered correspondingly, as

$$\boldsymbol{\mu} = [16/30 \;\; 4/30 \;\; 1/30 \;\; 9/30].$$

Thus, when we speak of the probability distribution $\boldsymbol{\mu}$ on the set $\mathbb{A}$, we need to specify the ordering of the elements of the set.

The way we have defined it above, a probability distribution associates a *weight* with each element of the set $\mathbb{A}$ of possible outcomes. Thus $\boldsymbol{\mu}$ can be thought of as a map from $\mathbb{A}$ into the interval $[0, 1]$. This notion of a weight of *individual elements* can be readily extended to define the weight of *each subset* of $\mathbb{A}$. This is called the probability **measure** P_μ associated with the distribution $\boldsymbol{\mu}$. Suppose $A \subseteq \mathbb{A}$. Then we define

$$P_\mu(A) := \Pr\{X \in A\} = \sum_{i=1}^{n} \mu_i I_A(x_i), \tag{1.2}$$

where $I_A(\cdot)$ is the so-called **indicator function** of the set A, defined by

$$I_A(x) = \begin{cases} 1 & \text{if} \quad x \in A, \\ 0 & \text{if} \quad x \notin A. \end{cases} \tag{1.3}$$

So (1.2) states that the **probability measure** of the set A, denoted by $P_\mu(A)$, is the sum of the probability weights of the individual elements of the set A. Thus, whereas $\boldsymbol{\mu}$ maps the set $\mathbb{A}$ into $[0, 1]$, the corresponding probability measure P_μ maps the "power set" $2^{\mathbb{A}}$ (that is, the collection of all subsets of $\mathbb{A}$) into the interval $[0, 1]$.

In this text, we need to deal with three kinds of objects:

- A probability distribution $\boldsymbol{\mu}$ on a finite set $\mathbb{A}$.
- A random variable X assuming values in $\mathbb{A}$, with the probability distribution $\boldsymbol{\mu}$.
- A probability measure P_μ on the power set $2^{\mathbb{A}}$, associated with the probability distribution $\boldsymbol{\mu}$.

We will use whichever interpretation is most convenient and natural in the given context. As for notation, throughout the text, boldface Greek letters such as $\boldsymbol{\mu}$ denote probability distributions. The probability measure corresponding to $\boldsymbol{\mu}$ is denoted by P_μ. Strictly speaking, we should write $P_{\boldsymbol{\mu}}$, but for reasons of aesthetics and appearance we prefer to use P_μ. Similar notation applies to all other boldface Greek letters.

From (1.2), it follows readily that the empty set $\emptyset$ has probability measure zero, while the complete set $\mathbb{A}$ has probability measure one. This is true irrespective of what the underlying probability distribution $\boldsymbol{\mu}$ is. Moreover, the following additional observations are easy consequences of (1.2):

Theorem 1.3 *Suppose $\mathbb{A}$ is a finite set and $\boldsymbol{\mu}$ is a probability distribution on $\mathbb{A}$, and let P_μ denote the corresponding probability measure on $\mathbb{A}$. Then*

1. $0 \leq P_\mu(A) \leq 1 \; \forall A \subseteq \mathbb{A}$.
2. $P_\mu(\emptyset) = 0$ *and* $P_\mu(\mathbb{A}) = 1$.
3. *If A, B are disjoint subsets of $\mathbb{A}$, then*

$$P_\mu(A \cup B) = P_\mu(A) + P_\mu(B). \tag{1.4}$$

In the next paragraph we give a brief glimpse of axiomatic probability theory in a general setting, where the set $\mathbb{A}$ of possible outcomes is not necessarily finite. This paragraph is *not needed* to understand the remainder of the book, and therefore the reader can skip it with no aftereffects. In axiomatic probability theory, one actually *begins* with generalizations of the two properties above. One starts with a collection of subsets $\mathcal{S}$ of $\mathbb{A}$ that has three properties:

1. Both the empty set $\emptyset$ and $\mathbb{A}$ itself belong to $\mathcal{S}$.

2. If A belongs to $\mathcal{S}$, so does its complement A^c.

3. If $\{A_1, A_2, \ldots\}$ is a *countable* collection of sets belonging to $\mathcal{S}$, then their union $B := \cup_{i=1}^{\infty} A_i$ also belongs to $\mathcal{S}$.

Such a collection $\mathcal{S}$ is called a **σ-algebra** of subsets of A, and the pair $(A, \mathcal{S})$ is called a **measurable space**. Note that, on the same set A, it is possible to define different σ-algebras. Given a measurable space $(A, \mathcal{S})$, a **probability measure** P is defined to be a function that maps the σ-algebra $\mathcal{S}$ into $[0, 1]$, or in other words a map that assigns a number $P(A) \in [0, 1]$ to each set A belonging to $\mathcal{S}$, such that two properties hold.

1. $P(\emptyset) = 0$ and $P(\mathbb{A}) = 1$.

2. If $\{A_1, A_2, \ldots\}$ is a *countable* collection of *pairwise disjoint* sets belonging to $\mathcal{S}$, then

$$P\left(\bigcup_{i=1}^{\infty} A_i\right) = \sum_{i=1}^{\infty} P(A_i).$$

Starting with just these two simple sets of axioms, together with the notion of independence (introduced later), it is possible to build a tremendously rich edifice of probability theory; see [81].

In the case where the set $\mathbb{A}$ is either finite or countably infinite, by tradition one takes $\mathcal{S}$ to be the collection of *all* subsets of $\mathbb{A}$, because any other σ-algebra $\mathcal{S}'$ of subsets of $\mathcal{A}$ must in fact be a subalgebra of $\mathcal{S}$, in the sense that every set that is contained in the collection $\mathcal{S}'$ must also belong to $\mathcal{S}$. Now suppose P is a probability measure on $\mathcal{S}$. Then P assigns a weight $P(\{a_i\}) =: \mu_i$ to each element $a_i \in \mathbb{A}$. Moreover, if A is a subset of $\mathbb{A}$, then the measure $P(A)$ is just the sum of the weights assigned to individual elements of A. So if we let $\boldsymbol{\mu}$ denote the sequence of nonnegative numbers $\boldsymbol{\mu} := (\mu_i, i = 1, 2, \ldots)$, then, in conformity with earlier notation, we can identify the probability measure P with P_μ. Conversely, if $\boldsymbol{\mu}$ is any sequence of nonnegative numbers such that $\sum_{i=1}^{\infty} \mu_i = 1$, then the associated probability measure P_μ is defined for every subset $A \subseteq \mathbb{A}$ by

$$P_\mu(A) = \sum_{a_i \in A} \mu_i = \sum_{i=1}^{\infty} \mu_i I_A(a_i),$$

where as before $I_A(\cdot)$ denotes the indicator function of the set A.

If $\mathbb{A}$ is finite, and if $\{A_1, A_2, \ldots\}$ is a countable collection of pairwise disjoint sets, then the only possibility is that all but finitely many sets are empty. So Property 3 above can be simplified to:

$$P(A \cup B) = P(A) + P(B) \text{ if } A \cap B = \emptyset,$$

which is precisely Property 2 from Theorem 1.3. In other words, the case where $\mathbb{A}$ is countably infinite is not any more complicated than the case where $\mathbb{A}$ is a finite set. This is why most "elementary" books on Markov chains assume that the underlying set $\mathbb{A}$ is countably infinite. But if $\mathbb{A}$ is an uncountable infinite set (such as the real numbers, for example), this approach based on assigning weights to individual elements of the set $\mathbb{A}$ does not work, and one requires the more general version of the theory of probability as introduced in [81].

At this point the reader can well ask: But what does it all *mean*? As with much of mathematics, probability theory exists at many distinct levels. It can be viewed as an exercise in pure reasoning, an intellectual pastime, a challenge to one's wits. While that may satisfy some persons, the theory would have very little by way of *application* to "real" situations unless the notion of probability is given a little more concrete interpretation. So we can think of the probability distribution $\boldsymbol{\mu}$ as arising in one of two ways. First, the distribution can be *postulated*, as in the previous subsection. Thus if we are drawing from an urn containing four balls that are identical in all respects save their color, it makes sense to *postulate* that each of the four outcomes is equally likely. Similarly, if the balls are identical except for their diameter, and if we believe that the likelihood of drawing a ball is proportional to the surface area, then once again we can *postulate* that the four components of $\boldsymbol{\mu}$ are in proportion to the surface areas (or equivalently, to the diameter squared) of the four balls. Then the requirement that the components of $\boldsymbol{\mu}$ must add up to one gives the normalizing constant. Second, the distribution can be *estimated*, as with the adhesive-coated balls in Example 1.3. In this case there is a true but unknown probability vector $\boldsymbol{\mu}$, and our estimate of $\boldsymbol{\mu}$, based on 1,000 draws of balls, is $\hat{\boldsymbol{\mu}} = [0.451 \;\; 0.187 \;\; 0.174 \;\; 0.188]$. Then we can try to develop theories that allow us to say *how close* $\hat{\boldsymbol{\mu}}$ *is to* $\boldsymbol{\mu}$*, and with what confidence we can make this statement.* This question is addressed in Section 1.3.3.

1.1.3 Function of a Random Variable, Expected Value

Suppose X is a random variable assuming values in a finite set $\mathbb{A} = \{a_1, \ldots, a_n\}$, with the probability measure P_μ and the probability distribution $\boldsymbol{\mu}$. Suppose f is a function mapping the set $\mathbb{A}$ into another set $\mathbb{B}$. Since $\mathbb{A}$ is finite, it is clear that the set $\{f(a_1), \ldots, f(a_n)\}$ is finite. So there is no loss of generality in assuming that the set $\mathbb{B}$ (the range of the function f) is also a finite set. Moreover, it is *not* assumed that the values $f(a_1), \ldots, f(a_n)$ are distinct. Thus the image of the set $\mathbb{A}$ under the function f can have fewer than n elements. Now $f(X)$ is itself a random variable. Moreover, the distribution of $f(X)$ can be computed readily from the distribution of X. Suppose $\boldsymbol{\mu} \in \mathcal{S}_n$ is the distribution of X. Thus $\mu_i = \Pr\{X = x_i\}$. To compute the distribution of $f(X)$, we need to address the possibility that $f(a_1), \ldots, f(a_n)$ may not be distinct elements. Let $\mathbb{B} = \{b_1, \ldots, b_m\}$ denote the set of all possible outcomes of $f(X)$, and note that $m \leq n$. Then

$$\Pr\{f(X) = b_j\} = \sum_{a_i \in f^{-1}(b_j)} \mu_i.$$

In other words, the probability that $f(X) = b_j$ is the sum of the probabilities of all the preimages of b_j under the function f.

Example 1.4 Suppose that in Example 1.1, we receive a payoff of \$2 if we draw a green ball, we pay a penalty of \$1 if we draw a red ball, and we neither pay a penalty nor receive a payment if we draw a white ball or a blue ball. The situation can be represented by defining $X = \{R, B, W, G\}$, $Y = \{-1, 0, 2\}$, and $f(R) = -1$, $f(G) = 2$, and $f(W) = f(B) = 0$. Moreover, if the balls are identical except for color, then

$$\Pr\{Y = -1\} = \Pr\{X = R\} = 0.25,$$

$$\Pr\{Y = 0\} = \Pr\{X = B\} + \Pr\{X = W\} = 0.5,$$

$$\Pr\{Y = 2\} = \Pr\{X = G\} = 0.25.$$

Example 1.5 We give an example from biology, which is discussed in greater detail in Chapter 8. Let $\mathcal{D}$ equal $\{A, C, G, T\}$, the set of DNA nucleotides, and let $\mathcal{R}$ equal $\{A, C, G, U\}$, the set of RNA nucleotides. As explained in Chapter 8, during the transcription phase of DNA replication thymine (T) gets replaced by uracil (U). Then, during the translation phase of DNA replication, each triplet of RNA nucleotides, known as a "codon," gets converted into one of 20 amino acids, or into the STOP symbol. The map that associates each codon with the corresponding amino acid (or the STOP symbol) is called the "genetic code." The complete genetic code was discovered by Hargobind mhorana, building upon earlier work by Marshall Nirenberg, and is shown in Figure 1.1. For this pioneering effort, Khorana and Nirenberg were awarded the Nobel Prize in medicine in 1968, along with Robert Holley who discovered tRNA (translation RNA). Further details can be found in Section 8.1.2.

Now let $\mathbb{A} = \mathcal{R}^3$, the set of codons, which has cardinality $4^3 = 64$, and let $\mathbb{B}$ denote the set of amino acids plus the STOP symbol, which has cardinality 21. Then the genetic code can be thought of as a function $f : \mathbb{A} \rightarrow \mathbb{B}$. From Figure 1.1 it is clear that the number of preimages $f^{-1}(b)$ varies quite considerably for each of the 21 elements of $\mathbb{B}$, ranging from a high of 6 for leucine (Leu) down to 1 for tryptophan (Trp). Thus, if we know the frequency of distribution of codons in a particular stretch of genome (a frequency distribution on $\mathbb{A}$), we can convert this into a corresponding frequency distribution of amino acids and stop codons.

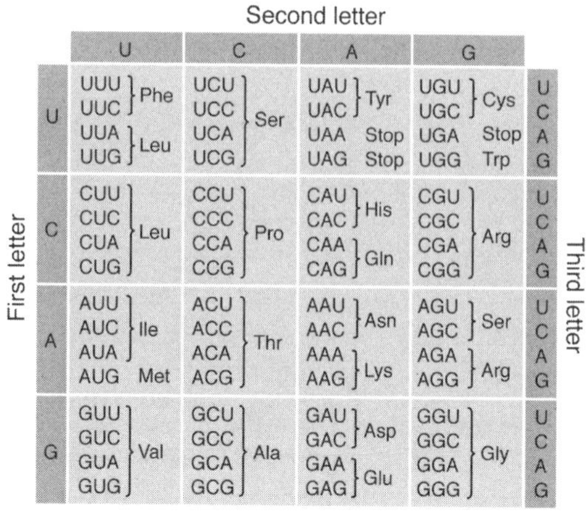

Second letter

First letter	U	C	A	G	Third letter
U	UUU, UUC } Phe UUA, UUG } Leu	UCU, UCC, UCA, UCG } Ser	UAU, UAC } Tyr UAA Stop UAG Stop	UGU, UGC } Cys UGA Stop UGG Trp	U C A G
C	CUU, CUC, CUA, CUG } Leu	CCU, CCC, CCA, CCG } Pro	CAU, CAC } His CAA, CAG } Gln	CGU, CGC, CGA, CGG } Arg	U C A G
A	AUU, AUC, AUA } Ile AUG Met	ACU, ACC, ACA, ACG } Thr	AAU, AAC } Asn AAA, AAG } Lys	AGU, AGC } Ser AGA, AGG } Arg	U C A G
G	GUU, GUC, GUA, GUG } Val	GCU, GCC, GCA, GCG } Ala	GAU, GAC } Asp GAA, GAG } Glu	GGU, GGC, GGA, GGG } Gly	U C A G

Figure 1.1 The Genetic Code in Tabular Form. Reproduced with permission from Stephen Carr.

Definition 1.4 *Suppose X is a* real-valued *random variable assuming values $\mathbb{A} = \{a_1, \ldots, a_n\} \subseteq \mathbb{R}$, with the probability distribution $\boldsymbol{\mu}$, and associated probability measure P_μ. Then the* **expected value** *of X is denoted by $E[X, P_\mu]$ and is defined by*

$$E[X, P_\mu] := \sum_{i=1}^{n} a_i \mu_i. \tag{1.5}$$

It is important to note that, while the notion of probability can be defined for *any* random variable (for example, the set of nucleotides or the set of amino acids), the notion of an expected value can be defined only for *real-valued* random variables. Also, if the underlying probability distribution P_μ is clear from the context, we sometimes omit it and write $E(X)$ or $E[X]$ instead of $E(X, P_\mu)$ or $E[X, P_\mu]$.

Suppose now that X is a random variable assuming values in some finite set $\mathbb{A}$ (not necessarily a subset of $\mathbb{R}$), and f is a function mapping the set $\mathbb{A}$ into the real numbers $\mathbb{R}$. Thus to each element $a_i \in \mathbb{A}$, the function f assigns a real number $f(a_i)$. Let $\boldsymbol{\mu}$ denote the distribution of X and let P_μ denote the associated probability measure.

Definition 1.5 *The* **expected value** *of the function f is denoted by $E[f, P_\mu]$ and is defined by*

$$E[f, P_\mu] := \sum_{i=1}^{n} f(a_i)\mu_i = \sum_{i=1}^{n} f(a_i) \Pr\{X = a_i\}. \tag{1.6}$$

The point to note in the above definition is that it is permissible for X to be an "abstract" random variable, not necessarily real-valued, so long as

$f(X)$ is real-valued. It is left to the reader to verify that the above equation is the same as the expected value of the real-valued random variable $f(X)$. The point to note is that the formula (1.6) is valid even if the real numbers $f(a_1), \ldots, f(a_n)$ are not all distinct.

Observe that the expected value is *linear* in the function f. Thus, if f, g are two functions of a random variable X with probability measure P_μ and the probability distribution $\boldsymbol{\mu}$, and α, β are two real numbers, then

$$E[\alpha f + \beta g, P_\mu] = \alpha E[f, P_\mu] + \beta E[g, P_\mu].$$

Suppose X is a *real-valued* random variable assuming values in $\mathbb{A} = \{a_1, \ldots, a_n\}$, with probability measure P_μ and the probability distribution $\boldsymbol{\mu}$. Then the quantity defined earlier as the expected value of X, namely

$$E[X, P_\mu] = \sum_{i=1}^{n} a_i \mu_i,$$

is also called the **mean value** or just simply the **mean** of X. The quantity $E[(X - E[X])^2, P_\mu]$ is called the **variance** of X. The square root of the variance is called the **standard deviation** of X. In many books, the symbols $\mu(X)$, $V(X)$, and $\sigma(X)$ are commonly used to denote the mean, the variance, and standard deviation respectively. This notation is used only rarely in this book.

Note that we can also define the variance of X as $E[X^2, P_\mu] - (E[X, P_\mu])^2$. This is because, by the linearity of the expected value operation, we have

$$\begin{aligned} E[(X - E[X])^2, P_\mu] &= E[X^2, P_\mu] - 2(E[X, P_\mu])^2 + (E[X, P_\mu])^2 \\ &= E[X^2, P_\mu] - (E[X, P_\mu])^2. \end{aligned} \tag{1.7}$$

The above argument also shows that, for every random variable X, we have

$$E[X^2, P_\mu] \geq (E[X, P_\mu])^2.$$

This is a special case of a very general result known as Schwarz's inequality.

In general, for every positive integer l, the quantity

$$E[X^l, P_\mu] = \sum_{i=1}^{n} a_i^l \mu_i$$

is called the ***l*-th moment** of the random variable X. Since the random variable assumes only finitely many values, the l-th moment is well-defined for every positive integer l. In this connection, one can define the so-called **moment generating function** $\mathrm{mgf}(\lambda; X)$ as follows:

$$\mathrm{mgf}(\lambda; X) = E[\exp(\lambda X), P_\mu] = \sum_{i=1}^{n} \exp(\lambda a_i) \mu_i.$$

Again, since X assumes only finitely many values, the moment generating function is well-defined for every real number λ. Moreover, it is a direct consequence of the definition that

$$\left[\frac{d^l \mathrm{mgf}(\lambda; X)}{d\lambda}\right]_{\lambda=0} = \left[\sum_{i=1}^{n} a_i^l \exp(\lambda a_i) \mu_i\right]_{\lambda=0} = \sum_{i=1}^{n} a_i^l \mu_i$$

is indeed the l-th moment of the random variable X.

1.1.4 Total Variation Distance

Suppose $\mathbb{A} = \{a_1, \ldots, a_n\}$ is a finite set, and $\boldsymbol{\mu}, \boldsymbol{\nu}$ are two probability distributions on X. Let P_μ, P_ν denote the corresponding probability measures. In this section, we show how to quantify the "difference" between the two measures.

Definition 1.6 *Let P_μ, P_ν be two probability measures on a finite set $\mathbb{A} = \{a_1, \ldots, a_n\}$, corresponding to the distributions $\boldsymbol{\mu}, \boldsymbol{\nu}$ respectively. Then the* **total variation distance** *between P_μ and P_ν (or between $\boldsymbol{\mu}$ and $\boldsymbol{\nu}$), denoted by $\rho(P_\mu, P_\nu)$ or $\rho(\boldsymbol{\mu}, \boldsymbol{\nu})$ is defined as*

$$\rho(P_\mu, P_\nu), \rho(\boldsymbol{\mu}, \boldsymbol{\nu}) := \max_{S \subseteq \mathbb{A}} |P_\mu(S) - P_\nu(S)|. \tag{1.8}$$

Now it is shown that $\rho(\cdot, \cdot)$ is indeed a proper metric or "distance" on $\mathbb{S}_n$, which can be identified with the set of all probability distributions on the set $\mathbb{A}$ of cardinality n.

Lemma 1.7 *The function $\rho(\cdot, \cdot)$ defined in (1.8) satisfies the following properties:*

1. *$\rho(P_\mu, P_\nu) \geq 0$ for all $\boldsymbol{\mu}, \boldsymbol{\nu} \in \mathbb{S}_n$.*
2. *$\rho(P_\mu, P_\nu) = 0$ if and only if $\boldsymbol{\mu} = \boldsymbol{\nu}$.*
3. *$\rho(P_\mu, P_\nu) = \rho(P_\nu, P_\mu)$ for all $\boldsymbol{\mu}, \boldsymbol{\nu} \in \mathbb{S}_n$.*
4. *The "triangle inequality" is satisfied, namely:*

$$\rho(P_\mu, P_\nu) \leq \rho(P_\mu, P_\phi) + \rho(P_\phi, P_\nu) \; \forall \boldsymbol{\mu}, \boldsymbol{\nu}, \boldsymbol{\phi} \in \mathbb{S}_n. \tag{1.9}$$

Proof. Property 1 is obvious. To prove Property 2, note that $\rho(P_\mu, P_\nu) = 0$ if $\boldsymbol{\mu} = \boldsymbol{\nu}$. Thus the key observation is the converse, or equivalently, $\rho(P_\mu, P_\nu) > 0$ if $\boldsymbol{\mu} \neq \boldsymbol{\nu}$. Note that if $\boldsymbol{\mu} \neq \boldsymbol{\nu}$, then $\mu_i \neq \nu_i$ for at least one index i (actually for at least two indices). Let $S = \{i\}$, where i is an index such that $\mu_i \neq \nu_i$. Then

$$P_\mu(S) = \mu_i \neq \nu_i = P_\nu(S).$$

Hence $\rho(P_\mu, P_\nu) > 0$. Property 3 is again obvious. Finally, Property 4 follows from the triangle inequality for real numbers, namely:

$$|x - y| \leq |x - z| + |z - y|, \; \forall x, y, z \in \mathbb{R}.$$

Now suppose $S \subseteq \mathbb{A}$ is arbitrary. Then the triangle inequality for real numbers implies that

$$|P_\mu(S) - P_\nu(S)| \leq |P_\mu(S) - P_\phi(S)| + |P_\phi(S) - P_\nu(S)|, \; \forall A \subseteq \mathbb{A}.$$

Taking the maximum over all $S \subseteq \mathbb{A}$ proves Property 4. □

As defined in (1.8), $\rho(P_\mu, P_\nu)$ is the maximum difference between $P_\mu(S)$ and $P_\nu(S)$ as S varies over the 2^n subsets of $\mathbb{A}$. Clearly, (1.8) is an impractical formula for actually *computing* the number $\rho(P_\mu, P_\nu)$. The next theorem gives a number of equivalent formulas for computing $\rho(P_\mu, P_\nu)$. Note that, given a real number $x \in \mathbb{R}$, the symbol x_+ denotes the positive part of x, that is, $\max\{x, 0\}$. Similarly x_- denotes $\min\{x, 0\}$.

Theorem 1.8 *Suppose* $\mathbb{A} = \{a_1, \ldots, a_n\}$ *is a finite set, and that* P_μ, P_ν *are two probability measures on* $\mathbb{A}$ *with associated distributions* $\boldsymbol{\mu}$ *and* $\boldsymbol{\nu}$ *respectively. Then*

$$\rho(P_\mu, P_\nu) = \sum_{i=1}^{n} (\mu_i - \nu_i)_+ \tag{1.10}$$

$$= -\sum_{i=1}^{n} (\mu_i - \nu_i)_- \tag{1.11}$$

$$= \frac{1}{2} \sum_{i=1}^{n} |\mu_i - \nu_i|. \tag{1.12}$$

Proof. Define $\delta_i := \mu_i - \nu_i$, for $i = 1, \ldots, n$. Then, since $\boldsymbol{\mu}, \boldsymbol{\nu} \in \mathbb{S}_n$, it follows that $\sum_{i=1}^n \delta_i = 0$. Moreover, for any set $S \subseteq \mathbb{A}$, we have that

$$P_\mu(S) - P_\nu(S) = \sum_{a_i \in S} \delta_i = \sum_{i=1}^{n} I_S(a_i)\delta_i,$$

where $I_S(\cdot)$ is the indicator function of the set S. Now, let us look at the 2^n numbers $P_\mu(S) - P_\nu(S)$ generated by varying S over all subsets of $\mathbb{A}$. (These numbers may not all be distinct.) Let $K \subseteq \mathbb{R}$ denote the set of all these numbers. The first point to note is that

$$P_\mu(S^c) - P_\nu(S^c) = [1 - P_\mu(S)] - [1 - P_\nu(S)] = -[P_\mu(S) - P_\nu(S)].$$

Hence the set K is symmetric: If $x \in K$ (corresponding to a set S), then $-x \in K$ (corresponding to the set S^c). Observe that $\rho(P_\mu, P_\nu)$ is the largest value of $|x|$ for all $x \in K$. However, because of the symmetry of the set K, $\rho(P_\mu, P_\nu)$ also equals the largest value of $x \in K$, and also

$$\rho(P_\mu, P_\nu) = -\min\{x \in K\}.$$

So if we can find the largest or the smallest element in K, then we would have found $\rho(P_\mu, P_\nu)$.

Next, let $\mathbb{N}_+ \subseteq \mathbb{N} := \{1, \ldots, n\}$ denote the set of indices i for which $\delta_i \geq 0$, and let $\mathbb{N}_-$ denote the set of indices i for which $\delta_i < 0$. Then

$$P_\mu(S) - P_\nu(S) = \sum_{i \in \mathbb{N}_+} I_S(a_i)\delta_i + \sum_{i \in \mathbb{N}_-} I_S(a_i)\delta_i.$$

Now the first summation consists of only nonnegative numbers, while the second summation consists only of nonpositive numbers. Therefore the largest possible value of $P_\mu(S) - P_\nu(S)$ is

$$\sum_{i \in \mathbb{N}_+} I_S(a_i)\delta_i = \sum_{i=1}^{n} (\delta_i)_+,$$

and corresponds to the choice $S = \{a_i : i \in \mathbb{N}_+\}$. By the discussion in the preceding paragraph, it follows that

$$\rho(P_\mu, P_\nu) = \sum_{i=1}^{n} (\delta_i)_+,$$

which is precisely (1.10). Similarly, the smallest value of $P_\mu(S) - P_\nu(S)$ is

$$\sum_{i \in \mathbb{N}_-} I_S(a_i)\delta_i = \sum_{i=1}^n (\delta_i)_-,$$

corresponding to the choice $S = \{a_i : i \in \mathbb{N}_-\}$. Again from the discussion in the previous paragraph, it follows that

$$\rho(P_\mu, P_\nu) = -\sum_{i=1}^n (\delta_i)_-,$$

which is precisely (1.11). Finally, observe that

$$\sum_{i=1}^n |\delta_i| = \sum_{i=1}^n (\delta_i)_+ - \sum_{i=1}^n (\delta_i)_- = 2\rho(P_\mu, P_\nu),$$

which establishes (1.12). □

From the definition (1.8), it is immediate that $\rho(P_\mu, P_\nu) \in [0,1]$. This is because both $P_\mu(S)$ and $P_\nu(S)$ lie in the range $[0,1]$, and so $P_\mu(S) - P_\nu(S) \in [-1,1]$. Now the proof of Theorem 1.8 shows that, for every pair of probability measures P_μ and P_ν, there actually exists a set S such that $P_\mu(S) - P_\nu(S) = \rho(P_\mu, P_\nu)$; one such choice is $S = \{a_i : i \in \mathbb{N}_+\}$. Now, if $\rho(P_\mu, P_\nu)$ actually equals one, this implies that $P_\mu(S) = 1$ and $P_\nu(S) = 0$ (and also that $P_\mu(S^c) = 0$ and $P_\nu(S^c) = 1$). In such a case the two measures P_μ and P_ν are said to be **mutually singular**, because their weights are supported on disjoint sets: The weights μ_i are concentrated on the set S whereas the weights ν_i are concentrated on the set S^c.

Lemma 1.9 *Suppose $\mathbb{A} = \{a_1, \ldots, a_n\}$ is a finite set, and P_μ, P_ν are two probability measures on $\mathbb{A}$. Suppose $f : \mathbb{A} \to [-1,1]$. Then*

$$|E[f, P_\mu] - E[f, P_\nu]| \leq 2\rho(P_\mu, P_\nu). \tag{1.13}$$

Proof. The proof follows by direct substitution. We have that

$$\begin{aligned} |E[f, P_\mu] - E[f, P_\nu]| &= \left| \sum_{i=1}^n f(a_i)(\mu_i - \nu_i) \right| \\ &\leq \sum_{i=1}^n |f(a_i)| \cdot |\mu_i - \nu_i| \\ &\leq \sum_{i=1}^n |\mu_i - \nu_i| \text{ since } |f(a_i)| \leq 1 \\ &= 2\rho(P_\mu, P_\nu). \end{aligned}$$

□

Lemma 1.10 *Suppose $\mathbb{A} = \{a_1, \ldots, a_n\}$ is a finite set, and P_μ, P_ν are two probability measures on $\mathbb{A}$. Suppose $f : \mathbb{A} \to [0,1]$. Then*

$$|E[f, P_\mu] - E[f, P_\nu]| \leq \rho(P_\mu, P_\nu). \tag{1.14}$$

Proof. This lemma can be derived as a corollary of Lemma 1.9 by observing that $f(X) - 0.5$ assumes values in $[-0.5, 0.5]$; but we will give an alternate proof. We have that

$$\begin{aligned} E[f, P_\mu] - E[f, P_\nu] &= \sum_{i=1}^{n} f(a_i)(\mu_i - \nu_i) \\ &\leq \sum_{i=1}^{n} f(a_i) \cdot (\mu_i - \nu_i)_+ \text{ since } 0 \leq f(a_i)\ \forall i \\ &\leq \sum_{i=1}^{n} (\mu_i - \nu_i)_+ \text{ since } f(a_i) \leq 1\ \forall i \\ &= \rho(P_\mu, P_\nu). \end{aligned}$$

By entirely analogous reasoning, it follows that

$$E[f, P_\mu] - E[f, P_\nu] \geq \sum_{i=1}^{n} (\mu_i - \nu_i)_- = -\rho(P_\mu, P_\nu).$$

The desired bound now follows by combining these two inequalities. □

Problem 1.1 Let X be a random variable assuming values in the four-color alphabet $\mathbb{A} = \{R, B, W, G\}$ as in Examples 1.1 and 1.2, and let $f : \mathbb{A} \to \mathbb{R}$ be the payoff function defined in Example 1.4. First let X have the probability distribution in Example 1.1. Compute the mean, variance, and standard deviation of X. Repeat when X has the probability distribution in Example 1.2.

Problem 1.2 Suppose X is a "binary" random variable, assuming just two real values, namely 0 and 1, with $\Pr\{X = 1\} = \alpha \in (0, 1)$. We denote this by $X = \mathcal{B}(1, \alpha)$. Compute the mean and standard deviation of X.

Problem 1.3 Suppose an urn contains both white and black balls in the proportion α to $1 - \alpha$.[2] Let X be the associated binary random variable as defined in Problem 1.2 above. Now suppose we draw n balls from the urn, one after the other, replacing the ball drawn after each trial. Let Y_n denote the number of white balls drawn after n trials. Then Y_n is a "binomially distributed" random variable, whereby

$$\Pr\{Y_n = i\} = \binom{n}{i} \alpha^i (1 - \alpha)^{n-i},$$

where

$$\binom{n}{i} = \frac{n!}{i!(n-i)!}$$

[2]This suggests that α is a rational number, but the problem makes sense even without this assumption.

is called the combinatorial parameter. Y_n is the number of different sequences of n draws containing precisely i white balls. We denote this by $Y_n = \mathcal{B}(n, \alpha)$. Compute the mean and standard deviation of Y_n.

In case α is an irrational number, the urn and balls interpretation is not meaningful. Instead we should think in terms of generating n independent outcomes $X_1, \ldots, X_n$ where each X_i is binary with $\Pr\{X_i = 1\} = \alpha$. Such a sequence of random variables is known as a "Bernoulli process" or a set of "Bernoulli trials." If we equate an outcome of 1 with "success" and 0 with "failure," then Y_n is the number of successes in n Bernoulli trials.

Problem 1.4 A counterpart of the binomial distribution is the hypergeometric distribution. Suppose an urn contains N balls, out of which M are white. Now suppose we draw n balls one after the other, but this time *without replacing* the ball drawn. Let Z denote the resulting number of white balls. Show that

$$\Pr\{Z = i\} = \frac{\binom{M}{i}\binom{N-M}{n-i}}{\binom{N}{n}},$$

with the convention that

$$\binom{M}{i} = 0 \text{ if } M < i.$$

We denote this by $Z = \mathcal{H}(n, \alpha, N)$ where $\alpha = M/N$ is the fraction of white balls in the urn.

Problem 1.5 Suppose $Z = \mathcal{H}(n, \alpha, N)$ have the hypergeometric distribution. Show that, as $N \to \infty$, the hypergeometric distribution approaches the binomial distribution. Can you explain why?

Problem 1.6 Suppose $\boldsymbol{\mu} = [0.4 \;\; 0.6], \boldsymbol{\nu} = [0.6 \;\; 0.4]$. Compute the total variation distance $\rho(\boldsymbol{\mu}, \boldsymbol{\nu})$.

Problem 1.7 Let $n = 10$, and let Y, Z be binomially distributed random variables, with $Y_n = \mathcal{B}(10, 0.6), Z = \mathcal{B}(10, 0.4)$. Compute the total variation distance between the probability distributions of Y and Z.

Problem 1.8 Given a finite set $\mathbb{A}$, let $\mathcal{F}(\mathbb{A})$ denote all functions mapping $\mathbb{A}$ into the interval $[0, 1]$. Suppose $\boldsymbol{\mu}, \boldsymbol{\nu}$ are two probability distributions on $\mathbb{A}$. Show that

$$\max_{f \in \mathcal{F}(\mathbb{A})} |E[f, P_\mu] - E[f, P_\nu]| = \rho(P_\mu, P_\nu).$$

Problem 1.9 Prove the following generalization of Lemma 1.9: Suppose $f : \mathbb{A} \to [a, b]$ where a, b are real numbers, and that $\boldsymbol{\mu}, \boldsymbol{\nu}$ are probability distributions on $\mathbb{A}$. Then

$$|E[f, P_\mu] - E[f, P_\nu]| \leq (b - a)\rho(P_\mu, P_\nu).$$

Is this the best possible bound?

1.2 MULTIPLE RANDOM VARIABLES

1.2.1 Joint and Marginal Distributions

Up to now we have discussed only one random variable. It is also possible to have more than one random variable, each assuming values in its own set. Suppose $\mathbb{A} = \{a_1, \ldots, a_n\}$ and $\mathbb{B} = \{b_1, \ldots, b_m\}$ are finite sets. Then the Cartesian product $\mathbb{A} \times \mathbb{B}$ consists of all pairs of elements from the two sets; specifically

$$\mathbb{A} \times \mathbb{B} = \{(a_i, b_j) : a_i \in \mathbb{A}, b_j \in \mathbb{B}\}.$$

Moreover, the "joint random variable" $Z = (X, Y)$ assumes values in $\mathbb{A} \times \mathbb{B}$. Define

$$\phi_{ij} = \Pr\{X = a_i \wedge Y = b_j\},$$

where the wedge symbol $\wedge$ denotes "and." Then clearly the vector $\boldsymbol{\phi} \in \mathbb{S}_{nm}$. We can think of $\boldsymbol{\phi}$ as the probability distribution of some random variable that assumes values in a set of cardinality nm. But the fact that the range of values of Z is a cartesian product of two sets allows us to carry out a refined analysis. The probability distribution $\boldsymbol{\phi}$ on the set $\mathbb{A} \times \mathbb{B}$ is called the **joint distribution** of the variables X and Y. Thus

$$\Pr\{Z = (a_i, b_j)\} = \Pr\{X = a_i \wedge Y = b_j\} = \phi_{ij}, \ \forall a_i \in X, b_j \in Y.$$

So we can arrange the nm elements of ϕ in an array, as shown below.

$$\Phi = \begin{bmatrix} \phi_{11} & \ldots & \phi_{1m} \\ \vdots & \vdots & \vdots \\ \phi_{n1} & \ldots & \phi_{nm} \end{bmatrix}.$$

Up to now we have gained nothing by arranging the nm values of the probability distribution in an array. But now can take the analysis to another level.

Let us define the vectors $\boldsymbol{\phi}_X$ and $\boldsymbol{\phi}_Y$ as follows.

$$(\boldsymbol{\phi}_X)_i := \sum_{j=1}^{m} \phi_{ij}, i = 1, \ldots, n, \tag{1.15}$$

$$(\boldsymbol{\phi}_Y)_j := \sum_{i=1}^{n} \phi_{ij}, j = 1, \ldots, m. \tag{1.16}$$

Then it follows that $\boldsymbol{\phi}_X \in \mathbb{S}_n$ and $\boldsymbol{\phi}_Y \in \mathbb{S}_m$. This is because $\boldsymbol{\phi}$ is a probability distribution and as a result we have

$$\left(\sum_{i=1}^{n} \sum_{j=1}^{m} \phi_{ij} = 1 \right) \Rightarrow \left(\sum_{i=1}^{n} \left[\sum_{j=1}^{m} \phi_{ij} \right] = 1 \right) \Rightarrow \left(\sum_{i=1}^{n} (\boldsymbol{\phi}_X)_i = 1 \right).$$

Similarly

$$\left(\sum_{i=1}^{n} \sum_{j=1}^{m} \phi_{ij} = 1 \right) \Rightarrow \left(\sum_{j=1}^{m} (\boldsymbol{\phi}_Y)_j = 1 \right).$$

So $\boldsymbol{\phi}_X \in \mathbb{S}_n$ and $\boldsymbol{\phi}_Y \in \mathbb{S}_m$. The distribution $\boldsymbol{\phi}_X$ is called the **marginal distribution** of the random variable X. Similarly $\boldsymbol{\phi}_Y$ is called the marginal distribution of the random variable Y. Depending on the context, it may be more natural to write $\boldsymbol{\phi}_{\mathbb{A}}$ in the place of $\boldsymbol{\phi}_X$ and $\boldsymbol{\phi}_{\mathbb{B}}$ in the place of $\boldsymbol{\phi}_Y$. Mimicking earlier notation, we refer to the measure corresponding to the distribution $\boldsymbol{\phi} \in \mathbb{S}_{nm}$ as the **joint measure** P_ϕ of the joint random variable (X, Y). We refer to the measure corresponding to the distribution $\boldsymbol{\phi}_X \in \mathbb{S}_n$ as the **marginal measure** of X (or the marginal measure on the set $\mathbb{A}$), and denote it by $P_{\phi,X}$ or $P_{\phi,\mathbb{A}}$. The symbols $P_{\phi,Y}$ and $P_{\phi,\mathbb{B}}$ are defined analogously.

Now we proceed to show that indeed it is the case that

$$\Pr\{X = a_i\} = (\boldsymbol{\phi}_X)_i, \ \forall i = 1, \ldots, n.$$

To see this, fix an index i and observe that the singleton sets $\{(a_i, b_1)\}$ through $\{(a_i, b_m)\}$ are all pairwise disjoint subsets of $\mathbb{A} \times \mathbb{B}$. Moreover, it is clear that

$$\{(X, Y) \in \mathbb{A} \times \mathbb{B} : X = a_i\} = \bigcup_{j=1}^{m} \{(a_i, b_j)\}.$$

Hence from Property 2 of Theorem 1.3, we can conclude that

$$\begin{aligned} \Pr\{X = a_i\} &= \Pr\{(X, Y) \in \mathbb{A} \times \mathbb{B} : X = a_i\} \\ &= P_\phi \left(\bigcup_{j=1}^{m} \{(a_i, b_j)\} \right) \\ &= \sum_{j=1}^{m} P_\phi\{(a_i, b_j)\} \\ &= \sum_{j=1}^{m} \phi_{ij} = (\boldsymbol{\phi}_X)_i. \end{aligned}$$

By entirely analogous reasoning, it follows that

$$\Pr\{Y = b_j\} = \sum_{i=1}^{n} \phi_{ij} = (\boldsymbol{\phi}_Y)_j, \ \forall j = 1, \ldots, m.$$

1.2.2 Independence and Conditional Distributions

Up to now we have introduced the notion of a joint distribution of the two random variables X and Y, as well as their individual distributions, which can be obtained as the marginal distributions of the joint distribution. The next notion is among the most fundamental notions of probability theory.

Definition 1.11 *Suppose X, Y are random variables assuming values in finite sets $\mathbb{A}$ and $\mathbb{B}$ respectively. Let P_ϕ denote their joint probability measure, and let $\boldsymbol{\phi} \in \mathbb{S}_{nm}$ denote their joint distribution. Then the two random variables are said to be* **independent** *under the measure P_ϕ (or the distribution $\boldsymbol{\phi}$) if*

$$\phi_{ij} = (\boldsymbol{\phi}_X)_i \cdot (\boldsymbol{\phi}_Y)_j, \ \forall i = 1, \ldots, n, j = 1, \ldots, m. \tag{1.17}$$

An equivalent way of stating (1.17) is:

$$\Pr\{X = a_i \wedge Y = b_j\} = \Pr\{X = a_i\} \cdot \Pr\{Y = b_j\}, \ \forall a_i \in \mathbb{A}, b_j \in \mathbb{B}. \quad (1.18)$$

The above definition can be made a little more intuitive by introducing the notion of a "product" distribution. Suppose $\boldsymbol{\mu} \in \mathbb{S}_n, \boldsymbol{\nu} \in \mathbb{S}_m$ are distributions on the sets $\mathbb{A}, \mathbb{B}$ respectively. Then their **product distribution** $\boldsymbol{\mu} \times \boldsymbol{\nu}$ on the set $\mathbb{A} \times \mathbb{B}$ is defined by

$$(\boldsymbol{\mu} \times \boldsymbol{\nu})_{ij} = \mu_i \cdot \nu_j, \ \forall i, j. \quad (1.19)$$

With this definition, it follows that the two random variables X, Y are independent under the distribution $\boldsymbol{\phi}$ if and only if $\boldsymbol{\phi} = \boldsymbol{\phi}_X \times \boldsymbol{\phi}_Y$.

In the sequel we will often have occasion to speak about "independent and identically distributed" random variables. This notion can be made to fit into the above frame work by using product distributions where each of the marginals is the same. Thus if $\boldsymbol{\mu} \in \mathbb{S}_n$, then the product distribution $\boldsymbol{\mu}^2 \in \mathbb{S}_{n^2}$ is defined by

$$(\boldsymbol{\mu}^2)_{ij} = \mu_i \cdot \mu_j, \ \forall i, j.$$

The associated probability measure is often denoted by P_μ^2. Higher "powers" of $\boldsymbol{\mu}$ and P_μ are defined in an entirely analogous fashion.

Theorem 1.12 *Suppose X, Y are random variables assuming values in finite sets $\mathbb{A}$ and $\mathbb{B}$ respectively. Suppose $\boldsymbol{\phi} \in \mathbb{S}_{nm}$ is their joint distribution, that P_ϕ is their joint measure, and that X, Y are independent under the measure P_ϕ. Suppose further that $f : \mathbb{A} \to \mathbb{R}, g : \mathbb{B} \to \mathbb{R}$ are functions on $\mathbb{A}$ and $\mathbb{B}$ respectively. Then*

$$E[f(X)g(Y), P_\phi] = E[f(X), P_{\phi,X}] \cdot E[g(Y), P_{\phi,Y}]. \quad (1.20)$$

The point of the theorem is this: Even if f is a function of X alone and g is a function of Y alone, the function fg depends on both X and Y, and the pair (X, Y) has the joint measure P_ϕ. For an arbitrary joint probability measure P_ϕ, we cannot say anything about the expected value of the function $f(X)g(Y)$ under the measure P_ϕ. However, if the two random variables are *independent* under the measure P_ϕ, then the expected value factors neatly into the product of two different expected values, namely the expected value of f under the marginal measure $P_{\phi,X}$, and the expected value of g under the marginal measure $P_{\phi,Y}$.

Proof. The proof is a ready consequence of (1.17). We have

$$\begin{aligned} E[f(X)g(Y), P_\phi] &= \sum_{i=1}^{n} \sum_{j=1}^{m} f(a_i)g(b_j)\phi_{ij} \\ &= \sum_{i=1}^{n} \sum_{j=1}^{m} f(a_i)g(b_j)(\boldsymbol{\phi}_X)_i(\boldsymbol{\phi}_Y)_j \\ &= \left[\sum_{i=1}^{n} f(a_i)(\boldsymbol{\phi}_X)_i \right] \cdot \left[\sum_{j=1}^{m} g(b_j)(\boldsymbol{\phi}_Y)_j \right] \\ &= E[f(X), P_{\phi,X}] \cdot E[g(Y), P_{\phi,Y}]. \end{aligned}$$

This is precisely the desired conclusion. □

The above observation motivates the notion of the correlation coefficient between two real-valued random variables.

Definition 1.13 *Suppose X, Y are real-valued random variables assuming values in finite sets $\mathbb{A}, \mathbb{B} \subseteq \mathbb{R}$ respectively. Let $\boldsymbol{\phi}$ denote their joint distribution, and $\boldsymbol{\phi}_X, \boldsymbol{\phi}_Y$ the two marginal distributions. Let $E[XY, \boldsymbol{\phi}]$, $E[X, \boldsymbol{\phi}_X]$, $E[Y, \boldsymbol{\phi}_Y]$ denote expectations, and let $\sigma(X), \sigma(Y)$ denote the standard deviations of X, Y under their respective marginal distributions. Then the quantity*

$$C(X,Y) := \frac{E[XY, \boldsymbol{\phi}] - E[X, \boldsymbol{\phi}_X] E[Y, \boldsymbol{\phi}_Y]}{\sigma(X)\sigma(Y)} \tag{1.21}$$

is called the **correlation coefficient** *between X and Y.*

Note that some authors refer to $C(X, Y)$ as the "Pearson" correlation coefficient after its inventor. It can be shown that the correlation coefficient $C(X, Y)$ always lies in the interval $[-1, 1]$. It is often said that X, Y are **uncorrelated** if $C(X, Y) = 0$, **positively correlated** if $C(X, Y) > 0$, and **negatively correlated** if $C(X, Y) < 0$. One of the advantages of the correlation coefficient is that it is invariant under both scaling and centering. In other words, if $\alpha, \beta, \gamma, \delta$ are any real numbers, then

$$C(\alpha X + \beta, \gamma Y + \delta) = C(X, Y). \tag{1.22}$$

If two random variables are independent, then they are uncorrelated. However, the converse statement is most definitely not true; see Problem 1.12.

The next definition is almost as important as the notion of independence.

Definition 1.14 *Suppose X, Y are random variables assuming values in finite sets $\mathbb{A}$ and $\mathbb{B}$ respectively, and let $\boldsymbol{\phi} \in \mathbb{S}_{nm}$ denote their joint distribution. The* **conditional probability** *of X given an observation $Y = b_j$ is defined as*

$$\Pr\{X = a_i | Y = b_j\} := \frac{\Pr\{X = a_i \wedge Y = b_j\}}{\Pr\{Y = b_j\}} = \frac{\phi_{ij}}{\sum_{i'=1}^{n} \phi_{i'j}}. \tag{1.23}$$

In case $\Pr\{Y = b_j\} = 0$, we define $\Pr\{X = a_i | Y = b_j\} = \Pr\{X = a_i\} = (\boldsymbol{\phi}_X)_i$.

Let us use the notation $\phi_{\{a_i | b_j\}}$ as a shorthand for $\Pr\{X = a_i | Y = b_j\}$. Then the vector

$$\boldsymbol{\phi}_{\{X|Y=b_j\}} := [\phi_{\{a_1|b_j\}} \cdots \phi_{\{a_n|b_j\}}] \in \mathbb{S}_n. \tag{1.24}$$

This is obvious from (1.23). So $\boldsymbol{\phi}_{\{X|Y=b_j\}}$ is a probability distribution on the set $\mathbb{A}$; it is referred to as the **conditional distribution** of X, given that $Y = b_j$. The corresponding probability measure is denoted by $P_{\boldsymbol{\phi},\{X|Y=b_j\}}$ and is referred to as the **conditional measure** of X, given that $Y = b_j$.

We also use the simplified notation $\phi_{X|b_j}$ and $P_{\phi,X|b_j}$ if the variable Y is clear from the context.

Now we briefly introduce the notion of convex combinations of vectors; we will discuss this idea in greater detail in Section 2.1. If $\mathbf{x}, \mathbf{y} \in \mathbb{R}^n$ are n-dimensional vectors and $\lambda \in [0, 1]$, then the vector $\lambda\mathbf{x} + (1 - \lambda)\mathbf{y}$ is called a **convex combination** of $\mathbf{x}$ and $\mathbf{y}$. More generally, if $\mathbf{x}_1, \ldots, \mathbf{x}_l \in \mathbb{R}^n$ and $\boldsymbol{\lambda} \in \mathbb{S}_l$, then the vector $\sum_{i=1}^{l} \lambda_i \mathbf{x}_i$ is called a **convex combination** of the vectors $\mathbf{x}_1$ through $\mathbf{x}_l$. In the present context, it is easy to see that

$$(\phi_X)_i = \sum_{j=1}^{m} (\phi_Y)_j \phi_{\{X|Y=b_j\}}. \tag{1.25}$$

Thus, the marginal distribution ϕ_X is a convex combination of the m conditional distributions $\phi_{\{X|Y=b_j\}}, j = 1, \ldots, m$. The proof of (1.25) is a straightforward consequence of the definitions and is left as an exercise.

Thus far we have introduced a lot of terminology and notation, so let us recapitulate. Suppose X and Y are random variables, assuming values in finite sets $\mathbb{A} = \{a_1, \ldots, a_n\}$ and $\mathbb{B} = \{b_1, \ldots, b_m\}$ respectively. Then they have a *joint probability measure* P_ϕ, defined on the product set $\mathbb{A} \times \mathbb{B}$. Associated with P_ϕ is a *marginal probability* $P_{\phi,X}$, which is a measure on $\mathbb{A}$, and a *marginal probability* $P_{\phi,Y}$, which is a measure on $\mathbb{B}$. Finally, for each of the m possible values of Y, there is an associated *conditional probability* $P_{\phi,\{X|Y=b_j\}}$, which is a measure on $\mathbb{A}$. Similarly, for each of the n possible values of X, there is an associated *conditional probability* $P_{\phi,\{Y|X=a_i\}}$, which is a measure on $\mathbb{B}$.

Example 1.6 Let us return to the problem studied earlier of an urn containing four uniform balls with colors red, blue, white and green. Suppose we draw two balls from the urn, one after the other, but *without* replacing the first ball before drawing the second ball. Let X denote the color of the first ball, and Y the color of the second ball. We can ask: What is the probability of drawing a red ball the second time? The answer is somewhat counter-intuitive because, as shown below, the answer is 0.25. We know that, when we make the second draw, there are only three balls in the urn, and which three colors they represent depends on X, the outcome of the first draw. Nevertheless, the probability of drawing a red ball (or any other colored ball) turns out to be 0.25, as is shown next.

Let us first compute the marginal or "unconditional" distribution of X, the outcome of the first draw. Since the balls are assumed to be uniform and there are four balls when we draw for the first time, we can define $\mathbb{A} = \{R, B, W, G\}$ and with this definition the distribution ϕ_X of X is given by

$$\phi_X = [0.25 \;\; 0.25 \;\; 0.25 \;\; 0.25].$$

Now let us compute the conditional probability of Y given X. If $X = R$, then at the second draw there are only B, G, Y in the urn. So we can say that

$$\phi_{\{Y|X=R\}} = [0 \;\; 1/3 \;\; 1/3 \;\; 1/3].$$

Similarly,

$$\phi_{\{Y|X=B\}} = [1/3 \;\; 0 \;\; 1/3 \;\; 1/3],$$

$$\phi_{\{Y|X=W\}} = [1/3 \;\; 1/3 \;\; 0 \;\; 1/3],$$

$$\phi_{\{Y|X=G\}} = [1/3 \;\; 1/3 \;\; 1/3 \;\; 0].$$

Therefore

$$\Pr\{Y = R\} = \Pr\{Y = R|X = R\} \cdot \Pr\{X = R\} + \cdots \\ + \Pr\{Y = R|X = G\} \cdot \Pr\{X = G\},$$

summing over all possible outcomes for Y. Doing this routine calculation shows that

$$\phi_Y = [0.25 \;\; 0.25 \;\; 0.25 \;\; 0.25].$$

This somewhat counter-intuitive result can be explained as follows: When we make the second draw to determine Y, there are indeed only three balls in the urn. However, *which three* they are depends on X, the outcome of the first draw. There are four possible sets of three colors, and each of them is equally likely. Hence the probability of getting a red ball the first time is exactly the same as the probability of getting a red ball the second time, even though we are not replacing the first ball drawn.

Example 1.7 The purpose of this example is to show that it is necessary to verify the condition (1.26) for *every* possible value b_j. Suppose $\mathbb{A} = \{a_1, a_2\}$, $\mathbb{B} = \{b_1, b_2, b_3\}$, and that the joint probability distribution is

$$[\phi_{ij}] = \begin{bmatrix} 0.12 & 0.08 & 0.20 \\ 0.20 & 0.10 & 0.30 \end{bmatrix}.$$

Then it follows from (1.15) and (1.16) that

$$\phi_X = [0.4 \;\; 0.6] \text{ and } \phi_Y = [0.32 \;\; 0.18 \;\; 0.50].$$

It can be readily checked that the condition (1.26) holds when $j = 3$ but not when $j = 1$ or $j = 2$. Hence the variables X and Y are not independent.

Lemma 1.15 *Suppose X, Y are random variables assuming values in finite sets $\mathbb{A}$ and $\mathbb{B}$ respectively, and let $\phi \in \mathbb{S}_{nm}$ denote their joint distribution. Then X and Y are independent if and only if*

$$\phi_{\{X|Y=b_j\}} = \phi_X, \; \forall b_j \in Y. \tag{1.26}$$

There is an apparent asymmetry in the statement of Lemma 1.15. It appears as though we should say "X is independent of Y if (1.26) holds" as opposed to "X and Y are independent if (1.26) holds." It is left as an exercise to show that (1.26) is equivalent to the statement

$$\phi_{\{Y|X=a_i\}} = \phi_Y, \; \forall a_i \in X. \tag{1.27}$$

The following observation follows readily from Definition 1.11.

Lemma 1.16 *Suppose X, Y are random variables assuming values in finite sets $\mathbb{A}$ and $\mathbb{B}$ respectively, and let $\boldsymbol{\phi} \in \mathbb{S}_{nm}$ denote their joint distribution. Then X and Y are independent if and only if the matrix*

$$\Phi := \begin{bmatrix} \phi_{11} & \dots & \phi_{1m} \\ \vdots & \vdots & \vdots \\ \phi_{n1} & \dots & \phi_{nm} \end{bmatrix}$$

has rank one.

The proof is easy and is left as an exercise.

In the preceding discussion, there is nothing special about having *two* random variables—we can have any finite number of them. We can also condition the probability distribution on multiple events, and the results are consistent. To illustrate, suppose X, Y, Z are random variables assuming values the finite sets $\mathbb{A} = \{a_1, \dots, a_n\}$, $\mathbb{B} = \{b_1, \dots, b_m\}$, $\mathbb{C} = \{c_1, \dots, c_l\}$ respectively. Let P_θ denote their joint probability measure and $\boldsymbol{\theta} = [\theta_{ijk}] \in \mathbb{S}_{nml}$ their joint probability distribution. Then

$$\Pr\{X = a_i | Y = b_j \wedge Z = c_k\} = \frac{\Pr\{X = a_i \wedge Y = b_j \wedge Z = c_k\}}{\Pr\{Y = b_j \wedge Z = c_k\}}. \quad (1.28)$$

In the shorthand notation introduced earlier, this becomes

$$\theta_{\{a_i | b_j \wedge c_k\}} = \left[\frac{\theta_{ijk}}{\sum_{i'=1}^{n} \theta_{i'jk}}, i = 1, \dots, n \right] \in \mathbb{S}_n. \quad (1.29)$$

When there are three random variables, the "law of iterated conditioning" applies, namely:

$$\boldsymbol{\theta}_{\{X | Y = b_j \wedge Z = b_k\}} = \boldsymbol{\theta}_{\{\{X \wedge Y | Z = c_k\} | Y = b_j\}}. \quad (1.30)$$

In other words, in order to compute the conditional distribution of X given that $Y = b_j$ and $Z = c_k$, we can think of two distinct approaches. First, we can directly apply (1.28). Second, we can begin by computing the joint conditional distribution of $X \wedge Y$ given that $Z = c_k$, and then condition this distribution of $Y = b_j$. Both approaches give the same answer.

The proof of (1.30) is straightforward. To begin with, we have

$$\boldsymbol{\theta}_{\{X \wedge Y | Z = c_k\}} = \left[\frac{\theta_{ijk}}{\sum_{i'=1}^{n} \sum_{j'=1}^{m} \theta_{i'j'k}}, i = 1, \dots, n, j = 1, \dots, m \right] \in \mathbb{S}_{nm}. \quad (1.31)$$

To make this formula less messy, let us define

$$\zeta_k := \sum_{i'=1}^{n} \sum_{j'=1}^{m} \theta_{i'j'k}.$$

Then

$$\boldsymbol{\theta}_{\{X \wedge Y | Z = c_k\}} = \left[\frac{\theta_{ijk}}{\zeta_k}, i = 1, \dots, n, j = 1, \dots, m \right].$$

If we now condition this joint distribution of $X \wedge Y$ on $Y = b_j$, we get

$$\theta_{\{\{X \wedge Y | Z = c_k\} | Y = b_j\}} = \left[\frac{\theta_{ijk}/\zeta_k}{\sum_{i'=1}^{n} \theta_{i'jk}/\zeta_k}, \mathbf{i} = 1, \ldots, n \right]$$
$$= \left[\frac{\theta_{ijk}}{\sum_{i'=1}^{n} \theta_{i'jk}}, i = 1, \ldots, n \right],$$

which is the same as (1.29).

The next notion introduced is conditional independence, which is very important in the case of hidden Markov processes, which are a central theme of this book.

Definition 1.17 *Suppose X, Y, Z are random variables assuming values in finite sets $\mathbb{A} = \{a_1, \ldots, a_m\}$, $\mathbb{B} = \{b_1, \ldots, b_m\}$ and $\mathbb{C} = \{c_1, \ldots, c_l\}$ respectively. Then we say that X, Y are* **conditionally independent given** *Z if, for all $c_k \in \mathbb{C}, b_j \in \mathbb{B}, a_i \in \mathbb{A}$, we have*

$$\Pr\{X = a_i \wedge Y = b_j | Z = c_k\} = \Pr\{X = a_i | Z = c_k\} \cdot \Pr\{Y = b_j | Z = c_k\} \tag{1.32}$$

Example 1.8 Consider three random variables X, Y, Z, each assuming values in $\{0, 1\}$. Suppose the joint distribution of the three variables is given by

$$\begin{bmatrix} \theta_{000} & \theta_{001} \\ \theta_{010} & \theta_{011} \end{bmatrix} = \begin{bmatrix} 0.018 & 0.056 \\ 0.042 & 0.084 \end{bmatrix}, \begin{bmatrix} \theta_{100} & \theta_{101} \\ \theta_{110} & \theta_{111} \end{bmatrix} = \begin{bmatrix} 0.096 & 0.192 \\ 0.224 & 0.288 \end{bmatrix},$$

where θ_{ijk} denotes $\Pr\{X = i \wedge Y = j \wedge Z = k\}$. It is now shown that X and Y are conditionally independent given Z. This is achieved by verifying (1.32).

From the given data, we can compute the joint distributions of $X \wedge Z$, and of $Y \wedge Z$. This gives

$$\begin{array}{ccc} X \wedge Z & 0 & 1 \\ 0 & \theta_{000} + \theta_{010} & \theta_{001} + \theta_{011} \\ 1 & \theta_{100} + \theta_{110} & \theta_{101} + \theta_{111} \end{array} = \begin{array}{ccc} X \wedge Z & 0 & 1 \\ 0 & 0.06 & 0.14 \\ 1 & 0.32 & 0.48 \end{array}$$

Hence, if $\boldsymbol{\psi}$ denotes the joint distribution of X and Z, then

$$\boldsymbol{\psi}_{\{X|Z=0\}} = \frac{1}{0.38} \begin{bmatrix} 0.06 \\ 0.32 \end{bmatrix}, \boldsymbol{\psi}_{\{X|Z=1\}} = \frac{1}{0.62} \begin{bmatrix} 0.14 \\ 0.48 \end{bmatrix}.$$

Next, let $\boldsymbol{\eta}$ denote the joint distribution of Y and Z. Then an entirely similar computation yields that

$$\begin{array}{ccc} Y \wedge Z & 0 & 1 \\ 0 & \theta_{000} + \theta_{100} & \theta_{001} + \theta_{101} \\ 1 & \theta_{010} + \theta_{110} & \theta_{011} + \theta_{111} \end{array} = \begin{array}{ccc} Y \wedge Z & 0 & 1 \\ 0 & 0.114 & 0.248 \\ 1 & 0.266 & 0.372 \end{array}$$

Hence

$$\boldsymbol{\eta}_{\{Y|Z=0\}} = \frac{1}{0.38}[0.114 \;\; 0.266] = [0.3 \;\; 0.7],$$

$$\eta_{\{Y|Z=1\}} = \frac{1}{0.62}[0.248 \;\; 0.372] = [0.4 \;\; 0.6].$$

Next, let us compute the joint distribution of $X \wedge Y$ conditioned on Z. From either of the above computations, it is clear that the marginal distribution of Z is given by

$$\psi_Z = \eta_Z = [0.38 \;\; 0.62].$$

Therefore the joint distribution of $X \wedge Y$ conditioned on Z can be computed using (1.31). This gives

$$\begin{array}{ccc} X \wedge Y|Z=0 & 0 & 1 \\ 0 & 0.018/0.38 & 0.042/0.38 \\ 1 & 0.096/0.38 & 0.224/0.38 \end{array} = \frac{1}{0.38}\begin{bmatrix} 0.06 \\ 0.32 \end{bmatrix}[0.3 \;\; 0.7]$$

$$= \psi_{\{X|Z=0\}} \times \eta_{\{Y|Z=0\}}.$$

Similarly,

$$\begin{array}{ccc} X \wedge Y|Z=1 & 0 & 1 \\ 0 & 0.056/0.62 & 0.084/0.62 \\ 1 & 0.192/0.62 & 0.288/0.62 \end{array} = \frac{1}{0.62}\begin{bmatrix} 0.14 \\ 0.48 \end{bmatrix}[0.4 \;\; 0.6]$$

$$= \psi_{\{X|Z=1\}} \times \eta_{\{Y|Z=1\}}.$$

Hence X and Y are conditionally independent given Z.

Note that, if we are not "given Z," then X and Y are *not* independent. From earlier discussion, it follows that the joint distribution of X and Y is given by

$$\Pr\{X = a_i \wedge Y = b_j\} = \sum_{k=1}^{l} \Pr\{X = a_i \wedge Y = b_j | Z = c_k\} \cdot \Pr\{Z = c_k\}.$$

So if we were to write the joint distribution of X and Y in a matrix, then it would be

$$\begin{bmatrix} 0.018 & 0.042 \\ 0.096 & 0.224 \end{bmatrix} + \begin{bmatrix} 0.056 & 0.084 \\ 0.192 & 0.288 \end{bmatrix} = \begin{bmatrix} 0.074 & 0.126 \\ 0.288 & 0.512 \end{bmatrix},$$

where the columns correspond to the values of Y and the row corresponds to the values of X. Since this matrix does not have rank one, X and Y are not independent. The point is that a convex combination of rank one matrices need not be of rank one.

Once we have the notion of a conditional distribution, the notion of conditional expected value is natural. Suppose X, Y are random variables assuming values in $\mathbb{A}$ and $\mathbb{B}$ respectively, and suppose $f : \mathbb{A} \to \mathbb{R}$ is some real-valued function. Let ϕ denote the joint distribution of X and Y. Then the "unconditional" expected value of f is denoted by $E[f, P_{\phi,X}]$ or less cumbersomely by $E[f, \phi_X]$, and is defined as

$$E[f, \phi_X] = \sum_{i=1}^{n} f(a_i)(\phi_X)_i.$$

The "conditional" expected value of f is denoted by $E[f, P_{\phi,\{X|Y=b_j\}}]$ or less cumbersomely by $E[f, \phi_{\{X|Y=b_j\}}]$, and is defined as

$$E[f, \phi_{\{X|Y=b_j\}}] = \sum_{i=1}^{n} f(a_i)\phi_{\{a_i|b_j\}}.$$

We conclude this subsection by introducing another notion called the conditional expectation of a random variable. The dual usage of the adjective "conditional" is a source of endless confusion to students. The conditional *expected value* of a random variable (or a function of a random variable) is a *real number*, whereas the conditional *expectation* of a random variable is another *random variable*. Unfortunately, this dual usage is too firmly entrenched in the probability literature for the present author to deviate from it.

Suppose X, Y are random variables assuming values in finite sets $\mathbb{A} = \{a_1, \ldots, a_n\}$, $\mathbb{B} = \{b_1, \ldots, b_m\}$ respectively. Let $\phi \in \mathbb{S}_{nm}$ denote their joint distribution. Now suppose $h : \mathbb{A} \times \mathbb{B} \to \mathbb{R}$ is a function of both X and Y. Then we can think of $h(X, Y)$ as a real-valued random variable. Now one can ask: What is the best approximation of $h(X, Y)$ in terms of a real-valued function of X alone? In other words, we seek a function $f : \mathbb{A} \to \mathbb{R}$ such that f best approximates h. A natural error criterion is the "least-squares error," namely,

$$J(f) = E[f - h, P_\phi] = \sum_{i=1}^{n}\sum_{j=1}^{m}(f_i - h_{ij})^2\phi_{ij},$$

where we use the shorthand $f_i = f(a_i), h_{ij} = h(a_i, b_j)$. The choice of f that minimizes J is easy to compute. Note that

$$\frac{\partial J}{\partial f_i} = 2\sum_{j=1}^{m}(f_i - h_{ij})\phi_{ij}.$$

Hence the optimal choice of f_i is obtained by setting this partial derivative to zero, that is,

$$f_i = \frac{\sum_{j=1}^{m} h_{ij}\phi_{ij}}{\sum_{j=1}^{m}\phi_{ij}} = \frac{\sum_{j=1}^{m} h_{ij}\phi_{ij}}{(\phi_X)_i} = \frac{E[h(X,Y), \phi]}{(\phi_X)_i}. \tag{1.33}$$

Hence if we define a function $f : \mathbb{A} \to \mathbb{R}$ by $f(a_i) = f_i$, then $f(X)$ is the best approximation to $h(X, Y)$ that depends on X alone. We formalize this idea through a definition.

Definition 1.18 *Suppose X, Y are random variables assuming values in finite sets $\mathbb{A} = \{a_1, \ldots, a_n\}$, $\mathbb{B} = \{b_1, \ldots, b_m\}$ respectively. Let $\phi \in \mathbb{S}_{nm}$ denote their joint distribution. Suppose $h : \mathbb{A} \times \mathbb{B} \to \mathbb{R}$. Then the* **conditional expectation of h with respect to X** *is the function $f : \mathbb{A} \to \mathbb{R}$ defined by (1.33), and is denoted by $h|\mathbb{A}$ or $h|X$.*

In the above definition, if $(\phi_X)_i = 0$ for some index i, then the corresponding value f_i can be assigned arbitrarily. This is because, if $(\phi_X)_i = 0$

for some index i, then $\phi_{ij} = 0$ for all j. As a result, we can actually just drop the corresponding element a_i from the set $\mathbb{A}$ and carry on without affecting anything.

Lemma 1.19 *Suppose* $h : \mathbb{A} \times \mathbb{B} \rightarrow \mathbb{R}_+$. *Then* $h|X : \mathbb{A} \rightarrow \mathbb{R}_+$. *Suppose* $h : \mathbb{A} \times \mathbb{B} \rightarrow [\alpha, \beta]$ *for some finite numbers* $\alpha < \beta$. *Then* $h|X : \mathbb{A} \rightarrow [\alpha, \beta]$.

Proof. The first part of the lemma says that if the original function h assumes only nonnegative values, then so does its conditional expectation $h|X$. This fact is obvious from the definition (1.33). The second part follows readily upon observing that if $h : \mathbb{A} \times \mathbb{B} \rightarrow [\alpha, \beta]$, then both $h - \alpha$ and $\beta - h$ are nonnegative-valued functions. □

A very useful property of the conditional expectation is given next.

Theorem 1.20 *Suppose* X, Y *are random variables assuming values in finite sets* $\mathbb{A} = \{a_1, \ldots, a_n\}$, $\mathbb{B} = \{b_1, \ldots, b_m\}$ *respectively. Let* $\phi \in \mathbb{S}_{nm}$ *denote their joint distribution. Suppose* $h : \mathbb{A} \times \mathbb{B} \rightarrow \mathbb{R}$ *and that* $g : \mathbb{A} \rightarrow \mathbb{R}$. *Then*

$$E[g(X)h(X,Y), \phi] = E[g \cdot h|X, \phi_{\mathbb{A}}]. \tag{1.34}$$

The point of the theorem is that $g(X)h(X,Y)$ is a random variable that depends on both X and Y, whereas $g \cdot h|X$ is a random variable that depends on X alone. The theorem states that both random variables have the same expected value (with respect to their corresponding probability distributions).

Proof. This follows from just writing out the expected value as a summation. We have

$$\begin{aligned} E[g(X)h(X,Y), \phi] &= \sum_{i=1}^{n} \sum_{j=1}^{m} g_i h_{ij} \phi_{ij} \\ &= \sum_{i=1}^{n} g_i \sum_{j=1}^{m} h_{ij} \phi_{ij} \\ &= \sum_{i=1}^{n} g_i (h_{\mathbb{A}})_i (\phi_{\mathbb{A}})_i \\ &= E[g \cdot h|X, \phi_{\mathbb{A}}]. \end{aligned}$$

This is the desired result. □

1.2.3 Bayes' Rule

The next result, known as **Bayes' rule**, is widely used.

Lemma 1.21 *Suppose* X *and* Y *are random variables assuming values in finite sets* $\mathbb{A}$ *and* $\mathbb{B}$ *of cardinality* n *and* m *respectively. Then*

$$\Pr\{X = a_i | Y = b_j\} = \frac{\Pr\{Y = b_j | X = a_i\} \cdot \Pr\{X = a_i\}}{\Pr\{Y = b_j\}}. \tag{1.35}$$

Proof. An equivalent way of writing (1.32) is:

$$\Pr\{X = a_i | Y = b_j\} \cdot \Pr\{Y = b_j\} = \Pr\{Y = b_j | X = a_i\} \cdot \Pr\{X = a_i\}.$$

But this statement is clearly true, since each side is equal to $\Pr\{X = a_i \wedge Y = b_j\}$. □

Example 1.9 A typical use of Bayes' rule is when we try to invert the hypothesis and conclusion, and assess the probability of the resulting statement. To illustrate, suppose there is a diagnostic test for HIV, which is accurate 98% of the time on HIV-positive patients and 99% of the time on HIV-negative patients. In other words, the probability that the test is positive when the patient has HIV is 0.98, while the probability that the test is negative when the patient does not have HIV is 0.99. We may therefore be lulled into thinking that we have a very good test at hand. But the question that really interests us is this: What is the probability that a patient who tests positive actually has HIV?

Let us introduce two random variables: X for a patient's actual condition, and Y for the outcome of a test. Thus X assumes values in the set $\mathbb{X} = \{H, F\}$, where H denotes that the patient has HIV, while F denotes that the patient is free from HIV. Similarly, Y assumes values in the set $\mathbb{Y} = \{P, N\}$, where P denotes that the test is positive, while N denotes that the test is negative. The data given thus far can be summarized as follows:

$$\Pr\{Y = P | X = H\} = 0.98, \Pr\{Y = N | X = F\} = 0.99. \tag{1.36}$$

But what we really want to know is the value of

$$\Pr\{X = H | Y = P\},$$

that is, the probability that the patient really has HIV when the test is positive.

To compute this quantity, suppose the fraction of the population that has HIV is 1%. Thus the marginal probability distribution of X is

$$[\Pr\{X = H\} \;\; \Pr\{X = F\}] = [0.01 \;\; 0.99].$$

With this information and (1.36), we can compute the joint distribution of the variables X and Y. We get

$$\begin{bmatrix} \phi_{X=H \wedge Y=P} & \phi_{X=H \wedge Y=N} \\ \phi_{X=F \wedge Y=P} & \phi_{X=F \wedge Y=N} \end{bmatrix} = \begin{bmatrix} 0.0098 & 0.0002 \\ 0.0099 & 0.9801 \end{bmatrix}.$$

So by adding up the two columns, we get

$$[\Pr\{Y = P\} \;\; \Pr\{Y = N\}] = [0.0197 \;\; 0.9803].$$

Hence, by Bayes' rule, we can compute that

$$\Pr\{X = H | Y = P\} = \frac{0.0098}{0.0197} \approx 0.5.$$

So actually the diagnostic is quite unreliable, because the likelihood of a patient who tests positive *not* having HIV is just about equal to the likelihood of a patient who tests positive actually having HIV.

This apparent paradox is easily explained: For the sake of simplicity, assume that the test is equally accurate both for patients actually having HIV and for patients not having HIV. Let β denote the inaccuracy of the test. Thus

$$\Pr\{Y = P|X = H\} = \Pr\{Y = N|X = F\} = 1 - \beta.$$

Let α denote the fraction of the population that actually has HIV. We can carry through all of the above computations in symbolic form and obtain

$$\begin{bmatrix} \phi_{X=H \wedge Y=P} & \phi_{X=H \wedge Y=N} \\ \phi_{X=F \wedge Y=P} & \phi_{X=F \wedge Y=N} \end{bmatrix} = \begin{bmatrix} \alpha(1-\beta) & \alpha\beta \\ (1-\alpha)\beta & (1-\alpha)(1-\beta) \end{bmatrix}.$$

So

$$\Pr\{X = H|Y = P\} = \frac{\alpha(1-\beta)}{\alpha(1-\beta) + (1-\alpha)\beta}.$$

If, as is reasonable, both α and β are small, we can approximate both $1 - \alpha$ and $1 - \beta$ by 1, which leads to

$$\Pr\{X = H|Y = P\} \approx \frac{\alpha}{\alpha + \beta}.$$

So, unless $\beta \ll \alpha$, we get a test that is pretty useless. On the other hand, if $\beta \ll \alpha$, then $\Pr\{X = H|Y = P\}$ is very close to one and we have an excellent diagnostic test. The point to note is that the error of the diagnostic test must be small, not in comparison with 1, but with the likelihood of occurrence of the condition that we are trying to detect.

1.2.4 MAP and Maximum Likelihood Estimates

In the previous subsections, we have discussed the issue of computing the probability distribution of one random variable, given an observation of another random variable. Now let us make the question a little more specific, and ask: What is the *most likely value* of one random variable, given an observation of another random variable? It is shown below that there are two distinct ways of formalizing this notion, and each is reasonable in its own way.

Definition 1.22 *Suppose X and Y are random variables assuming values in finite sets $\mathbb{A} = \{a_1, \ldots, a_n\}$ and $\mathbb{B} = \{b_1, \ldots, b_m\}$ respectively. Let ϕ denote their joint distribution. Then the* **maximum a posteriori (MAP)** *estimate of X given an observation $Y = b_j$ is the a_{i^*} such that*

$$\phi_{\{a_{i^*}|b_j\}} = \max_i \phi_{\{a_i|b_j\}}. \tag{1.37}$$

Thus the MAP estimate of X given an observation $Y = b_j$ is the most likely value of X using the conditional distribution $\phi_{\{X|Y=b_j\}}$. Since

$$\phi_{\{a_i|b_j\}} = \frac{\phi_{ij}}{(\phi_Y)_j},$$

and the denominator is independent of i, we can see that

$$i^* = \operatorname*{argmin}_i \phi_{ij}.$$

So computing the MAP estimate is very easy. Given an observation $Y = b_j$, we simply scan down the j-th column of the joint distribution matrix, and pick the row i where the element ϕ_{ij} is the largest. (If there is a tie, we can use any sensible tie-breaking rule.)

The next definition gives an alternate way of defining the "most likely" value of X.

Definition 1.23 *Suppose X and Y are random variables assuming values in finite sets $\mathbb{A} = \{a_1, \ldots, a_n\}$ and $Y = \{b_1, \ldots, b_m\}$ respectively. Let $\boldsymbol{\phi}$ denote their joint distribution. Then the* **maximum likelihood estimate (MLE)** *of X given an observation $Y = b_j$ is defined as the index i^* such that $\Pr\{Y = b_j | X = a_i\}$ is maximized when $i = i^*$.*

Thus the MLE of X given the observation $Y = b_j$ is the choice of a_i that makes the observed value the most likely one.

The choice between MAP and MLE is essentially dictated by whether we believe that X "causes" Y, or vice versa. The joint distribution $\boldsymbol{\phi}$ is strictly neutral, and does not at all address the issue of what causes what. If we believe that Y causes X, then we should believe that, following the observation $Y = b_j$, the probability distribution of X has shifted from the marginal distribution $\boldsymbol{\phi}_X$ to the conditional distribution $\boldsymbol{\phi}_{\{X|Y=b_j\}}$. Thus MAP is the most logical way to estimate X. If on the other hand we believe that X causes Y, we should choose the MLE of X, because that estimate makes the observation most likely.

Example 1.10 To show that MAP and MLE can lead to diametrically opposite conclusions, consider the case where $n = m = 2$ and the joint distribution of X, Y is given by

$$\boldsymbol{\phi} = \begin{bmatrix} 0.1 & 0.2 \\ 0.4 & 0.3 \end{bmatrix},$$

where the rows correspond to the value of X and the columns to the values of Y. Suppose we observe $Y = b_2$. Then, by examining the second column of $\boldsymbol{\phi}$, we see that the MAP estimate of X is a_2, because $\phi_{22} > \phi_{12}$. On the other hand, to compute the MLE of X, we compute

$$\boldsymbol{\phi}_{\{Y|X=x_1\}} = [1/3 \;\; 2/3], \quad \boldsymbol{\phi}_{\{Y|X=x_2\}} = [4/7 \;\; 3/7].$$

Since $2/3 > 3/7$, b_2 is the most likely value of Y if $X = a_1$. So the MLE of X given the observation $Y = y_2$ is a_1.

Problem 1.10 Prove (1.22).

Problem 1.11 Show that if X, Y are independent real-valued random variables, then their correlation coefficient is zero.

Problem 1.12 Suppose the joint distribution of two random variables X and Y, each of them assuming one of the five values $\{1, 2, 3, 4, 5\}$, is as shown in the table below.

$$\Phi = \left[\begin{array}{c|ccccc} X \setminus Y & 1 & 2 & 3 & 4 & 5 \\ \hline 1 & 0.0800 & 0.0260 & 0.0280 & 0.0320 & 0.0340 \\ 2 & 0.0280 & 0.0900 & 0.0300 & 0.0270 & 0.0250 \\ 3 & 0.0260 & 0.0200 & 0.0800 & 0.0340 & 0.0400 \\ 4 & 0.0340 & 0.0300 & 0.0290 & 0.0800 & 0.0270 \\ 5 & 0.0320 & 0.0340 & 0.0330 & 0.0270 & 0.0740 \end{array} \right].$$

Compute the following:

1. The five conditional probability distributions $\phi_{X|Y=n\}}$, for $n = 1, \ldots, 5$.
2. The five conditional probability distributions $\phi_{Y|X=n\}}$, for $n = 1, \ldots, 5$.
3. The five conditional expected values $E[X|Y = n]$ for $n = 1, \ldots, 5$.
4. The five conditional expected values $E[Y|X = n]$ for $n = 1, \ldots, 5$.
5. The MAP estimates of X given that $Y = n$ for $n = 1, \ldots, 5$.
6. The MAP estimates of Y given that $X = n$ for $n = 1, \ldots, 5$.
7. The correlation coefficient $C(X, Y)$.

Problem 1.13 Prove (1.25).

Problem 1.14 Show that (1.26) and (1.27) are equivalent conditions.

Problem 1.15 Prove Lemma 1.16.

Problem 1.16 Suppose, as in (1.33), that $h : \mathbb{A} \times \mathbb{B} \to \mathbb{R}$. For each $a_i \in \mathbb{A}$, define the function $h_{i\cdot} : \mathbb{B} \to \mathbb{R}$ by

$$h_{i\cdot} = h_{ij}.$$

Show that the expression (1.33) for the conditional expectation of h with respect to X can be defined as

$$(h_X)_i = E[h_{i\cdot}, \phi_{|X=a_i}].$$

1.3 RANDOM VARIABLES ASSUMING INFINITELY MANY VALUES

1.3.1 Some Preliminaries

Until now we have steadfastly restricted ourselves to random variables that assume values in a *finite* set. However, there are situations in which it is desirable to relax this assumption, and examine situations in which the range of the random variable under study is *infinite*. Within this, we make a further distinction between two situations: where the range is a *countable* set and where the range is an *uncountable* set. Recall that a set is said to be **countable** if it can be placed in one-to-one correspondence with the set of natural numbers $\mathbb{N} = \{1, 2, \ldots\}$.[3] For example, the set of integers and the set of rational numbers are both countable sets. Next, suppose $\mathbb{M}$ is a finite set, such as $\{H, T\}$, the set of possible outcomes of a coin toss experiment, or $\{A, C, G, T\}$, the set of nucleotides. Let $\mathbb{M}^*$ denote the set of all *finite* sequences taking values in $\mathbb{M}$. Thus $\mathbb{M}^*$ consists of all sequences $\{u_1, \ldots, u_n\}$ where $u_i \in \mathbb{M}$ for all i. Then $\mathbb{M}^*$ is countable. But uncountably infinite sets are also relevant, such as with real-valued random variables. A less familiar example is that, if $\mathbb{M}$ is a finite set, then the set of *all* sequences (not just finite sequences) taking values in $\mathbb{M}$ is an uncountably infinite set.

It turns out that the method adopted thus far to define probabilities over finite sets, namely just to assign nonnegative "weights" to each element in such a way that the weights add up to one, works perfectly well on countable sets. However, the approach breaks down when the range of the random variable is an uncountably infinite set. The great Russian mathematician A. N. Kolmogorov introduced the axiomatic foundations of probability theory precisely to cope with this situation; see [81]. Though this theory is very beautiful and comprehensive, for the most part in the present book we will not be using this more advanced theory.

We first discuss the case of random variables assuming values in a countable set $\mathcal{X} = \{x_i, i \in \mathbb{N}\}$, where $\mathbb{N}$ denotes the set of natural numbers. Let $\mu_i \geq 0$ be chosen such that $\sum_{i=1}^{\infty} \mu_i = 1$. Then for every subset $A \subseteq \mathcal{X}$, we can define the corresponding probability measure P_μ in analogy with (1.2), namely

$$P_\mu(A) := \sum_{i=1}^{\infty} I_A(x_i)\mu_i.$$

We can think of $P_\mu(A)$ as the probability $\Pr\{X \in A\}$ that the random variable X belongs to the set A. With this definition, the "axiomatic" properties described just after Theorem 1.3 continue hold, namely:

1. $0 \leq P_\mu(A) \leq 1$ for all subsets $A \subseteq X$.
2. $P_\mu(\emptyset) = 0$ and $P_\mu(\mathcal{X}) = 1$.

---

[3]Some authors also include 0 in $\mathbb{N}$.

3. If $\{A_i\}_{i\geq 1}$ is a countable collection of pairwise disjoint subsets of X, then

$$P_\mu\left(\bigcup_{i=1}^{\infty} A_i\right) = \sum_{i=1}^{\infty} P_\mu(A_i).$$

If X is a *real-valued* random variable assuming values in the countable set $\mathcal{X} = \{x_i, i \in \mathbb{N}\} \subseteq \mathbb{R}$, with the associated probability measure P_μ defined above, then we define the mean, variance, and standard deviation of X just as we did earlier for finite-valued random variables, namely

$$E[X, P_\mu] := \sum_{i=1}^{\infty} x_i \mu_i,$$

$$V[X, P_\mu] = E[(X - E[X, P_\mu])^2, P_\mu],$$

$$\sigma(X, P_\mu) = \sqrt{V[X, P_\mu]}.$$

The major potential difficulty is that, because we are dealing with *infinite sums*, the above summations are not guaranteed to converge. Example 1.11 gives one such random variable. However, Schwarz's inequality (of which we got only a glimpse earlier) guarantees that, if X has finite variance, then it also has finite mean.

Example 1.11 This example is sometimes referred to as the "St. Petersberg paradox." Suppose a gambler visits a casino where he plays a coin-tossing game. At each step, both the gambler and the casino put up equal stakes, after which the gambler predicts the outcome of the coin toss. If he calls correctly, he gets the entire stake, whereas if he calls incorrectly, the casino gets the entire stake. The game is fair, with the coin turning up heads or tails with equal probability. Moreover, each coin toss is independent of all previous coin tosses. To simplify the notation, let us suppose that the gambler always calls heads. In view of the independence and the fairness assumptions, this strategy has just as good a chance of winning as any other.

Now the following strategy is "guaranteed" to fetch a positive payoff to the gambler: At each step, he merely doubles his stake. Thus, at the "zeroth" step, he bets \$1. If he wins, he quits and goes home. If he loses, he bets \$2 at step 1. If he loses again, he bets \$4 at the next step, and so on. The game reaches n steps only if the gambler has lost all previous $n - 1$ times, meaning that his accumulated losses are $\$1 + 2 + \ldots 2^{n-1} = 2^n - 1$. At the n-th step, he bets 2^n. If he wins, his cumulative winning amount is precisely \$1, the initial bet.

One feels that there is something strange about this strategy; indeed the difficulty is that the random variable in this case is "heavy-tailed" and does not have a finite expectation. We can see that the game has a countable set of possible outcomes, namely $H, TH, T^2H, \ldots, T^nH, \ldots$, where T^nH denotes a sequence of n tails followed by a head. The probability of T^nH is obviously

$2^{-(n+1)}$ because the coin is fair. In this case, the accumulated losses before the n-th step are $2^n - 1$. Thus, in order to bet 2^n at the next step, the gambler must have an initial sum of $2^n - 1 + 2^n = 2^{n+1} - 1$. Therefore, the amount of money that the player must have to begin with, call it X, is an integer-valued random variable, and equals $2^{n+1} - 1$ with probability $2^{-(n+1)}$. If we try to compute the expected value of this random variable, we see that

$$\sum_{n=0}^{\infty}(2^{n+1} - 1) \cdot 2^{-(n+1)} = \sum_{n=0}^{\infty}[1 - 2^{-(n+1)}].$$

Now the second summation converges nicely to -1. Unfortunately, the first summation blows up. Hence, unless one has an infinite amount of money to begin with, the above "strategy" will not work.

Next we give a very cursory discussion of random variables whose range may consist of an *uncountable* subset of $\mathbb{R}$. With each real-valued random variable X we associate a so-called **cumulative distribution function (cdf)** P_X, defined as follows:

$$P_X(a) = \Pr\{X \leq a\}, \ \forall a \in \mathbb{R}.$$

The cdf is monotonically nondecreasing, as is obvious from the definition; thus

$$a \leq b \Rightarrow P_X(a) \leq P_X(b).$$

The cdf also has a property known as "cadlag," which is an abbreviation of the French expression "continué à droite, limité à gauche." In English this means "continuous from the right, and limit exists from the left." In other words, the function P_X has the property that, for each real number a,

$$\lim_{x \to a^+} P_X(x) = P_X(a),$$

while $\lim_{x \to a^-} P_X(x)$ exists, but may or may not equal $P_X(a)$. Due to the monotonicity of the cdf, it is clear that

$$\lim_{x \to a^-} P_X(x) \leq P_X(a).$$

If the above holds with equality, then P_X is continuous at a. Otherwise it has a positive jump equal to the difference between $P_X(a)$ and the limit on the left side. For every $a \in \mathbb{R}$, it is the case that

$$\Pr\{X = a\} = P_X(a) - \lim_{x \to a^-} P_X(x).$$

So if P_X is continuous at a, then the probability that X *exactly* equals a is zero. However, if P_X has a jump at a, then the magnitude of the jump is the probability that X exactly equals a.

In general the function P_X need not be differentiable, or even continuous. However, for the purposes of the present discussion, it is sufficient to consider the case where P_X is continuously differentiable everywhere, except for a countable set of points $\{x_i\}_{i=1}^{\infty}$, where the function has a jump. Thus

$$\lim_{x \to x_i^-} P_X(x) < P_X(x_i),$$

but $P_X(\cdot)$ is continuously differentiable at all $x \neq x_i$. In such a case, the difference

$$P_X(x_i) - \lim_{x \to x_i^-} P_X(x) =: \mu_i$$

can be interpreted as the (nonzero) probability that the random variable X *exactly equals* x_i, as discussed above. For all other values of x, it is interpreted that the probability of the random variable X *exactly* equaling x is zero. Moreover, if $a < b$, then the probability of the random variable X lying in the interval $(a, b]$ is taken as $P_X(b) - P_X(a)$.

To define the expected value of the random variable X, we adapt the earlier formulation to the present situation. To simplify notation, let $P(\cdot)$ denote the cdf of X. Then we define

$$E[X, P] = \int_{-\infty}^{\infty} x P(dx),$$

where the integral is a so-called Riemann-Stiltjes integral. If P is continuously differentiable over some interval $[a, b]$, we define

$$\int_a^b f(x) P(dx) = \int_a^b f(x) \frac{dP}{dx} dx,$$

and add the term $f(x_i)\mu_i$ whenever the interval $[a, b]$ contains one of the points x_i where P_X has a jump discontinuity. As before, the existence of the expected value is not guaranteed.

1.3.2 Markov and Chebycheff Inequalities

In this subsection, we introduce two very useful inequalities known as "Markov's inequality" and "Chebycheff's inequality." We use the formalism of a random variable assuming real values in order to state the result. Since the set of real numbers is uncountably infinite, the earlier approach of assigning a weight to each possible outcome does not work, and we need to adopt a different approach. What follows is a very superficial introduction to the subject, and a reader interested in a serious discussion of the subject is referred to any of the classic texts on the subject, such as [23, 19] for example.

Theorem 1.24 (Markov's Inequality) *Suppose X is a real-valued random variable with the property that $|X|$ has finite expected value. Then, for every real number a, we have*

$$\Pr\{|X| > a\} \leq \frac{E[|X|, P]}{a}. \tag{1.38}$$

Proof. By definition, we have

$$\begin{aligned} E[|X|, P] &= \int_{|x| \leq a} |x| P(dx) + \int_{|x| > a} |x| P(dx) \\ &\geq \int_{|x| > a} |x| P(dx) \\ &\geq \int_{|x| > a} a P(dx) \\ &= a \Pr\{|X| > a\}. \end{aligned}$$

The desired inequality follows by dividing both sides by a. □

Corollary 1.25 *Suppose X is a nonnegative-valued random variable with finite expected value. Then, for every real number a, we have*

$$\Pr\{X > a\} \leq \frac{E[X, P]}{a}. \tag{1.39}$$

The proof is entirely analogous to that of Theorem 1.24.

Markov's inequality in the above form is not particularly useful. However, we get a more useful version if we examine a function of X.

Corollary 1.26 *Suppose X is a real-valued random variable. Then*

$$\Pr\{X > \epsilon\} \leq \exp(-\gamma\epsilon) E[\exp(\gamma X), P], \ \forall \gamma \geq 0 \ \forall \epsilon > 0, \tag{1.40}$$

provided only that $\exp(\gamma X)$ *has finite expected value.*

Proof. Note that, whenever $\gamma \geq 0$, the function $x \mapsto \exp(\gamma x)$ is nonnegative-valued and nondecreasing. Hence, for every $\epsilon > 0$, we have

$$X > \epsilon \Leftrightarrow \exp(\gamma X) > \exp(\gamma\epsilon).$$

Now apply (1.39) with X replaced by $\exp(\gamma X)$ and a replaced by $\exp(\gamma\epsilon)$. □

There is a variant of Markov's inequality for random variables that have not only finite expected value but also finite variance. As before, we define the variance of X as

$$V(X) := E[(X - E(X))^2],$$

assuming it exists of course, and the standard deviation $\sigma(X)$ as $(V(X))^{1/2}$. With this notation, we now state the next result. Recall that if a random variable has finite variance, it automatically has a finite expected value.

Theorem 1.27 (Chebycheff's Inequality) *Suppose X is a real-valued random variable with finite variance $\sigma^2(X)$ and finite expected value $E(X)$. Then for each $\epsilon > 0$, we have*

$$\Pr\{|X - E(X)| > \epsilon\} \leq \frac{\sigma^2(X)}{\epsilon^2}. \tag{1.41}$$

Proof. We reason as follows:

$$\begin{aligned}\Pr\{|X - E(X)| > \epsilon\} &= \Pr\{(X - E(X))^2 > \epsilon^2\} \\ &\leq \frac{E[(X - E(X))^2]}{\epsilon^2} \text{ by (1.39)} \\ &= \frac{\sigma^2(X)}{\epsilon^2}.\end{aligned}$$

□

Chebycheff's inequality can be used to derive a quick, but rather conservative, estimate of the rate at which empirical estimates of the mean, derived from independent samples, converge to the true value.

Theorem 1.28 *Suppose X is a real-valued random variable with finite variance $\sigma^2(X)$ and finite expected value $E(X)$. Let $X_1, \ldots, X_l$ be independent copies of X, and define the l-fold average*

$$A_l := \frac{1}{l} \sum_{i=1}^{l} X_i. \tag{1.42}$$

Then

$$\Pr\{|A_l - E(X)| > \epsilon\} \leq \frac{\sigma^2(X)}{l\epsilon^2}. \tag{1.43}$$

Proof. Note that, by the linearity of the expected value, it follows that

$$E(A_l) = E(X), \ \forall l.$$

Also, the independence of the X_i leads to the observation that if we define S_l as $S_l = \sum_{i=1}^{l} X_i$, then $V(S_l) = l\sigma^2(X)$. In turn this implies that $V(A_l) = \sigma^2(X)/l$. Now apply Theorem 1.27, and specifically (1.41), to A_l. This readily leads to the desired inequality (1.43). □

Chebycheff's inequality can be used to estimate the probability that an empirical mean differs from its true mean. Specifically, suppose X is a real-valued random variable with finite variance $\sigma^2(X)$, and it is desired to estimate its expected value $E(X)$. One way to do this is to generate independent samples $x_1, \ldots, x_l$ of X, and define

$$\hat{E}(\mathbf{x}_1^l) := \frac{1}{l} \sum_{i=1}^{l} x_i, \tag{1.44}$$

where $\mathbf{x}_1^l := (x_1, \ldots, x_l) \in \mathbb{R}^n$. Then $\hat{E}(\mathbf{x}_1^l)$ is called the **empirical mean** of X based on the samples $x_1, \ldots, x_l =: \mathbf{x}_1^l$. At this point the reader might wonder what the difference is between the average A_l and the empirical mean $\hat{E}(\mathbf{x}_1^l)^l$. Think of A_l as a random variable and $\hat{E}(\mathbf{x}_1^l)$ as a realization of the random variable A_l. Both notations prove useful in different situations, so both are introduced. Now Chebycheff's inequality tells us that

$$\Pr\{|\hat{E}(\mathbf{x}_1^l) - E(X)| > \epsilon\} \leq \frac{\sigma^2(X)}{l\epsilon^2}. \tag{1.45}$$

Therefore it follows that

$$\lim_{l \to \infty} \Pr\{|\hat{E}(\mathbf{x}_1^l) - E(X)| > \epsilon\} = 0, \ \forall \epsilon > 0.$$

Note that we can also express this as

$$\lim_{l \to \infty} \Pr\{|A_l - E(X)| > \epsilon\} = 0, \ \forall \epsilon > 0.$$

This says that the sequence of random variables $\{A_l\}$ "converges in probability" to the true mean $E(X)$ as $l \to \infty$.

1.3.3 Hoeffding's Inequality

In this subsection we present a very powerful result known as Hoeffding's inequality, which is applicable only to *bounded* real-valued random variables. The advantage of Hoeffding's inequality over Chebycheff's inequality is that the estimates converge to zero much more quickly.

Theorem 1.29 (Hoeffding's Inequality) *Suppose $Y_1, \ldots, Y_l$ are independent random variables, where Y_i assumes values in the bounded interval $[a_1, b_i]$. Then for every $\epsilon > 0$, we have*

$$\Pr\{\sum_{i=1}^{l} [Y_i - E(Y_i)] > \epsilon\} \leq \exp\left[-2\epsilon^2 / \sum_{i=1}^{l} (b_i - a_i)^2\right], \tag{1.46}$$

where $E(Y_i)$ denotes the expected value of Y_i.

The proof of Hoeffding's inequality uses the following auxiliary lemma.

Lemma 1.30 *Suppose X is a zero-mean random variable assuming values in the interval $[a, b]$. Then for any $s > 0$, we have*

$$E[\exp(sX)] \leq \exp(s^2 (b-a)^2 / 8).$$

Proof. **(Of Lemma 1.30)**: We are forced to invoke Theorem 2.6, which has not yet been proven. Since the exponential is a convex function, the value of e^{sx} is bounded by the corresponding convex combination of its extreme values; that is,

$$\exp(sx) \leq \frac{x-a}{b-a} e^{sb} + \frac{b-x}{b-a} e^{sa}, \ \forall x \in [a, b].$$

Now take the expected values of both sides, and use the fact that $E(X) = 0$. This gives

$$\begin{aligned} E[\exp(sX)] &\leq \frac{b}{b-a} e^{sa} - \frac{a}{b-a} e^{sb} \\ &= (1 - p + p e^{s(b-a)}) e^{-ps(b-a)} \\ &=: \exp(\phi(u)), \end{aligned}$$

where

$$p := -a/(b-a), u := s(b-a), \phi(u) := -pu + \ln(1 - p + p e^u).$$

Clearly $\phi(u) = 0$. Moreover, a routine calculation shows that

$$\phi'(u) = -p + \frac{p}{p + (1-p)e^{-u}},$$

whence $\phi'(u) = 0$ as well. Moreover,

$$\phi''(u) = \frac{p(1-p)e^{-u}}{[p + (1-p)e^{-u}]^2} \leq 0.25 \; \forall u > 0.$$

Hence by Taylor's theorem, there exists a $\theta \in [0, u]$ such that

$$\phi(u) = \frac{\phi''(\theta)u^2}{2} \leq \frac{u^2}{8} = \frac{s^2(b-a)^2}{8}.$$

This completes the proof. □

Proof. **(Of Theorem 1.29)**: For any nonnegative random variable, we have from Corollary 1.26 that

$$\Pr\{X > \epsilon\} \leq e^{-s\epsilon} E[\exp(sX)].$$

Now apply this inequality to the random variable

$$Z_l := \sum_{i=1}^{l} (Y_i - E(Y_i)),$$

which has zero mean by the linearity of the expected value. Thus

$$\begin{aligned}
\Pr\{Z_l > \epsilon\} &\leq e^{-s\epsilon} \, E\left[\exp\left(s \sum_{i=1}^{l} (Y_i - E(Y_i))\right)\right] \\
&= e^{-s\epsilon} \prod_{i=1}^{l} E[\exp(s(Y_i - E(Y_i)))] \text{ by independence} \\
&\leq e^{-s\epsilon} \prod_{i=1}^{l} \exp[s^2(b_i - a_i)^2/8] \text{ by Lemma 1.30} \\
&= \exp\left[-s\epsilon + s^2 \sum_{i=1}^{l} \frac{(b_i - a_i)^2}{8}\right] \\
&= \exp\left[\frac{-2\epsilon^2}{\sum_{i=1}^{l} (b_i - a_i)^2}\right],
\end{aligned} \tag{1.47}$$

where the last step follows by choosing

$$s = \frac{4\epsilon}{\sum_{i=1}^{l} (b_i - a_i)^2}.$$

This completes the proof. □

A useful (and widely used) "corollary" of Hoeffding's inequality is obtained when we take repeated and independent measurements of *the same* random variable. Because of its importance, we state the "corollary" as a theorem.

Theorem 1.31 (Hoeffding's Inequality for Empirical Means) *Suppose X is a random variable assuming values in a bounded interval $[a, b]$, and that $x_1, \ldots, x_l$ are independent samples of X. Define the empirical mean $\hat{E}(\mathbf{x}_1^l)$ as in (1.44). Then for each $\epsilon > 0$, we have*

$$\Pr\{\hat{E}(\mathbf{x}_1^l) - E(X) > \epsilon\} \leq \exp[-2l\epsilon^2/(b-a)^2], \tag{1.48}$$

$$\Pr\{\hat{E}(\mathbf{x}_1^l) - E(X) < -\epsilon\} \leq \exp[-2l\epsilon^2/(b-a)^2], \tag{1.49}$$

$$\Pr\{|\hat{E}(\mathbf{x}_1^l) - E(X)| > \epsilon\} \leq 2\exp[-2l\epsilon^2/(b-a)^2]. \tag{1.50}$$

Proof. To prove (1.48), apply (1.46) with ϵ replaced by $l\epsilon$, and $a_i = a, b_i = b$ for all i. To prove (1.49), apply (1.48) with Y replaced by $-Y$. Finally (1.50) is a direct consequence of (1.48) and (1.49). □

Comparing (1.45) and (1.50) brings out clearly the superiority of Hoeffding's inequality over Chebycheff's inequality. In (1.45) the right side is $O(1/l)$ and thus decays very slowly with respect to the number of samples. In contrast, the right side of (1.50) decays as $\exp(-2l\epsilon^2)$.

Hoeffding's inequality was proved in 1963; see [65]. Since then various researchers had attempted to improve the bound, but could not succeed in doing so. And it is no wonder. In 1990, Massart [95] proved that Hoeffding's inequality is, in a very precise sense, the "best possible" inequality.

Note that Hoeffding's inequality is applicable to real-valued random variables. So how can it be applied to random variables that assume values in some discrete set that has no obvious interpretation as a subset of the real numbers (e.g., the set of nucleotides)? The next result gives the answer.

Theorem 1.32 *Suppose X is a random variable that assumes values in a finite set $\mathbb{A} = \{a_1, \ldots, a_n\}$, and let $\boldsymbol{\phi} \in \mathbb{S}_n$ denote the "true but unknown" probability distribution of X. Suppose $x_1, \ldots, x_l$ are independent samples of X. For each $j \in \{1, \ldots, n\}$, define the empirical frequency*

$$\hat{\phi}_{l,j} := \frac{1}{l} \sum_{i=1}^{l} I_{\{x_i = a_j\}},$$

and the vector $\hat{\boldsymbol{\phi}}_l \in \mathbb{S}_n$ as

$$\hat{\boldsymbol{\phi}}_l := [\hat{\phi}_{l,j}, j = 1, \ldots, n].$$

Then

$$\Pr\{\|\hat{\boldsymbol{\phi}}_l - \boldsymbol{\phi}\|_\infty > \epsilon\} \leq 2n\exp(-2l\epsilon^2). \tag{1.51}$$

Equivalently,

$$\Pr\{|\hat{\phi}_{l,j} - \phi_j| \leq \epsilon \ \forall j\} \geq 1 - 2n\exp(-2l\epsilon^2). \tag{1.52}$$

Proof. For each index $j \in \{1, \ldots, n\}$, we can define an associated binary-valued random variable Y_j as follows: $Y_j = 1$ if $X = a_j$, and $Y_j = 0$

otherwise. With this association, it is clear that the expected value of Y_j is precisely $\Pr\{X = a_j\} =: \phi_j$. Now define associated subsets $S_{l,j}(\epsilon)$ as follows:

$$S_{l,j}(\epsilon) := \{\mathbf{x} \in \mathbb{A}^l : |\hat{\phi}_{l,j} - \phi_j| > \epsilon\}.$$

Then it is clear that

$$\{\mathbf{x} \in \mathbb{A}^l : \|\hat{\boldsymbol{\phi}}_l - \boldsymbol{\phi}\|_\infty > \epsilon\} = \bigcup_{j=1}^{n} S_{l,j}(\epsilon).$$

Therefore

$$\Pr\{\|\hat{\boldsymbol{\phi}}_l - \boldsymbol{\phi}\|_\infty > \epsilon\} \leq \sum_{j=1}^{n} P_\phi^l(S_{l,j}(\epsilon)).$$

However, by (1.50), each of the summands on the right side is bounded by $2\exp(-2l\epsilon^2)$. □

Since n, the cardinality of the set $\mathbb{A}$, appears explicitly on the right side of the above equation, this approach is not useful for infinite sets, and alternate approaches need to be devised. However, for random variables X assuming values in a finite set of cardinality n, (1.51) states that, after l independent trials, we can state with confidence $1 - 2n\exp(-2l\epsilon^2)$ that *every one of the n estimates $\hat{\phi}_{j,l}$* is within ϵ of its true value ϕ_j.

If X assumes just two values, say H and T to denote "heads" and "tails," then it is possible to get rid of the factor 2 on the right side of (1.52). This is because, if the empirical estimate $\hat{p}_H$ is within ϵ of the true value p_H, then so is $\hat{p}_T$. However, this argument does not work when there are more than two possible outcomes.

1.3.4 Monte Carlo Simulation

In this subsection we give a brief introduction to Monte Carlo simulation, which is one of the most widely used randomized methods. There is no universal agreement on what constitutes the (or even a) Monte Carlo method, but the method given below is certainly often referred to as such. Also, while the original applications of the method were to problems in atomic physics, the application given here is to the computation (or more precisely, the estimation) of complex integrals.

Suppose $\mathbf{X}$ is a random variable assuming values in $\mathbb{R}^d, d \geq 1$. For such a random variable, the corresponding d-dimensional probability distribution function is defined in a manner analogous to the case of real-valued random variables. For each d-tuple $(a_1, \ldots, a_d) =: \mathbf{a}$, we define

$$P_{\mathbf{X}}(\mathbf{a}) = \Pr\{X_1 \leq a_1 \wedge \ldots \wedge X_d \leq a_d\}.$$

The reader is reminded that the wedge symbol $\wedge$ denotes "and." To make things simple, it is assumed that the distribution function is jointly differentiable with respect to all of its arguments, with the derivative denoted by

$p : \mathbb{R}^d \rightarrow \mathbb{R}_+$, which is called the **probability density function.** Therefore, for all $\mathbf{a} \in \mathbb{R}^d$, we have

$$\Pr\{X_1 \leq a_1 \wedge \ldots \wedge X_d \leq a_d\} = \int_{-\infty}^{a_1} \cdots \int_{-\infty}^{a_d} p(x_1, \ldots, x_d)\, dx_d \ldots dx_1.$$

Now suppose $f : \mathbb{R}^d \rightarrow [a, b]$ is a continuous function. The expected value of $f(\mathbf{X})$ can be defined as

$$E[f(\mathbf{X})] = \int_{-\infty}^{\infty} \cdots \int_{-\infty}^{\infty} f(x_1, \ldots, x_d)\, p(x_1, \ldots, x_d)\, dx_d \ldots dx_1,$$

where the existence of the integral is guaranteed by the fact that f assumes values in a bounded interval. Now, depending on the form of the function f and/or the density p, it might be difficult or impossible to compute the above integral exactly. So one possibility is to use "gridding," especially when $\mathbf{X}$ assumes values only in some hypercube in $\mathbb{R}^d$. Specifically, suppose

$$p(x_1, \ldots, x_d) = 0 \text{ if } (x_1, \ldots, x_d) \notin \prod_{i=1}^{d} [\alpha_i, \beta_i].$$

Thus the density is zero outside the hypercube $\prod_{i=1}^{d} [\alpha_i, \beta_i]$. The gridding technique consists of choosing some integer k, dividing each interval $[\alpha_i, \beta_i]$ into k equal intervals by choosing $k + 1$ grid points, such as

$$c_i^l = \alpha_i + l(\beta_i - \alpha_i)/k, l = 0, \ldots, k.$$

Then the set of points

$$\mathcal{G} = (c_1^{l_1}, \ldots, c_d^{l_d}), 0 \leq l_i \leq k$$

forms the grid. It is of course possible to choose different values of k for different indices i. Then one can evaluate f at each of these grid points, and then average them.

The main shortcoming of the gridding method is colorfully referred to as the "curse of dimensionality." Specifically, the grid $\mathcal{G}$ consists of $(k + 1)^d$ points. Clearly this number grows exponentially with respect to the dimension d. The method described next overcomes the curse of dimensionality by making the number of sample points independent of d.

Generate l independent samples $\mathbf{x}_1, \ldots, \mathbf{x}_l$ of $\mathbf{X}$, and define the empirical mean as

$$\hat{E}_l(f) := \frac{1}{l} \sum_{j=1}^{l} f(\mathbf{x}_j).$$

Then it readily follows from Hoeffding's inequality that

$$\Pr\{|\hat{E}(f) - E[f(\mathbf{X})]| > \epsilon\} \leq 2 \exp[-2l\epsilon^2/(b - a)^2].$$

Thus, in order to estimate the expected value to within a specified accuracy ϵ with confidence $1 - \delta$, it is enough to choose l samples provided

$$2 \exp[-2l\epsilon^2/(b - a)^2] \leq \delta, \text{ or } l \geq \frac{\log(2/\delta)}{2\epsilon^2}.$$

The key point is that the dimension d does not appear anywhere in the required number of samples!

The success of the Monte Carlo method depends on the ability to generate independent samples of the random variable $\mathbf{X}$ that follow the specified distribution function $P_{\mathbf{X}}$. There are many standard methods available for generating pseudorandom sequences, and the reader can find them through a routine literature search.

Just to look ahead, Markov Chain Monte Carlo (MCMC) refers to the situation where the samples $\mathbf{x}_j$ are not independent, but are instead the state sequence of a Markov chain. This is discussed in Section 4.3.

1.3.5 Introduction to Cramér's Theorem

In this subsection we give a brief introduction to a famous theorem by Cramér, which was one of the first results in "large deviation theory" for real-valued random variables. A fully rigorous treatment of the theory for real-valued random variables is beyond the scope of this book, and the interested reader is referred to an advanced text such as [41]. The simpler case of random variables assuming values in a finite set, which lends itself to analysis using combinatorial arguments, is studied in Chapter 5.

Suppose X is a real-valued random variable with the property that $e^{\lambda X}$ has finite expected value for every $\lambda \geq 0$. This means that if the probability distribution function $P_X(\cdot)$ is continuously differentiable everywhere, then the derivative $p(x) := dP_X(x)/dx$, known as the **probability density function**, has to decay faster than any exponential. The Gaussian random variable with mean μ and standard deviation σ, which has the density

$$p_{\text{Gauss}}(x;\mu,\sigma) = \frac{1}{\sqrt{2\pi}\sigma} \exp[-(x-\mu)^2/2\sigma^2],$$

satisfies this assumption; but other random variables might not.

For such a random variable, the function

$$\text{mgf}(\lambda; X) := E[e^{\lambda X}]$$

is called the **moment generating function** of X. Note that, once X is fixed, $\text{mgf}(\cdot; X) : \mathbb{R}_+ \to \mathbb{R}$. The rationale for the name is that, if we ignore the niceties associated with differentiating inside an integral, interchanging summation and integration, etc., then formally we have that

$$\left.\frac{d^n \text{mgf}(\lambda; X)}{d\lambda^n}\right|_{\lambda=0} = E\left[\left.\frac{d^n e^{\lambda X}}{d\lambda^n}\right|_{\lambda=0}\right] = E[X^n]$$

is the so-called n-th moment of X. So formally at least, we can write

$$\text{mgf}(\lambda; X) = \sum_{n=0}^{\infty} \frac{\lambda^n E[X^n]}{n!}.$$

The **logarithmic moment generating function (lmgf)** is defined as

$$\Lambda(\lambda; X) := \log \text{mgf}(\lambda; X) = \log E[e^{\lambda X}].$$

The problem studied in this subsection is the same as that addressed by the various inequalities presented thus far, namely: Suppose $X_1, \ldots, X_l$ are independent copies of X, and define the l-fold average A_l as in (1.42). Given a real number a, can we estimate $\Pr\{A_l > a\}$? Note that earlier we were estimating $\Pr\{A_l - E(X) > \epsilon\}$, but the difference is purely cosmetic–all we have to do is to replace a by $E(X) + \epsilon$ to get bounds similar to the earlier ones.

Since our aim is only to give an introduction to Cramér's theorem, and not to prove it rigorously, we deviate from the usual "theorem-proof" format wherein the end objective is stated up-front. Let us compute the mgf and lmgf of A_l. We have

$$\exp(\lambda A_l) = \exp\left[(\lambda/l)\sum_{i=1}^{l} X_i\right] = \prod_{i=1}^{l} \exp[(\lambda/l)X_i].$$

Therefore

$$E[\exp(\lambda A_l)] = \prod_{i=1}^{l} E[\exp((\lambda/l)X_i)] = (E[\exp((\lambda/l)X)])^l,$$

where the first step follows from the independence of the X_i's, and the second step follows from the fact that all X_i's have the same distribution. Therefore

$$\begin{aligned} \text{mgf}(\lambda; A_l) &= [\text{mgf}(\lambda/l; X)]^l, \\ \Lambda(\lambda; A_l) &= l\Lambda(\lambda/l; X). \end{aligned} \tag{1.53}$$

Next, let us apply Markov's inequality as stated in (1.40) to A_l. This leads to

$$\Pr\{A_l > a\} \leq e^{-\gamma a} E[e^{\gamma A_l}] = e^{-\gamma a}\text{mgf}(\gamma; A_l).$$

If we set $\gamma = \lambda l$ and make use of (1.53), we get

$$\begin{aligned} \Pr\{A_l > a\} &\leq e^{-\lambda l a}[\text{mgf}(\lambda; X)]^l, \\ \log \Pr\{A_l > a\} &\leq -\lambda l a + l\Lambda(\lambda; X), \\ \frac{1}{l}\Pr\{A_l > a\} &\leq -[\lambda a - \Lambda(\lambda; X)]. \end{aligned} \tag{1.54}$$

Note that the above inequality holds for *every* $\lambda \geq 0$. So we can "optimize" this inequality by tuning λ for each a. Define

$$F(a) := \sup_{\lambda \geq 0} [\lambda a - \Lambda(\lambda; X)].$$

Then (1.54) implies that

$$\frac{1}{l}\Pr\{A_l > a\} \leq -F(a). \tag{1.55}$$

Now a famous theorem of Cramér states that as $l \to \infty$ the bound in (1.55) is asymptotically tight. In other words, under suitable conditions,

$$\lim_{l\to\infty} \frac{1}{l}\Pr\{A_l > a\} = -F(a).$$

To put it another way, as $l \to \infty$,

$$\Pr\{A_l > a\} \sim \text{const.} e^{-lF(a)}.$$

This gives the rate at which the "tail probability" $\Pr\{A_l > a\}$ decays as a function of a. Again, the reader is directed to advanced texts such as [41] for full details, including a formal statement and proof.

Chapter Two

Introduction to Information Theory

In this chapter we give a brief introduction to some elementary aspects of information theory, specifically entropy in its various forms. One can think of entropy as the level of uncertainty associated with a random variable (or more precisely, the probability distribution of the random variable). Entropy has several useful properties, and the ones relevant to subsequent chapters are brought out here. When there are two or more random variables, it is worthwhile to study the conditional entropy of one random variable with respect to another. The last concept is relative entropy, also known as the Kullback-Leibler divergence, named after the two statisticians who invented the notion. The Kullback-Leibler divergence measures the "disparity" between two probability distributions, and has a very useful interpretation in terms of the rate at which one can learn to discriminate between the correct and an incorrect hypothesis, when there are two competing hypotheses. The same divergence is also encountered again in Chapter 5 when we study large deviation theory. To lay the foundation for introducing these concepts, we begin with the notion of convexity, which has many applications that go far beyond the few that are discussed in this book. Readers desirous of learning more are referred to the classic text [110] for convexity and to an authoritative text such as [30] for information theory.

2.1 CONVEX AND CONCAVE FUNCTIONS

In this section we introduce a very useful "universal" concept known as convexity (and its mirror image, concavity). Though we make use of this concept in a very restricted setting (namely, to study the properties of the entropy function), the concept itself has many applications. We begin with the notion of a convex set, and then move to the notion of a convex (or concave) function.

If x, y are real numbers, and $\lambda \in [0, 1]$, the number $\lambda x + (1 - \lambda)y$ is called a **convex combination** of x and y. More generally, if $\mathbf{x}, \mathbf{y} \in \mathbb{R}^n$ are n-dimensional vectors and $\lambda \in [0, 1]$, the vector $\lambda \mathbf{x} + (1 - \lambda)\mathbf{y}$ is called a **convex combination** of $\mathbf{x}$ and $\mathbf{y}$. If $\lambda \in (0, 1)$ and $\mathbf{x} \neq \mathbf{y}$, then the vector $\lambda \mathbf{x} + (1 - \lambda)\mathbf{y}$ is called a **strict convex combination** of $\mathbf{x}$ and $\mathbf{y}$. Some authors also call this a "nontrivial" convex combination.

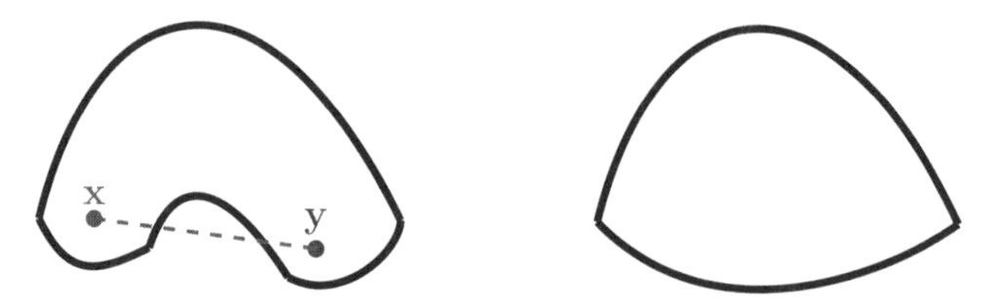

Figure 2.1 Examples of Convex and Nonconvex Sets

Definition 2.1 *A subset $S \subseteq \mathbb{R}^n$ is said to be a* **convex set** *if*

$$\lambda \mathbf{x} + (1 - \lambda)\mathbf{y} \in S \; \forall \lambda \in [0, 1], \; \forall \mathbf{x}, \mathbf{y} \in S. \tag{2.1}$$

Thus a set S is convex if every convex combination of two elements of S once again belongs to S. In two dimensions we can visualize a convex set very simply. If $\mathbf{x}, \mathbf{y} \in \mathbb{R}^2$, then the set $\{\lambda \mathbf{x} + (1 - \lambda)\mathbf{y} : \lambda \in [0, 1]\}$ is the line segment joining the two vectors $\mathbf{x}$ and $\mathbf{y}$. Thus a set $S \subseteq \mathbb{R}^2$ is convex if and only if the line segment joining any two points in the set S once again belongs to the set S. Therefore in Figure 2.1, the set on the left is not convex, because the line segment connecting $\mathbf{x}$ and $\mathbf{y}$ does not lie entirely in S; in contrast, the set on the right is convex. A similar interpretation also applies in higher dimensions, except that the "line" has to be imagined and cannot be drawn on a page.

Definition 2.1 is stated for a convex combination of *two* vectors, but can be easily extended to a convex combination of any finite number of vectors. Suppose $S \subseteq \mathbb{R}^n$ and $\mathbf{x}_1, \ldots, \mathbf{x}_k \in S$. Then a vector of the form

$$\mathbf{y} = \sum_{i=1}^{k} \lambda_i \mathbf{x}_i, \lambda_i \geq 0 \; \forall i, \sum_{i=1}^{k} \lambda_i = 1$$

is called a convex combination of the vectors $\mathbf{x}_1, \ldots, \mathbf{x}_k$. It is easy to show, by recursively applying Definition 2.1, that if $S \subseteq \mathbb{R}^n$ is a convex set then every convex combination of any finite number of vectors in S again belongs to S.

Example 2.1 The n-dimensional simplex $\mathbb{S}_n$, which can be identified with the set of probability distributions on a finite alphabet of cardinality n, is a convex set. Thus if $\boldsymbol{\mu}, \boldsymbol{\nu}$ are n-dimensional probability distributions, then so is the convex combination $\lambda \boldsymbol{\mu} + (1 - \lambda)\boldsymbol{\nu}$ for every λ in $[0, 1]$.

Example 2.2 An elaboration of the previous example comes from conditional distributions. Suppose X, Y are random variables assuming values in finite set $\mathbb{A} = \{a_1, \ldots, a_n\}$ and $\mathbb{B} = \{b_1, \ldots, b_m\}$ respectively, and let

$\boldsymbol{\phi} \in \mathbb{S}_{nm}$ denote their joint distribution. Recall from Section 1.2 that the marginal distributions $\boldsymbol{\phi}_X \in \mathbb{S}_n$ and $\boldsymbol{\phi}_Y \in \mathbb{S}_m$ are defined by

$$(\boldsymbol{\phi}_X)_i = \sum_{j=1}^{m} \phi_{ij}, (\boldsymbol{\phi}_Y)_j = \sum_{i=1}^{n} \phi_{ij},$$

while the m conditional distributions of X given the observations $Y = b_j$ are given by

$$\boldsymbol{\phi}_{\{X|Y=b_j\}} = \left[\frac{\phi_{ij}}{\sum_{i'=1}^{n} \phi_{i'j}}, i = 1, \ldots, n \right] \in \mathbb{S}_n.$$

Now it can be verified that the marginal distribution $\boldsymbol{\phi}_X$ is a convex combination of the m conditional distributions $\boldsymbol{\phi}_{\{X|Y=b_j\}}, j = 1, \ldots, m$, where the weights are the components of the marginal distribution $\boldsymbol{\phi}_Y$. In other words,

$$\phi_X = \sum_{j=1}^{m} \boldsymbol{\phi}_{\{X|Y=b_j\}} \cdot (\boldsymbol{\phi}_Y)_j.$$

This is a straightforward calculation and is left as an exercise.

Definition 2.2 *Suppose $S \subseteq \mathbb{R}^n$ is a convex set and $f : S \to \mathbb{R}$. We say that the function f is* **convex** *if*

$$f[\lambda \mathbf{x} + (1-\lambda)\mathbf{y}] \leq \lambda f(\mathbf{x}) + (1-\lambda) f(\mathbf{y}), \ \forall \lambda \in [0,1], \ \forall \mathbf{x}, \mathbf{y} \in S. \quad (2.2)$$

We say that the function f is **strictly convex** *if*

$$f[\lambda \mathbf{x} + (1-\lambda)\mathbf{y}] < \lambda f(\mathbf{x}) + (1-\lambda) f(\mathbf{y}), \ \forall \lambda \in (0,1), \ \forall \mathbf{x}, \mathbf{y} \in S, \mathbf{x} \neq \mathbf{y}. \quad (2.3)$$

We say that the function f is **concave** *if*

$$f[\lambda \mathbf{x} + (1-\lambda)\mathbf{y}] \geq \lambda f(\mathbf{x}) + (1-\lambda) f(\mathbf{y}), \ \forall \lambda \in [0,1], \ \forall \mathbf{x}, \mathbf{y} \in S. \quad (2.4)$$

Finally, we say that the function f is **strictly concave** *if*

$$f[\lambda \mathbf{x} + (1-\lambda)\mathbf{y}] > \lambda f(\mathbf{x}) + (1-\lambda) f(\mathbf{y}), \ \forall \lambda \in (0,1), \ \forall \mathbf{x}, \mathbf{y} \in S, \mathbf{x} \neq \mathbf{y}. \quad (2.5)$$

Equations (2.2) through (2.5) are stated for a convex combination of *two* vectors $\mathbf{x}$ and $\mathbf{y}$. But we can make repeated use of these equations and prove the following facts. If f is a convex function mapping a convex set S into $\mathbb{R}$, and $\mathbf{x}_1, \ldots, \mathbf{x}_k \in S$, then

$$f\left(\sum_{i=1}^{k} \lambda_i \mathbf{x}_i \right) \leq \sum_{i=1}^{k} \lambda_i f(\mathbf{x}_i), \text{ whenever } [\lambda_1 \ldots \lambda_k] =: \boldsymbol{\lambda} \in \mathbb{S}_k.$$

Analogous inequalities are valid for concave, strictly convex, and strictly concave functions.

The above definitions are all algebraic. But in the case where S is an interval $[a, b]$ in the real line (finite or infinite), the various inequalities can be given a simple pictorial interpretation. Suppose we plot the graph of the function f. This consists of all pairs $(x, f(x))$ as x varies over the interval

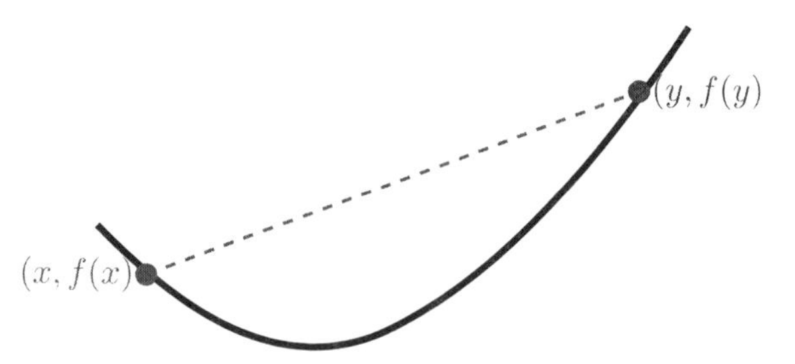

Figure 2.2 Graph Below Chord Interpretation of a Convex Function

$[a, b]$. Suppose $(x, f(x))$ and $(y, f(y))$ are two points on the graph. Then the straight line joining these two points is called the "chord" of the graph. (We can assume that the two points are distinct, because otherwise the various inequalities (2.2) through (2.5) all become trivial.) Equation (2.2) states that for any two points $x, y \in [a, b]$, the chord joining the two points $(x, f(x))$ and $(y, f(y))$ lies *above* the graph of the function $(z, f(z))$ whenever z lies between x and y. Equation (2.4) says exactly the opposite: It says that the chord joining the two points $(x, f(x))$ and $(y, f(y))$ lies *below* the graph of the function $(z, f(z))$ whenever z lies between x and y. Equation (2.3) says that, not only does the chord joining the two points $(x, f(x))$ and $(y, f(y))$ lie *above* the graph of the function $(z, f(z))$ whenever z lies between x and y, but in fact the chord does not even touch the graph, except at the two extreme points $(x, f(x))$ and $(y, f(y))$. Equation (2.5) says the opposite of the above. Finally, observe that f is (strictly) convex if and only if $-f$ is (strictly) concave.

Figure 2.2 depicts the interpretation of the definition of convexity. This figure depicts the fact that, for all z belonging to the chord connecting x and y, the value $f(z)$ lies below the chord value. The same figure also *suggests* that, if the chord is extended beyond the two original points x and y, then the value $f(z)$ actually lies *above* the chord value. This intuition is indeed correct, and not just in one dimension either!

Lemma 2.3 *Suppose* $f : \mathbb{R}^n \rightarrow \mathbb{R}$ *is convex, and suppose* $\mathbf{x}, \mathbf{y} \in \mathbb{R}^n$ *with* $\mathbf{x} \neq \mathbf{y}$. *Then, for every* $\lambda < 0$ *and every* $\lambda > 1$, *we have that*

$$f(\lambda\mathbf{x} + (1 - \lambda)\mathbf{y}) \geq \lambda f(\mathbf{x}) + (1 - \lambda) f(\mathbf{y}). \tag{2.6}$$

Proof. We begin with the case $\lambda < 0$. Let $\lambda = -\alpha$ where $\alpha > 0$, and define $\mathbf{z} = \lambda\mathbf{x} + (1 - \lambda)\mathbf{y}$. Then simple algebra shows that

$$\mathbf{y} = \frac{\alpha}{1 + \alpha}\mathbf{x} + \frac{1}{1 + \alpha}\mathbf{z},$$

so that $\mathbf{y}$ is a convex combination of $\mathbf{x}$ and $\mathbf{z}$. Now the convexity of $f(\cdot)$ implies that

$$f(\mathbf{y}) \leq \frac{\alpha}{1 + \alpha} f(\mathbf{x}) + \frac{1}{1 + \alpha} f(\mathbf{z}),$$

which can be rearranged as

$$f(\mathbf{z}) \geq -\alpha f(\mathbf{x}) + (1+\alpha) f(\mathbf{y}) = \lambda f(\mathbf{x}) + (1-\lambda) f(\mathbf{y}).$$

The case where $\lambda > 1$ is handled entirely similarly by interchanging the roles of $\mathbf{x}$ and $\mathbf{y}$. □

It can be shown that, for all practical purposes, a convex function (and thus a concave function) has to be continuous; see [110, Theorem 10.1]. But if a function is not merely continuous but also differentiable, then it is possible to give alternate characterizations of convexity (and of course concavity) that are more interesting. Recall that if $f : \mathbb{R}^n \to \mathbb{R}$ is differentiable at a point $\mathbf{x} \in \mathbb{R}^n$, then the **gradient** of f at $\mathbf{x}$ is an n-dimensional row vector denoted by $\nabla f(\mathbf{x})$ and equals

$$\nabla f(\mathbf{x}) = [\ \frac{\partial f}{\partial x_1} \ \ \cdots \ \ \frac{\partial f}{\partial x_n}\ ].$$

Lemma 2.4 *Suppose that $f : \mathbb{R}^n \to \mathbb{R}$ is continuously differentiable everywhere. If f is convex, then*

$$f(\mathbf{y}) \geq f(\mathbf{x}) + \nabla f(\mathbf{x})(\mathbf{y}-\mathbf{x}),\ \forall \mathbf{x}, \mathbf{y} \in \mathbb{R}^n, \mathbf{y} \neq \mathbf{x}. \tag{2.7}$$

If f is concave, then

$$f(\mathbf{y}) \leq f(\mathbf{x}) + \nabla f(\mathbf{x})(\mathbf{y}-\mathbf{x});\ \forall \mathbf{x}, \mathbf{y} \in \mathbb{R}^n, \mathbf{y} \neq \mathbf{x}. \tag{2.8}$$

If f is strictly convex, then

$$f(\mathbf{y}) > f(\mathbf{x}) + \nabla f(\mathbf{x})(\mathbf{y}-\mathbf{x}),\ \forall \mathbf{x}, \mathbf{y} \in \mathbb{R}^n, \mathbf{y} \neq \mathbf{x}. \tag{2.9}$$

If f is strictly concave, then

$$f(\mathbf{y}) < f(\mathbf{x}) + \nabla f(\mathbf{x})(\mathbf{y}-\mathbf{x}),\ \forall \mathbf{x}, \mathbf{y} \in \mathbb{R}^n, \mathbf{y} \neq \mathbf{x}. \tag{2.10}$$

For a proof of Lemma 2.4, see [110, Theorem 25.1].

Now we give interpretations of the various inequalities above in the case where $n = 1$, so that $f : \mathbb{R} \to \mathbb{R}$. Suppose f is continuously differentiable on some interval (a, b). Then for every $x \in (a, b)$, the function $y \mapsto f(x) + f'(x)(y - x)$ is the tangent to the graph of f at the point $(x, f(x))$. Thus (2.7) says that for a convex function, the tangent lies below the graph. This is to be contrasted with (2.2), which says that the chord lies above the graph. Equation (2.9) says that if the function is strictly convex, then not only does the tangent lie below the graph, but the tangent touches the graph only at the single point $(x, f(x))$. The interpretations of the other two inequalities are entirely similar. Figure 2.3 depicts the "graph above the tangent" property of a convex function, which is to be contrasted with the "graph below the chord" property depicted in Figure 2.2.

If the function is in fact *twice* continuously differentiable, then we can give yet another set of characterizations of the various forms of convexity.

Lemma 2.5 *Suppose $f : [a, b] \to \mathbb{R}$ is twice continuously differentiable on (a, b). Then*

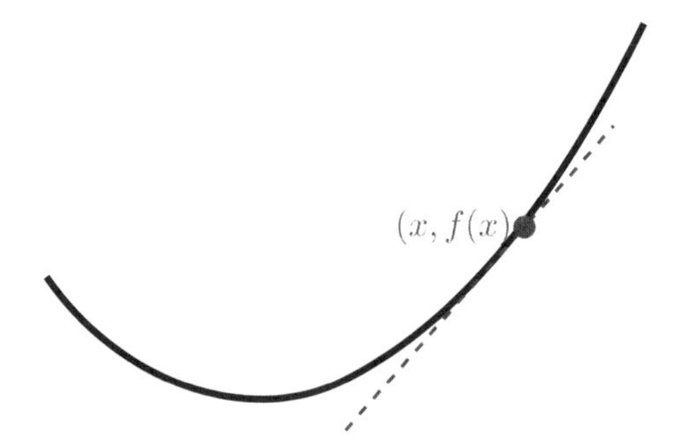

Figure 2.3 The Graph Above Tangent Property of a Convex Function

1. *f is convex if and only if $f''(x) \geq 0$ for all $x \in (a, b)$.*
2. *f is concave if and only if $f''(x) \leq 0$ for all $x \in (a, b)$.*
3. *f is strictly convex if and only if $f''(x) > 0$ for all $x \in (a, b)$.*
4. *f is strictly convex if and only if $f''(x) < 0$ for all $x \in (a, b)$.*

For a proof of this result, see [110, Theorem 4.5].

We now study two very specific functions that are very relevant to information theory.

Example 2.3 Consider the function $f(x) = \log x$, defined on $(0, \infty)$.[1] Since

$$f'(x) = 1/x,\ f''(x) = -1/x^2 < 0\ \forall x \in (0, \infty),$$

it follows that $\log x$ is strictly concave on $(0, \infty)$. As a result, if we substitute $x = 1$ in (2.10), we get

$$\log y < y - 1,\ \forall y > 0, y \neq 1. \tag{2.11}$$

Example 2.4 The function

$$h(p) = p \log(1/p) = -p \log p, p \in [0, 1] \tag{2.12}$$

plays a very central role in information theory. For $p \in (0, 1]$ the function is well-defined and can be differentiated as many times as one wishes. When $p = 0$ we can define $h(0) = 0$, since it is easy to verify using L'Hôpital's rule that $h(p) \to 0$ as $p \to 0^+$. Note that

$$h'(p) = -\log p - 1, h''(p) = -1/p < 0\ \forall p > 0.$$

Thus $h(\cdot)$ is strictly concave on $[0, 1]$ (and indeed on $(0, \infty)$, though values of p larger than one have no relevance to information theory). In particular, $h(p) > 0\ \forall p \in (0, 1)$, and $h(0) = h(1) = 0$. Figure 2.4 shows a plot of the function $h(\cdot)$.

Now we present a very useful result, known as **Jensen's inequality**.

---

[1] Here and elsewhere log denotes the natural logarithm, while lg denotes the binary logarithm, that is, logarithm to the base 2.

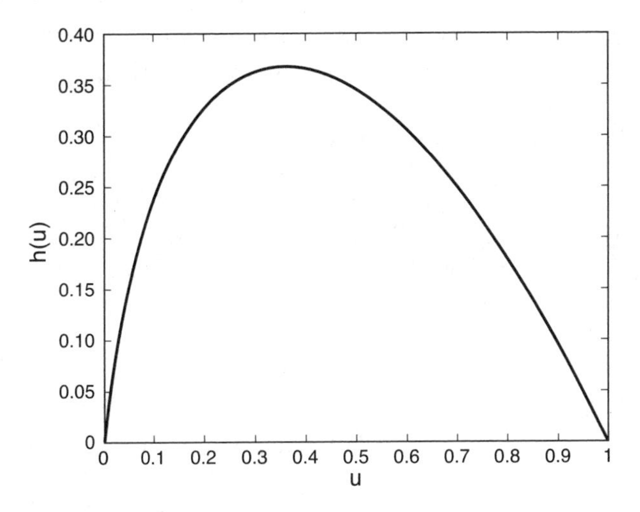

Figure 2.4 Plot of u vs. $h(u)$

Theorem 2.6 *Suppose X is a random variable assuming one of n real values $x_1, \ldots, x_n$ belonging an open interval (a, b), with probabilities $\mu_1, \ldots, \mu_n$. Suppose $f : (a, b) \to \mathbb{R}$ is convex. Then*

$$f(E[X, P_\mu]) \leq E[f(X), P_\mu] \tag{2.13}$$

Proof. The proof is a ready consequence of the definition of convexity. By definition, we have

$$E[X, P_\mu] = \sum_{i=1}^{n} \mu_i x_i,$$

$$f(E[X, P_\mu]) = f\left(\sum_{i=1}^{n} \mu_i x_i\right) \leq \sum_{i=1}^{n} \mu_i f(x_i) = E[f(X), P_\mu],$$

which is the desired conclusion. □

The above theorem statement and proof don't really do justice to Jensen's inequality, which actually holds in a far more abstract setting than the above.

Problem 2.1 Given a function $f : S \to \mathbb{R}$ where S is a convex subset of some Euclidean space $\mathbb{R}^d$, its **epigraph** is denoted by epi(f) and is defined by

$$\text{epi}(f) = \{(\mathbf{x}, y) : \mathbf{x} \in S, y \in \mathbb{R} \text{ and } y \geq f(\mathbf{x})\}.$$

Show that f is a convex *function* if and only if epi(f) is a convex *set* in $\mathbb{R}^{d+1}$. State and prove the analog of this statement for concave functions.

Problem 2.2 Suppose S is a convex set and $f : S \to \mathbb{R}$. Show that f is a convex function if and only if $-f$ is a concave function.

Problem 2.3 Suppose S is a convex subset of some Euclidean space $\mathbb{R}^d$ and $f : S \to \mathbb{R}$ is convex, while $g : \mathbb{R} \to \mathbb{R}$ is convex and nondecreasing. In other words, $\alpha \leq \beta$ implies that $g(\alpha) \leq g(\beta)$. Show that the function $\mathbf{x} \mapsto g(f(\mathbf{x}))$ is convex. Here the symbol $\mapsto$ is read as "mapsto," and $\mathbf{x} \mapsto g(f(\mathbf{x}))$ means the function that associates $g(f(\mathbf{x}))$ with each $\mathbf{x} \in S$. State and prove analogous statements for strictly convex, concave, and strictly concave functions.

Problem 2.4 Consider the following three probability distributions:

$$\boldsymbol{\phi} = [0.2 \;\; 0.3 \;\; 0.5], \boldsymbol{\psi} = [0.4 \;\; 0.5 \;\; 0.1], \boldsymbol{\theta} = [0.3 \;\; 0.4 \;\; 0.3].$$

Can you express any one of these three distributions as a convex combination of the other two? Justify your answer.

2.2 ENTROPY

The notion of entropy is very central to information theory. Its introduction by Claude Shannon in [117] led to the creation of an entirely new discipline that is still vibrant today, more than 60 years later. In this section we give a very brief introduction to this concept and derive several of its properties.

2.2.1 Definition of Entropy

Definition 2.7 *Suppose $\boldsymbol{\mu} \in \mathbb{S}_n$ is an n-dimensional probability distribution. The* **entropy** *of the distribution is denoted by $H(\boldsymbol{\mu})$ and is defined by*

$$H(\boldsymbol{\mu}) := \sum_{i=1}^{n} \mu_i \log(1/\mu_i) = -\sum_{i=1}^{n} \mu_i \log \mu_i. \tag{2.14}$$

Note that we can also write

$$H(\boldsymbol{\mu}) = \sum_{i=1}^{n} h(\mu_i),$$

where $h(\cdot)$ is the function defined in Example 2.4.

In (2.14), we are using the natural logarithm. It is of course possible to replace log by lg and take the logarithm to the base two. Some authors try to specify the base used to take the logarithm by saying that the entropy is measured in "bits" if lg is used, and "nats" if log is used. Clearly the two numbers will always differ by the constant factor log 2, so it does not really matter which base is used for the logarithm, so long as one is consistent.

If X is a random variable assuming values in a set $\mathbb{A} = \{a_1, \ldots, a_n\}$, and $\boldsymbol{\mu}$ is the probability distribution of X, then we can also refer to $H(\boldsymbol{\mu})$ as $H(X)$, the entropy of the random variable X, rather than as the entropy of the *probability distribution* of the random variable X. It is helpful to be able to switch back and forth between the two usages, and the two notations $H(\boldsymbol{\mu})$ and $H(X)$. However, the reader is cautioned that, strictly speaking, entropy is a property of probability distributions, and not of the random variables associated with those probability distributions. For example, consider two random variables: the outcome of tossing a fair coin, and drawing a ball from a box containing two identical balls of different colors. The underlying random variables are in some sense "different," but the entropies are the same. Also note that the entropy of a random variable X assuming values in a finite set $\mathbb{A}$ depends only on its probability distribution, but does not in any way depend on the elements of the set $\mathbb{A}$. Thus, if X assumes the values 0 and 1 with probabilities 0.3 and 0.7 respectively, while Y assumes the values -1 and $+1$ with probabilities 0.3 and 0.7 respectively, then X and Y have the same entropy.

2.2.2 Properties of the Entropy Function

Theorem 2.8 *We have the following properties of the entropy function.*

1. $H(\boldsymbol{\mu}) \geq 0$ *for all probability distributions* $\boldsymbol{\mu} \in \mathbb{S}_n$.
2. $H(\boldsymbol{\mu}) = 0$ *if and only if* $\boldsymbol{\mu}$ *is a degenerate distribution, i.e., there is an index* i *such that* $\mu_i = 1$ *and* $\mu_j = 0$ *for all* $j \neq i$.
3. $H(\boldsymbol{\mu}) \leq \log n \; \forall \boldsymbol{\mu} \in \mathbb{S}_n$, *with equality if and only if* $\boldsymbol{\mu}$ *is the uniform distribution, that is,* $\mu_i = 1/n$ *for all indices* i.

Proof. By definition we have

$$H(\boldsymbol{\mu}) = \sum_{i=1}^{n} h(\mu_i),$$

where the function $h(\cdot)$ is defined in (2.12). Since $h(\mu_i) \geq 0$ for all $\mu_i \in [0,1]$, it follows that $H(\boldsymbol{\mu}) \geq 0$. This proves the first statement. Moreover, $H(\boldsymbol{\mu}) = 0$ if and only if $h(\mu_i) = 0$ for all indices i, that is, if and only if $\mu_i = 0$ or 1 for every index i. But since the μ_i's must add up to one, we see that $H(\boldsymbol{\mu}) = 0$ if and only if all components of $\boldsymbol{\mu}$ are zero except for one component which must equal one. This proves the second statement. To prove the third statement, suppose $\boldsymbol{\mu} = (\mu_1, \ldots, \mu_n) \in \mathcal{S}_n$. Then, since $\mu_i \in [0,1] \; \forall i$ and $\sum_{i=1}^{n} \mu_i = 1$, it follows that

$$\frac{1}{n} = \sum_{i=1}^{n} \frac{\mu_i}{n}$$

is a convex combination of the numbers μ_1 through μ_n (with equal weights $1/n$). Now recall that the function $h(\cdot)$ in Example 2.4 is strictly concave.

As a result it follows that

$$h(1/n) = h\left(\frac{1}{n}\sum_{i=1}^{n}\mu_i\right) \geq \frac{1}{n}\sum_{i=1}^{n} h(\mu_i),$$

with equality if and only if all μ_i are equal (and thus equal $1/n$). Since $h(1/n) = (1/n)\log n$, the above equation can be rewritten as

$$H(\boldsymbol{\mu}) = \sum_{i=1}^{n} h(\mu_i) \leq nh(1/n) = \log n,$$

with equality holding if and only if all μ_i are equal to $1/n$. □

The motivation for the next theorem is that we can view the entropy function $H(\cdot)$ as a function mapping the convex set $\mathbb{S}_n$ of n-dimensional probability distributions into the set $\mathbb{R}_+$ of nonnegative numbers. The theorem states that the entropy function is strictly concave.

Theorem 2.9 *For each integer n, the function $H(\cdot) : \mathbb{S}_n \to \mathbb{R}_+$ is strictly concave. Thus if $\boldsymbol{\mu}, \boldsymbol{\nu} \in \mathbb{S}_n$ are probability distributions and $\lambda \in (0,1)$, then*

$$H[\lambda\boldsymbol{\mu} + (1-\lambda)\boldsymbol{\nu}] \geq \lambda H(\boldsymbol{\mu}) + (1-\lambda)H(\boldsymbol{\nu}). \tag{2.15}$$

Moreover, equality holds if and only if $\boldsymbol{\mu} = \boldsymbol{\nu}$.

Proof. Recall that the function $h(\cdot)$ defined in (2.12) is strictly concave. From the definition of $H(\cdot)$ it follows that

$$\begin{aligned} H[\lambda\boldsymbol{\mu} + (1-\lambda)\boldsymbol{\nu}] &= \sum_{i=1}^{n} h(\lambda\mu_i + (1-\lambda)\nu_i) \\ &\geq \sum_{i=1}^{n} \lambda h(\mu_i) + (1-\lambda)h(\nu_i) \\ &= \lambda\sum_{i=1}^{n} h(\mu_i) + (1-\lambda)\sum_{i=1}^{n} h(\nu_i) \\ &= \lambda H(\boldsymbol{\mu}) + (1-\lambda)H(\boldsymbol{\nu}). \end{aligned}$$

Moreover, if $\mu_i \neq \nu_i$ for even a single index i, then the corresponding inequality in the second step becomes strict. □

2.2.3 Conditional Entropy

Suppose X and Y are random variables assuming values in finite sets $\mathbb{A} = \{a_1, \ldots, a_n\}$ and $\mathbb{B} = \{b_1, \ldots, b_m\}$ respectively. In Section 1.2 we studied the notions of the joint distribution of X and Y, as well as the conditional distribution of X given an observed value of Y. Specifically, let $\boldsymbol{\phi}$ denote the joint distribution of (X,Y). Then $\boldsymbol{\phi} \in \mathbb{S}_{nm}$, but we can represent $\boldsymbol{\phi}$ as a doubly indexed set of numbers, as follows:

$$\Phi = \begin{bmatrix} \phi_{11} & \ldots & \phi_{1m} \\ \vdots & \vdots & \vdots \\ \phi_{n1} & \ldots & \phi_{nm} \end{bmatrix}.$$

We also defined the "marginal distributions" $\phi_X \in \mathbb{S}_n$ of X and $\phi_Y \in \mathbb{S}_m$ of Y as follows:

$$(\phi_X)_i = \sum_{j=1}^{m} \phi_{ij}, i = 1, \ldots, n, \text{ and } (\phi_Y)_j = \sum_{i=1}^{n} \phi_{ij}, j = 1, \ldots, m.$$

Given an observation $X = a_j$, we defined the conditional probability distribution $\phi_{\{Y|X=a_i\}}$ as[2]

$$\phi_{\{Y|X=a_i\}} = [\phi_{\{b_1|a_i\}} \ldots \phi_{\{b_m|a_i\}}] \in \mathbb{S}_m,$$

where

$$\phi_{\{b_j|a_i\}} := \frac{\phi_{ij}}{(\phi_X)_i} = \frac{\phi_{ij}}{\sum_{j'=1}^{m} \phi_{ij'}}.$$

Then $\phi_{\{Y|X=a_i\}} \in \mathbb{S}_m \ \forall i$. All of the above is just a reprise of previously discussed material for the convenience of the reader.

Definition 2.10 *Suppose X and Y are random variables assuming values in finite sets $\mathbb{A} = \{a_1, \ldots, a_n\}$ and $\mathbb{B} = \{b_1, \ldots, b_m\}$ respectively. Let $\phi \in \mathbb{S}_{nm}$ denote their joint probability distribution, and let $\phi_X \in \mathbb{S}_n$, $\phi_Y \in \mathbb{S}_m$ denote the marginal distributions of X and Y respectively. Then the quantity*

$$H(Y|X) := \sum_{i=1}^{n} (\phi_X)_i H(\phi_{\{Y|X=a_i\}})$$

is called the **conditional entropy** *of the random variable Y with respect to the random variable X.*

Theorem 2.11 *Suppose X and Y are random variables assuming values in finite sets $\mathbb{A} = \{a_1, \ldots, a_n\}$ and $\mathbb{B} = \{b_1, \ldots, b_m\}$ respectively. Then*

$$H(Y) \geq H(Y|X), \tag{2.16}$$

or equivalently

$$H(\phi_Y) \geq \sum_{i=1}^{n} (\phi_X)_i H(\phi_{\{Y|X=a_i\}}). \tag{2.17}$$

Proof. It is obvious from the various formulas that

$$\phi_Y = \sum_{i=1}^{n} (\phi_X)_i \cdot \phi_{\{Y|X=a_i\}}. \tag{2.18}$$

In other words, the marginal probability distribution ϕ_Y of the random variable Y is a *convex combination* of the various conditional probability distributions $\phi_{\{Y|X=a_i\}}$, weighted by the probabilities $(\phi_X)_i$. Thus the desired inequality (2.17) follows from Theorem 2.9 and repeated application of (2.15). □

---

[2] Just for variety's sake we are using the conditional distribution of Y given X, whereas in Chapter 1 we used the conditional distribution of X given Y.

The quantity $H(\phi_{\{Y|X=x_i\}})$ is the entropy of the conditional probability of Y, after we have observed that the value of X is a_i. One way of interpreting Theorem 2.11 is that, *on average*, the conditional entropy of Y following an observation of another variable X is no larger than the unconditional entropy $H(Y)$. But this statement applies *only* "on average." It is quite possible that *for some specific observations*, this entropy is in fact higher than the "unconditional entropy" $H(\phi_Y)$; see Example 2.5. However, Theorem 2.11 states that *on average*, one cannot be worse off by making an observation of X than by not observing X.

Example 2.5 Suppose $|\mathbb{A}| = |\mathbb{B}| = 2$, and that the joint probability distribution ϕ equals

$$\phi = \begin{bmatrix} 0.1 & 0.3 \\ 0.2 & 0.4 \end{bmatrix},$$

where the rows correspond to the values of X and the columns to values of Y. Thus

$$\phi_X = [0.4 \;\; 0.6], \; \phi_Y = [0.3 \;\; 0.7].$$

So, if we know nothing about X, we get

$$H(Y) = H(\phi_Y) = -0.3 \log 0.3 - 0.7 \log 0.7 \approx 0.6109.$$

Now suppose we measure X and it turns out that $X = a_1$. Then

$$\phi_{\{Y|X=a_1\}} = [0.1/0.4 \;\; 0.3/0.4] = [1/4 \;\; 3/4].$$

In this case we have

$$H(\phi_{\{Y|X=a_1\}}) = (1/4)\log 4 + (3/4)\log(4/3) \approx 0.5623,$$

which is *lower* than the unconditional entropy. On the other hand,

$$\phi_{\{Y|X=a_2\}} = [1/3 \;\; 2/3],$$

and

$$H(\phi_{\{Y|X=a_2\}}) \approx 0.6365,$$

which is *higher* than the unconditional entropy. Finally

$$H(Y|X) = (\phi_X)_1 H(\phi_{\{Y|X=a_1\}}) + (\phi_X)_2 H(\phi_{\{Y|X=a_2\}}) \approx 0.6068,$$

which is *lower than* $H(Y) = H(\phi_Y)$, as stated by Theorem 2.11.

Theorem 2.12 *Suppose X and Y are random variables assuming values in finite sets $\mathbb{A} = \{a_1, \ldots, a_n\}$ and $\mathbb{B} = \{b_1, \ldots, b_m\}$ respectively. Let $\phi \in \mathbb{S}_{nm}$ denote their joint probability distribution, and let $\phi_X \in \mathbb{S}_n$, $\phi_Y \in \mathbb{S}_m$ denote the marginal distributions. Then*

$$H((X,Y)) = H(X) + H(Y|X). \tag{2.19}$$

In the above theorem statement, we have used the convention of associating entropy with a random variable rather than with its probability distribution. Thus (2.19) says that the entropy of the *joint* random variable (X, Y) is the sum of the entropy of X by itself, and the conditional entropy of Y given X.

Proof. The desired equality (2.19) is equivalent to

$$H(Y|X) = H((X,Y)) - H(X). \tag{2.20}$$

To establish this relation, let us compute the right side, observing that

$$(\boldsymbol{\phi}_X)_i = \sum_{j=1}^{n} \phi_{ij} \ \forall i, \phi_{ij} = (\boldsymbol{\phi}_X)_i \cdot \phi_{b_j|a_i} \ \forall i, j.$$

Thus

$$\begin{aligned}
H((X,Y)) - H(X) &= -\sum_{i=1}^{n}\sum_{j=1}^{m} \phi_{ij} \log \phi_{ij} + \sum_{i=1}^{n} (\boldsymbol{\phi}_X)_i \log(\boldsymbol{\phi}_X)_i \\
&= -\sum_{i=1}^{n}\sum_{j=1}^{m} \phi_{ij} \log \phi_{ij} + \sum_{i=1}^{n}\sum_{j=1}^{m} \phi_{ij} \log(\boldsymbol{\phi}_X)_i \\
&= -\sum_{i=1}^{n}\sum_{j=1}^{m} \phi_{ij} \log \left(\frac{\phi_{ij}}{(\boldsymbol{\phi}_X)_i} \right) \\
&= -\sum_{i=1}^{n}\sum_{j=1}^{m} (\boldsymbol{\phi}_X)_i \cdot \phi_{b_j|a_i} \log \phi_{b_j|a_i} \\
&= -\sum_{i=1}^{n} (\boldsymbol{\phi}_X)_i \sum_{j=1}^{m} \phi_{b_j|a_i} \log \phi_{b_j|a_i} \\
&= \sum_{i=1}^{n} (\boldsymbol{\phi}_X)_i H(\boldsymbol{\phi}_{\{Y|X=a_i\}}) = H(Y|X).
\end{aligned}$$

This establishes (2.20) and completes the proof. □

Corollary 2.13 *Let all notation be as in Theorem 2.12. Then*

$$H((X,Y)) = H(X) + H(Y)$$

if and only if X and Y are independent.

Proof. The statement $H((X,Y)) = H(X) + H(Y)$ is equivalent to $H(Y|X) = H(Y)$. Now we know from (2.18) that $\boldsymbol{\phi}_Y$ is a convex combination of the n conditional probability distributions $\boldsymbol{\phi}_{\{Y|X=a_i\}}$. Since the entropy function $H(\cdot)$ is strictly concave, we have that

$$H(Y) = H(\boldsymbol{\phi}_Y) \geq \sum_{i=1}^{n} (\boldsymbol{\phi}_X)_i H(\boldsymbol{\phi}_{\{Y|X=a_i\}}) = H(Y|X),$$

with equality if and only if

$$\boldsymbol{\phi}_{\{Y|X=a_i\}} = \boldsymbol{\phi}_Y \ \forall i,$$

which is the same as saying that X and Y are independent. □

Corollary 2.14 *Let X, Y be as in Theorem 2.12. Then*

$$H((X,Y)) = H(Y) + H(X|Y).$$

Proof. From the definition it is obvious that $H((X,Y)) = H((Y,X))$. The conclusion now follows from Corollary 2.13 by interchanging X and Y. □

Definition 2.15 *Suppose X and Y are random variables assuming values in finite sets $\mathbb{A} = \{a_1, \ldots, a_n\}$ and $\mathbb{B} = \{b_1, \ldots, b_m\}$ respectively. Let $\phi \in \mathbb{S}_{nm}$ denote their joint probability distribution, and let $\phi_X \in \mathbb{S}_n$, $\phi_Y \in \mathbb{S}_m$ denote the marginal distributions. Then the quantity*

$$I(X,Y) := H(X) + H(Y) - H((X,Y)) = H(\phi_X) + H(\phi_Y) - H(\phi)$$

is called the **mutual information** *between X and Y.*

From the above definition, it is clear that the mutual information is "symmetric," that is, $I(X,Y) = I(Y,X)$. However, as the next lemma shows, there are other, equivalent, definitions that do not appear to be symmetric.

Lemma 2.16 *Suppose X and Y are random variables assuming values in finite sets $\mathbb{A} = \{a_1, \ldots, a_n\}$ and $\mathbb{B} = \{b_1, \ldots, b_m\}$ respectively. Then*

$$I(X,Y) = H(Y) - H(Y|X) = H(X) - H(X|Y).$$

Thus $I(X,Y) \geq 0$, and $I(X,Y) = 0$ if and only if X, Y are independent.

Proof. From Theorem 2.12, we have that

$$H(Y|X) = H((X,Y)) - H(X), H(X|Y) = H((X,Y)) - H(Y).$$

So

$$\begin{aligned} I(X,Y) &= H(X) + H(Y) - H((X,Y)) \\ &= H(X) + H(Y) - [H(Y|X) + H(X)] \\ &= H(Y) - H(Y|X), \end{aligned}$$

and similarly for the other equation. The second statement is obvious. □

2.2.4 Uniqueness of the Entropy Function

Up to now the concept of entropy has been defined, and several of its useful properties have been derived. In this subsection we present a result due to Shannon [117] that the entropy function as defined in Definition 2.7 is *the only function* (within a scale factor) that has these desirable properties.

Theorem 2.17 *Suppose $H_n : \mathbb{S}_n \to \mathbb{R}_+, n \geq 1$ is a family of maps with the following properties:*[3]

1. *H is continuous for each fixed integer n.*

---

[3] For reasons of aesthetics the argument n is suppressed and we just write $H : \mathbb{S}_n \to \mathbb{R}_+$, the value of the integer n being obvious from the context.

2. *Let* $\mathbf{u}_n \in \mathbb{S}_n$ *denote the uniform distribution. Then* $H(\mathbf{u}_n)$ *is a nondecreasing function of* n.

3. *Suppose* $\boldsymbol{\phi} \in \mathbb{S}_n$*, and let* $I_1, \ldots, I_m, m < n$ *denote a partition of* $\mathbb{N} = \{1, \ldots, n\}$ *into disjoint nonempty sets. Define* $\boldsymbol{\psi} \in \mathbb{S}_m$*,* $\boldsymbol{\phi}_{|j} \in \mathbb{S}_{|I_j|}$ *by*

$$\psi_j = \sum_{i \in I_j} \phi_i, j = 1, \ldots, m, \tag{2.21}$$

$$(\phi_{|j})_i = \frac{\phi_i}{\psi_j}, \ \forall i \in I_j. \tag{2.22}$$

Then

$$H(\boldsymbol{\phi}) = H(\boldsymbol{\psi}) + \sum_{j=1}^{m} \psi_j H(\boldsymbol{\phi}_{|j}). \tag{2.23}$$

Under these assumptions, there is a constant λ*, independent of* n*, such that for every* n *and every* $\boldsymbol{\phi} \in \mathbb{S}_n$*, we have*

$$H(\boldsymbol{\phi}) = -\lambda \sum_{i=1} \phi_i \log \phi_i. \tag{2.24}$$

Proof. For convenience let $L(n)$ denote $H(\mathbf{u}_n)$. Suppose $n = lm$ where l and m are integers greater or equal to two. Partition $\mathbb{N} = \{1, \ldots, n\}$ into m disjoint sets $I_1, \ldots, I_m$ consisting of l elements each. Let $\boldsymbol{\phi} = \mathbf{u}_n$. Then it is clear that $\boldsymbol{\psi} = \mathbf{u}_m$ and $\boldsymbol{\phi}_{|j} = \mathbf{u}_l$ for each j. So (2.23) implies that

$$L(n) = L(m) + L(l).$$

In particular

$$L(m^2) = 2L(m), \text{ and}$$

$$L(m^{s+1}) = L(m^s) + L(m) = (s+1)L(m), \tag{2.25}$$

where the last step follows by induction.

Next, suppose l, m, n are arbitrary integers ≥ 2, and determine a unique integer s according to

$$m^s \leq l^n < m^{s+1}.$$

This implies that

$$s \log m \leq n \log l \leq (s+1) \log m, \text{ or}$$

$$\frac{s}{n} \leq \frac{\log l}{\log m} \leq \frac{s}{n} + \frac{1}{n}. \tag{2.26}$$

Now since $L(\cdot)$ is an increasing function of its integer argument, it follows that

$$L(m^s) \leq L(l^n) \leq L(m^{s+1}).$$

Applying (2.25) now shows that

$$sL(m) \leq nL(l) \leq (s+1)L(m),$$

or

$$\frac{s}{n} \leq \frac{L(l)}{L(m)} \leq \frac{s}{n} + \frac{1}{n}. \tag{2.27}$$

Combining (2.26) and (2.27) shows that

$$\left| \frac{\log l}{\log m} - \frac{L(l)}{L(m)} \right| \leq \frac{1}{n}.$$

Since the left side of the inequality does not contain n, we can let $n \to \infty$, which shows that

$$\frac{\log l}{\log m} = \frac{L(l)}{L(m)},$$

or

$$L(m) = \frac{L(l)}{l} \log m, \ \forall l, m.$$

However, since l, m are arbitrary, we can define

$$\lambda := \frac{L(2)}{2},$$

and conclude that

$$L(m) = \lambda \log m, \ \forall m. \tag{2.28}$$

Next, suppose $\boldsymbol{\psi} \in \mathbb{S}_m$ where each ψ_j is a rational number, say

$$\psi_j = \frac{n_j}{n}, j = 1, \ldots, m, \text{ and } n = \sum_{j=1}^{m} n_j.$$

Now define $\boldsymbol{\phi} = \mathbf{u}_n$, and sets $I_1, \ldots, I_m$ as disjoint subsets of $\mathbb{N} = \{1, \ldots, n\}$ where I_j contains n_j elements, for each j. Define $\boldsymbol{\psi}$ and $\boldsymbol{\phi}_{|j}$ as in (2.21) and (2.22) respectively. Then indeed $\psi_j = n_j/n$, and $\boldsymbol{\phi}_{|j} = \mathbf{u}_{n_j}$. It is already known from (2.28) that

$$H(\boldsymbol{\phi}) = H(\mathbf{u}_n) = \lambda \log n, H(\boldsymbol{\phi}_{|j}) = H(\mathbf{u}_{n_j}) = \lambda \log n_j \ \forall j.$$

Substituting these into (2.23) leads to

$$\lambda \log n = H(\boldsymbol{\psi}) + \sum_{j=1}^{m} \lambda \psi_j \log n_j.$$

Rearranging gives

$$\begin{aligned}
H(\boldsymbol{\psi}) &= \lambda \log n - \lambda \sum_{j=1}^{m} \psi_j \log n_j \\
&= -\lambda \sum_{j=1}^{m} \psi_j (\log n_j - \log n) \\
&= -\lambda \sum_{j=1}^{m} \psi_j \log(n_j/n) \\
&= -\lambda \sum_{j=1}^{m} \psi_j \log \psi_j,
\end{aligned}$$

where the second step follows from the fact that

$$\sum_{j=1}^{m} \psi_j = 1.$$

This proves that the function H has the desired form (2.24) whenever all components of $\boldsymbol{\psi}$ are rational. The general case now follows from the fact that rational numbers are dense in the real numbers, and H is continuous. □

For an alternate derivation of the uniqueness of the entropy function, see [79].

2.3 RELATIVE ENTROPY AND THE KULLBACK-LEIBLER DIVERGENCE

In this section we introduce a very important notion known as relative entropy. It is also called the Kullback-Leibler divergence after the two persons who invented this notion.

Definition 2.18 *If $\boldsymbol{\mu}, \boldsymbol{\nu} \in \mathbb{S}_n$ are probability vectors with n components each, define $\boldsymbol{\mu} \ll \boldsymbol{\nu}$ or $\boldsymbol{\nu} \gg \boldsymbol{\mu}$ (read as "$\boldsymbol{\mu}$ is dominated by $\boldsymbol{\nu}$," or "$\boldsymbol{\nu}$ dominates $\boldsymbol{\mu}$") if $\nu_i = 0 \Rightarrow \mu_i = 0$. Then*

$$D(\boldsymbol{\mu}\|\boldsymbol{\nu}) := \sum_{i=1}^{n} \mu_i \log(\mu_i/\nu_i) \tag{2.29}$$

is called the **relative entropy** *of $\boldsymbol{\mu}$ with respect to $\boldsymbol{\nu}$, or the* **Kullback-Leibler (K-L) divergence** *of $\boldsymbol{\mu}$ with respect to $\boldsymbol{\nu}$.*

Note that if $\nu_i = 0$ for some index i, the assumption that $\boldsymbol{\mu} \ll \boldsymbol{\nu}$ guarantees that μ_i also equals zero. So in defining the K-L divergence, we take $0 \log(0/0)$ to equal 0. If $\boldsymbol{\mu} \not\ll \boldsymbol{\nu}$, then there exists an index i such that $\nu_i = 0$ but $\mu_i \neq 0$. In this case we can take $D(\boldsymbol{\mu}\|\boldsymbol{\nu}) = \infty$.

Actually, the quantity defined by Kullback and Leibler in their original paper [87] would correspond to $D(\boldsymbol{\mu}\|\boldsymbol{\nu}) + D(\boldsymbol{\nu}\|\boldsymbol{\mu})$, which would result in a quantity that is symmetric in $\boldsymbol{\mu}$ and $\boldsymbol{\nu}$. However, subsequent researchers have recognized the advantages of defining the divergence as in (2.29).

From the definition (2.29), it is not even clear that the K-L divergence is nonnegative. Since both $\boldsymbol{\mu}$ and $\boldsymbol{\nu}$ are probability distributions, it is clear that the components of both vectors add up to one. Hence, if $\boldsymbol{\mu} \neq \boldsymbol{\nu}$, then for some i the ratio μ_i/ν_i exceeds one and for other i it is less than one. So $\log(\mu_i/\nu_i)$ is positive for some i and negative for other i. Why then should the divergence be nonnegative? The question is answered next.

Theorem 2.19 *Suppose $\boldsymbol{\mu}, \boldsymbol{\nu} \in \mathbb{S}_n$ with $\boldsymbol{\mu} \ll \boldsymbol{\nu}$. Then $D(\boldsymbol{\mu}\|\boldsymbol{\nu}) \geq 0$, and $D(\boldsymbol{\mu}\|\boldsymbol{\nu}) > 0$ if $\boldsymbol{\mu} \neq \boldsymbol{\nu}$.*

Proof. Observe that log is a strictly concave function, whence $-\log$ is a strictly convex function. Therefore

$$\begin{aligned} D(\boldsymbol{\mu}\|\boldsymbol{\nu}) &= \sum_{i=1}^{n} \mu_i \log \frac{\mu_i}{\nu_i} \\ &= -\sum_{i=1}^{n} \mu_i \log \frac{\nu_i}{\mu_i} \\ &\geq -\log \left(\sum_{i=1}^{n} \mu_i \frac{\nu_i}{\mu_i} \right) \\ &= -\log \left(\sum_{i=1}^{n} \nu_i \right) \\ &= -\log(1) = 0. \end{aligned}$$

To prove the second part of the conclusion, note that if $\boldsymbol{\mu} \neq \boldsymbol{\nu}$, then there exist at least two indices i, j such that $\mu_i \neq \nu_i$, $\mu_j \neq \nu_j$. In this case, the quantity

$$\sum_{i=1}^{n} \mu_i \frac{\nu_i}{\mu_i}$$

is a *nontrivial convex combination* of the number $\nu_1/\mu_1, \ldots, \nu_n/\mu_n$. Thus the inequality in the above becomes strict, because $\log(\cdot)$ is a strictly concave function. □

A ready corollary of the above argument is the following.

Corollary 2.20 *Suppose $\boldsymbol{\mu}, \boldsymbol{\nu} \in \mathbb{S}_n$ are probability distributions with n components, and we define the "loss" function*

$$J(\boldsymbol{\mu}, \boldsymbol{\nu}) := \sum_{i=1}^{n} \mu_i \log(1/\nu_i). \tag{2.30}$$

Then, viewed as a function of $\boldsymbol{\nu}$, the loss function assumes its minimum value when $\boldsymbol{\nu} = \boldsymbol{\mu}$, and the minimum value is $H(\boldsymbol{\mu})$. In other words,

$$J(\boldsymbol{\mu}, \boldsymbol{\nu}) \geq J(\boldsymbol{\mu}, \boldsymbol{\mu}) = H(\boldsymbol{\mu}) \ \forall \boldsymbol{\nu} \in \mathbb{S}_n. \tag{2.31}$$

Moreover, equality holds if and only if $\boldsymbol{\nu} = \boldsymbol{\mu}$.

Proof. Routine algebra shows that

$$D(\boldsymbol{\mu}\|\boldsymbol{\nu}) = J(\boldsymbol{\mu}, \boldsymbol{\nu}) - J(\boldsymbol{\mu}, \boldsymbol{\mu}).$$

Since we already know that $D(\boldsymbol{\mu}\|\boldsymbol{\nu}) \geq 0$ and that $D(\boldsymbol{\mu}\|\boldsymbol{\nu}) = 0$ if and only if $\boldsymbol{\mu} = \boldsymbol{\nu}$, it follows that $J(\boldsymbol{\mu}, \boldsymbol{\nu}) \geq J(\boldsymbol{\mu}, \boldsymbol{\mu})$ with equality if and only if $\boldsymbol{\nu} = \boldsymbol{\mu}$. It is routine to verify that $J(\boldsymbol{\mu}, \boldsymbol{\mu}) = H(\boldsymbol{\mu})$. □

The proof of Theorem 2.19 makes it clear that if we define the "divergence" between two probability distributions $\boldsymbol{\mu}, \boldsymbol{\nu} \in \mathbb{S}_n$ by replacing the logarithm function by *any* strictly convex function $\eta(\cdot)$ such that $\eta(1) = 0$,[4] the conclusions of Theorem 2.19 would still hold. In other words, if $\eta(\cdot)$ is strictly convex and satisfies $\eta(1) = 0$, and we define

$$D_\eta(\boldsymbol{\mu}\|\boldsymbol{\nu}) := -\sum_{i=1}^{n} \mu_i \eta(\nu_i/\mu_i),$$

then we would have $D_\eta(\boldsymbol{\mu}\|\boldsymbol{\nu}) \geq 0 \; \forall \boldsymbol{\mu}, \boldsymbol{\nu} \in \mathbb{S}_n$, and $D_\eta(\boldsymbol{\mu}\|\boldsymbol{\nu}) = 0$ if and only if $\boldsymbol{\mu} = \boldsymbol{\nu}$. The paper [90] contains a survey of various "divergences" that can be obtained for different choices of the function η, and their properties. However, Corollary 2.20 depends crucially on the fact that $\log(\nu_i/\mu_i) = \log \nu_i - \log \mu_i$ and thus does not hold for more general "divergences."

The K-L divergence has a very straightforward interpretation in terms of the likelihood of misclassifying an observation. Suppose X is a random variable taking values in the set $\mathbb{A} = \{a_1, \ldots, a_n\}$. Suppose we are told that the random variable X has the distribution $\boldsymbol{\mu}$ or $\boldsymbol{\nu}$, and our challenge is to decide between the two competing hypotheses (namely "X is distributed according to $\boldsymbol{\mu}$" and "X is distributed according to $\boldsymbol{\nu}$"). For this purpose, we make a series of independent observations of X, resulting in a "sample path" $u_1^l := (u_1, u_2, \ldots, u_l)$ of length l, where each u_k belongs to $\mathbb{A}$.

In order to discriminate between the two competing hypotheses, we compute the **likelihood** of the observed sample path under each of the two hypotheses. Now, given the observation u_1^l of length l, let l_i denote the number of occurrences of a_i in the sample path. Then the likelihood of this sample path, in case the underlying distribution is $\boldsymbol{\mu}$, equals

$$L_{\boldsymbol{\mu}}(u_1^l) = \prod_{i=1}^{n} \mu_i^{l_i}.$$

Similarly, the likelihood of this sample path in case the underlying distribution is $\boldsymbol{\nu}$ equals

$$L_{\boldsymbol{\nu}}(u_1^l) = \prod_{i=1}^{n} \nu_i^{l_i}.$$

We choose the hypothesis that is more likely. Thus, we say that the underlying distribution is $\boldsymbol{\mu}$ if $L_{\boldsymbol{\mu}}(u_1^l) > L_{\boldsymbol{\nu}}(u_1^l)$, and that the underlying distribution is $\boldsymbol{\nu}$ if $L_{\boldsymbol{\mu}}(u_1^l) < L_{\boldsymbol{\nu}}(u_1^l)$. For the time being, let us not worry about a "tie," because the probability of a tie approaches zero as $l \to \infty$. Now, instead of comparing $L_{\boldsymbol{\mu}}(u_1^l)$ and $L_{\boldsymbol{\nu}}(u_1^l)$, it is more customary to compute the **log likelihood ratio** $\log(L_{\boldsymbol{\mu}}(u_1^l)/L_{\boldsymbol{\nu}}(u_1^l))$. There are many reasons for this. From the standpoint of analysis, taking the logarithm turns products into sums, and thus the log likelihood function is often easier to analyze. From the standpoint of computation, the true likelihoods $L_{\boldsymbol{\mu}}(\mathbf{u})$ and $L_{\boldsymbol{\nu}}(\mathbf{u})$ often approach zero at a geometric rate as $l \to \infty$. So if l is at all a large

[4] This by itself is not a limitation, because if $\eta(\cdot)$ is a strictly convex function, so is $\eta(\cdot) - \eta(1)$.

number, then $L_{\boldsymbol{\mu}}(u_1^l)$ and $L_{\boldsymbol{\nu}}(u_1^l)$ rapidly fall below the machine zero, thus making computation difficult.

Now we can compute that

$$\log(L_{\boldsymbol{\mu}}(u_1^l)/L_{\boldsymbol{\nu}}(u_1^l)) = \sum_{i=1}^{n} l_i \log(\mu_i/\nu_i).$$

However, the indices l_i are themselves random numbers, as they correspond to the number of times that the outcome a_i appears in the sample path. Suppose now that "the truth" is $\boldsymbol{\mu}$. In other words, the underlying probability distribution really is $\boldsymbol{\mu}$. Then the expected value of l_i equals $\mu_i l$. Thus the expected value of the log likelihood ratio, when "the truth" is $\boldsymbol{\mu}$, equals

$$E[\log(L_{\boldsymbol{\mu}}(u_1^l)/L_{\boldsymbol{\nu}}(u_1^l)), P_\mu] = l \sum_{i=1}^{n} \mu_i \log(\mu_i/\nu_i) = lD(\boldsymbol{\mu}\|\boldsymbol{\nu}).$$

In other words, the expected value of the log likelihood ratio equals $D(\boldsymbol{\mu}\|\boldsymbol{\nu})$ multiplied by the length of the observation. For this reason, we can also interpret $D(\boldsymbol{\mu}\|\boldsymbol{\nu})$ as the "per sample" contribution to the log likelihood ratio. This computation shows us two things. First, if we are trying to choose between two competing hypotheses where one of them is the "truth" and the other is not, then "in the long run" we will choose the truth. Moreover, "the farther" the other competing hypothesis is from the "truth," the more quickly the log likelihood classifier will zero in on the "truth."

We shall return to this argument again in Chapter 5, when we discuss large deviation theory.

This interpretation can be extended to a more general situation. The interpretation of the K-L divergence given above assumes that one of the two competing hypotheses is in fact the truth. What if this assumption does not hold? Suppose the samples are generated by an underlying probability distribution $\boldsymbol{\mu}$, and the competing hypotheses are that the distribution is $\boldsymbol{\nu}$ or $\boldsymbol{\theta}$. Then

$$\begin{aligned} E[\log(L_{\boldsymbol{\nu}}(u_1^l)/L_{\boldsymbol{\theta}}(u_1^l)), P_\mu] &= l \sum_{i=1}^{n} \mu_i \log(\nu_i/\theta_i) \\ &= l \left[\sum_{i=1}^{n} \mu_i \log\left(\frac{\nu_i}{\mu_i}\right) - \sum_{i=1}^{n} \mu_i \log\left(\frac{\theta_i}{\mu_i}\right) \right] \\ &= l[D(\boldsymbol{\mu}\|\boldsymbol{\theta}) - D(\boldsymbol{\mu}\|\boldsymbol{\nu})]. \end{aligned}$$

Hence, in the long run, the maximum likelihood classifier would choose $\boldsymbol{\nu}$ if $D(\boldsymbol{\mu}\|\boldsymbol{\nu}) < D(\boldsymbol{\mu}\|\boldsymbol{\theta})$, and choose $\boldsymbol{\theta}$ if the inequality is reversed. In other words, the K-L divergence induces an ordering on the set of probability distributions. Given a fixed $\boldsymbol{\mu} \in \mathbb{S}_n$, all the other $\boldsymbol{\nu} \in \mathbb{S}_n$ can be ranked in terms of the divergence $D(\boldsymbol{\mu}\|\boldsymbol{\nu})$. If we try to choose between two competing hypotheses $\boldsymbol{\nu}$ and $\boldsymbol{\theta}$ when the "truth" is $\boldsymbol{\mu}$, in the long run we would choose the one that is "closer" to the "truth."

Unfortunately the K-L divergence is *not* symmetric in general. That is, in general

$$D(\boldsymbol{\mu}\|\boldsymbol{\nu}) \neq D(\boldsymbol{\nu}\|\boldsymbol{\mu}).$$

Moreover, in general

$$D(\boldsymbol{\mu}\|\boldsymbol{\nu}) + D(\boldsymbol{\nu}\|\boldsymbol{\theta}) \not\geq D(\boldsymbol{\mu}\|\boldsymbol{\theta}).$$

Hence we cannot use the K-L divergence to define any kind of a "metric" distance between probability measures. Nevertheless, the K-L divergence is very useful, because it quantifies the probability of misclassification using the log likelihood criterion. Again, it is shown in [90] that if the log function is replaced by some other strictly convex function, then it is possible for the resulting "divergence" to satisfy the triangle inequality. However, all the other nice features such as the interpretation in terms of the log likelihood function would be lost.

In Definition 2.15, we introduced the notion of "mutual information" between two random variables. It is now shown that the mutual information can also be defined in terms of the K-L divergence.

Theorem 2.21 *Suppose X, Y are random variables assuming values in finite sets $\mathbb{A}, \mathbb{B}$ respectively, and let $\boldsymbol{\phi}$ denote their joint distribution. Then the mutual information $I(X, Y)$ is given by*

$$I(X, Y) = D(\boldsymbol{\phi}\|\boldsymbol{\phi}_X \times \boldsymbol{\phi}_Y).$$

Proof. By definition we have that

$$I(X, Y) = H(X) + H(Y) - H((X, Y)).$$

Now substitute directly in terms of $\boldsymbol{\phi}$, and observe that

$$(\boldsymbol{\phi}_X)_i = \sum_{j=1}^{m} \phi_{ij}, (\boldsymbol{\phi}_Y)_j = \sum_{i=1}^{n} \phi_{ij}.$$

This gives

$$\begin{aligned} I(X, Y) &= -\sum_{i=1}^{n} \left[\sum_{j=1}^{m} \phi_{ij} \right] \log(\boldsymbol{\phi}_X)_i - \sum_{j=1}^{m} \left[\sum_{i=1}^{n} \phi_{ij} \right] \log(\boldsymbol{\phi}_Y)_j \\ &\quad + \sum_{i=1}^{n} \sum_{j=1}^{m} \phi_{ij} \log \phi_{ij} \\ &= \sum_{i=1}^{n} \sum_{j=1}^{m} \phi_{ij} \log \left[\frac{\phi_{ij}}{(\boldsymbol{\phi}_X)_i (\boldsymbol{\phi}_Y)_j} \right] \\ &= D(\boldsymbol{\phi}\|\boldsymbol{\phi}_X \times \boldsymbol{\phi}_Y). \end{aligned}$$

□

The above line of reasoning is extended in the next result, which is sometimes referred to as the "information inequality."

Theorem 2.22 *Suppose $\mathbb{A}, \mathbb{B}$ are finite sets, that $\boldsymbol{\mu}_{\mathbb{A}}, \boldsymbol{\mu}_{\mathbb{B}}$ are distributions on $\mathbb{A}, \mathbb{B}$ respectively, and that $\boldsymbol{\nu}$ is a distribution on the product set $\mathbb{A} \times \mathbb{B}$. Then*

$$D(\boldsymbol{\nu}\|\boldsymbol{\mu}_{\mathbb{A}} \times \boldsymbol{\mu}_{\mathbb{B}}) \geq D(\boldsymbol{\nu}_{\mathbb{A}}\|\boldsymbol{\mu}_{\mathbb{A}}) + D(\boldsymbol{\nu}_{\mathbb{B}}\|\boldsymbol{\mu}_{\mathbb{B}}). \tag{2.32}$$

with equality if and only if $\boldsymbol{\nu}$ is a product distribution, i.e., $\boldsymbol{\nu} = \boldsymbol{\nu}_{\mathbb{A}} \times \boldsymbol{\nu}_{\mathbb{B}}$.

Proof. Let us write the K-L divergence in terms of the loss function, and note that $H(\boldsymbol{\nu}) \leq H(\boldsymbol{\nu}_{\mathbb{A}}) + H(\boldsymbol{\nu}_{\mathbb{B}})$. This gives

$$\begin{aligned} D(\boldsymbol{\nu}\|\boldsymbol{\mu}_{\mathbb{A}} \times \boldsymbol{\mu}_{\mathbb{B}}) &= J(\boldsymbol{\nu}\|\boldsymbol{\mu}_{\mathbb{A}} \times \boldsymbol{\mu}_{\mathbb{B}}) - H(\boldsymbol{\nu}) \\ &\geq J(\boldsymbol{\nu}\|\boldsymbol{\mu}_{\mathbb{A}} \times \boldsymbol{\mu}_{\mathbb{B}}) - H(\boldsymbol{\nu}_{\mathbb{A}}) - H(\boldsymbol{\nu}_{\mathbb{B}}) \\ &= J(\boldsymbol{\nu}_{\mathbb{A}}\|\boldsymbol{\mu}_{\mathbb{A}}) + J(\boldsymbol{\nu}_{\mathbb{B}}\|\boldsymbol{\mu}_{\mathbb{B}}) - H(\boldsymbol{\nu}_{\mathbb{A}}) - H(\boldsymbol{\nu}_{\mathbb{B}}) \\ &= D(\boldsymbol{\nu}_{\mathbb{A}}\|\boldsymbol{\mu}_{\mathbb{A}}) + D(\boldsymbol{\nu}_{\mathbb{B}}\|\boldsymbol{\mu}_{\mathbb{B}}). \end{aligned}$$

Here we make use of the easily proved fact

$$J(\boldsymbol{\nu}\|\boldsymbol{\mu}_{\mathbb{A}} \times \boldsymbol{\mu}_{\mathbb{B}}) = J(\boldsymbol{\nu}_{\mathbb{A}}\|\boldsymbol{\mu}_{\mathbb{A}}) + J(\boldsymbol{\nu}_{\mathbb{B}}\|\boldsymbol{\mu}_{\mathbb{B}}).$$

In order for the inequality to be an equality, we must have $H(\boldsymbol{\nu}) = H(\boldsymbol{\nu}_{\mathbb{A}}) + H(\boldsymbol{\nu}_{\mathbb{B}})$, which is the case if and only if $\boldsymbol{\nu}$ is a product distribution, i.e., $\boldsymbol{\nu} = \boldsymbol{\nu}_{\mathbb{A}} \times \boldsymbol{\nu}_{\mathbb{B}}$. □

The next theorem gives a nice expression for the K-L divergence between two probability distributions of joint random variables. In [30, p. 24], this result is referred to as the "chain rule" for the K-L divergence.

Theorem 2.23 *Suppose $\boldsymbol{\phi}, \boldsymbol{\theta}$ are probability distributions on a product set $\mathbb{A} \times \mathbb{B}$, where $|\mathbb{A}| = n$ and $|\mathbb{B}| = m$. Suppose $\boldsymbol{\phi} \ll \boldsymbol{\theta}$. Let $\boldsymbol{\phi}_X, \boldsymbol{\theta}_X$ denote the marginal distributions on $\mathbb{A}$. Then*

$$D(\boldsymbol{\phi}\|\boldsymbol{\theta}) = D(\boldsymbol{\phi}_X\|\boldsymbol{\theta}_X) + \sum_{i=1}^{n} (\boldsymbol{\phi}_X)_i D(\boldsymbol{\phi}_{\{Y|X=a_i\}}\|\boldsymbol{\theta}_{\{Y|X=a_i\}}). \tag{2.33}$$

Proof. To simplify the notation, let us use the symbols

$$f_i := \sum_{j=1}^{m} \phi_{ij}, 1 \leq i \leq n, \mathbf{f} := [f_1 \ldots f_n] \in \mathbb{S}_n,$$

$$g_i := \sum_{j=1}^{m} \theta_{ij}, 1 \leq i \leq n, \mathbf{g} := [g_1 \ldots g_n] \in \mathbb{S}_n.$$

Thus both $\mathbf{f}$ and $\mathbf{g}$ are probability distributions on $\mathbb{A}$. Next, for each i between 1 and n, define

$$c_{ij} := \frac{\phi_{ij}}{f_i}, d_{ij} := \frac{\theta_{ij}}{g_i}, 1 \leq j \leq m.$$

$$\mathbf{c}_i := [c_{i1} \ldots c_{im}], \mathbf{d}_i := [d_{i1} \ldots d_{im}].$$

Then clearly

$$\boldsymbol{\phi}_{\{Y|X=a_i\}} = \mathbf{c}_i, \text{ and } \boldsymbol{\theta}_{\{Y|X=a_i\}} = \mathbf{d}_i.$$

As per the conventions established earlier, if $f_i = 0$ for some i, we take $\mathbf{c}_i = \boldsymbol{\phi}_Y$. In other words, when conditioned on the "impossible" event $X = a_i$, the conditional distribution of Y is taken as its marginal distribution. Similar remarks apply in case $g_i = 0$ for some i. Next let us study the dominance between these conditional distributions. If $g_i = 0$ for some index i, then it follows that $\theta_{ij} = 0 \; \forall j$. Since $\boldsymbol{\phi} \ll \boldsymbol{\theta}$, this implies that $\phi_{ij} = 0 \; \forall j$, i.e., that $f_i = 0$. In other words, the dominance condition $\boldsymbol{\phi} \ll \boldsymbol{\theta}$ implies that $\boldsymbol{\phi}_X \ll \boldsymbol{\theta}_X$. The same dominance condition $\boldsymbol{\phi} \ll \boldsymbol{\theta}$ also implies that $\mathbf{c}_i \ll \mathbf{d}_i \; \forall i$.

Now, in terms of the new symbols, the desired conclusion (2.33) can be rewritten as

$$D(\boldsymbol{\phi}\|\boldsymbol{\theta}) = D(\mathbf{f}\|\mathbf{g}) + \sum_{i=1}^{n} f_i D(\mathbf{c}_i\|\mathbf{d}_i).$$

This relationship can be established simply by expanding $D(\boldsymbol{\phi}\|\boldsymbol{\theta})$. Note that

$$\begin{aligned} D(\boldsymbol{\phi}\|\boldsymbol{\theta}) &= \sum_{i=1}^{n}\sum_{j=1}^{m} \phi_{ij} \log\left(\frac{\phi_{ij}}{\theta_{ij}}\right) \\ &= \sum_{i=1}^{n}\sum_{j=1}^{m} \phi_{ij} \log\left(\frac{f_i c_{ij}}{g_i d_{ij}}\right) \\ &= \sum_{i=1}^{n}\left(\sum_{j=1}^{m} \phi_{ij}\right) \log\frac{f_i}{g_i} + \sum_{i=1}^{n}\sum_{j=1}^{m} f_i c_{ij} \log\left(\frac{c_{ij}}{d_{ij}}\right) \\ &= D(\mathbf{f}\|\mathbf{g}) + \sum_{i=1}^{n} f_i D(\mathbf{c}_i\|\mathbf{d}_i). \end{aligned}$$

This is the desired relationship. □

We conclude this chapter with a couple of useful properties of the K-L divergence.

Theorem 2.24 *The K-L divergence $D(\boldsymbol{\phi}\|\boldsymbol{\theta})$ is jointly convex in $\boldsymbol{\phi}, \boldsymbol{\theta}$; that is, if $\boldsymbol{\phi}_1, \boldsymbol{\phi}_2, \boldsymbol{\theta}_1, \boldsymbol{\theta}_2 \in \mathbb{S}_n$ with $\boldsymbol{\phi}_1 \ll \boldsymbol{\theta}_1$, $\boldsymbol{\phi}_2 \ll \boldsymbol{\theta}_2$ and $\lambda \in (0,1)$, we have*

$$D(\lambda\boldsymbol{\phi}_1 + (1-\lambda)\boldsymbol{\phi}_2 \| \lambda\boldsymbol{\theta}_1 + (1-\lambda)\boldsymbol{\theta}_2) \leq \lambda D(\boldsymbol{\phi}_1\|\boldsymbol{\theta}_1) + (1-\lambda) D(\boldsymbol{\phi}_2\|\boldsymbol{\theta}_2). \quad (2.34)$$

The proof of Theorem 2.24 depends on an auxiliary lemma called the "log sum inequality" that is of independent interest.

Lemma 2.25 *Suppose $\alpha_1, \ldots, \alpha_m, \beta_1, \ldots, \beta_m$ are nonnegative, and that at least one α_i and at least one β_j are positive. Assume further that $\beta_i = 0 \Rightarrow \alpha_i = 0$. Then*

$$\sum_{i=1}^{m} \alpha_i \log\left(\frac{\alpha_i}{\beta_i}\right) \geq \left(\sum_{i=1}^{m} \alpha_i\right) \cdot \log\left(\frac{\sum_{i=1}^{m} \alpha_i}{\sum_{j=1}^{m} \beta_j}\right). \quad (2.35)$$

Remark: Note that neither the α_i's nor the β_i's need to add up to one in order for (2.35) to hold.

Proof. (Of Lemma 2.25): Define

$$A = \sum_{i=1}^{m} \alpha_i, B = \sum_{i=1}^{m} \beta_i, a_i = \frac{\alpha_i}{A}, b_i = \frac{\beta_i}{B}, i = 1, \ldots, n.$$

Then the vectors $\mathbf{a} = [a_1 \ldots a_m], \mathbf{b} = [b_1 \ldots b_m] \in \mathbb{S}_m$. Thus it follows from Theorem 2.19 that

$$\sum_{i=1}^{m} a_i \log\left(\frac{a_i}{b_i}\right) \geq 0.$$

Now let us substitute $\alpha_i = Aa_i, \beta_i = Bb_i$ for all i. This leads to

$$\begin{aligned}
\sum_{i=1}^{m} \alpha_i \log\left(\frac{\alpha_i}{\beta_i}\right) &= A\left[\sum_{i=1}^{n} a_i \log\left(\frac{a_i}{b_i}\right) + \sum_{i=1}^{n} a_i \log\frac{A}{B}\right] \\
&= A\left[\sum_{i=1}^{n} a_i \log\left(\frac{a_i}{b_i}\right) + \log\frac{A}{B}\right] \text{ because } \sum_{i=1}^{n} a_i = 1 \\
&\geq A \log\frac{A}{B}.
\end{aligned}$$

This is precisely (2.35). □

Proof. (of Theorem 2.24): The left side of (2.34) is the sum over $i = 1, \ldots, n$ of the term

$$g_i := [\lambda\phi_{1i} + (1-\lambda)\phi_{2i}] \cdot \log\left(\frac{\lambda\phi_{1i} + (1-\lambda)\phi_{2i}}{\lambda\theta_{1i} + (1-\lambda)\theta_{2i}}\right).$$

Now apply the log sum inequality with $m = 2$ and

$$\alpha_{1i} = \lambda\phi_{1i}, \alpha_{2i} = (1-\lambda)\phi_{2i}, \beta_{1i} = \lambda\theta_{1i}, \beta_{2i} = (1-\lambda)\theta_{2i}.$$

Note that since $\boldsymbol{\phi}_1 \ll \boldsymbol{\theta}_1$, $\boldsymbol{\phi}_2 \ll \boldsymbol{\theta}_2$, it follows that $\beta_{ji} = 0 \Rightarrow \alpha_{ji} = 0$ for $j = 1, 2$. Then it follows from (2.35) that, for each i, we have

$$\begin{aligned}
g_i &= (\alpha_{1i} + \alpha_{2i}) \log\frac{\alpha_{1i} + \alpha_{2i}}{\beta_{1i} + \beta_{2i}} \\
&\leq \alpha_{1i} \log\frac{\alpha_{1i}}{\beta_{1i}} + \alpha_{2i} \log\frac{\alpha_{2i}}{\beta_{2i}} \\
&= \lambda\phi_{1i} \log\frac{\phi_{1i}}{\theta_{1i}} + (1-\lambda)\phi_{2i} \log\frac{\phi_{2i}}{\theta_{2i}}.
\end{aligned}$$

Summing over $i = 1, \ldots, n$ shows that

$$\sum_{i=1}^{n} g_i \leq \lambda D(\boldsymbol{\phi}_1 \| \boldsymbol{\theta}_1) + (1-\lambda) D(\boldsymbol{\phi}_2 \| \boldsymbol{\theta}_2),$$

which is the desired inequality. □

The last result of this chapter relates the total variation metric and the Kullback-Leibler divergence.

Theorem 2.26 *Suppose $\boldsymbol{\nu}, \boldsymbol{\mu} \in \mathbb{S}_n$, and that $\boldsymbol{\mu} \ll \boldsymbol{\nu}$. Then*

$$\rho(\boldsymbol{\mu}, \boldsymbol{\nu}) \leq [(1/2)D(\boldsymbol{\mu}\|\boldsymbol{\nu})]^{1/2}. \tag{2.36}$$

Proof. Let us view $\boldsymbol{\mu}, \boldsymbol{\nu}$ as distributions on some finite set $\mathbb{A}$. Define $\mathbb{A}_+ := \{i : \mu_i \geq \nu_i\}$, and $\mathbb{A}_- := \{i : \mu_i < \nu_i\}$. Then together $\mathbb{A}_+, \mathbb{A}_-$ partition $\mathbb{A}$. Next, let us define

$$p := \sum_{i \in \mathbb{A}_+} \mu_i, q := \sum_{i \in \mathbb{A}_+} \nu_i,$$

and observe that $p \geq q$ by the manner in which we have defined the two sets. Moreover, it follows from (1.12) that

$$\rho(\boldsymbol{\mu}, \boldsymbol{\nu}) = p - q.$$

Next, by definition we have

$$\begin{aligned} D(\boldsymbol{\mu}\|\boldsymbol{\nu}) &= \sum_{i \in \mathbb{A}_+} \mu_i \log\left(\frac{\mu_i}{\nu_i}\right) + \sum_{i \in \mathbb{A}_-} \mu_i \log\left(\frac{\mu_i}{\nu_i}\right) \\ &\geq p \log \frac{p}{q} + (1-p) \log \frac{1-p}{1-q}. \end{aligned}$$

The last step follows from the log sum inequality because

$$\sum_{i \in \mathbb{A}_+} \mu_i \log\left(\frac{\mu_i}{\nu_i}\right) \geq \left(\sum_{i \in \mathbb{A}_+} \mu_i\right) \log\left(\frac{\sum_{i \in \mathbb{A}_+} \mu_i}{\sum_{i \in \mathbb{A}_+} \nu_i}\right) = p \log \frac{p}{q},$$

and similarly for the other term.

Thus the proof of the inequality (2.36) can be achieved by studying only distributions on a set of cardinality two. Define $\boldsymbol{\phi} = (p, 1-p), \boldsymbol{\psi} = (q, 1-q)$ and suppose $p \geq q$. The objective is to prove that

$$p - q \leq [(1/2)D(\boldsymbol{\phi}\|\boldsymbol{\psi})]^{1/2},$$

or equivalently that

$$2(p-q)^2 \leq p \log \frac{p}{q} + (1-p) \log \frac{1-p}{1-q}, \ \forall q \in (0, p], \ \forall p \in (0, 1). \tag{2.37}$$

To prove (2.37), define

$$f(q) := p \log \frac{p}{q} + (1-p) \log \frac{1-p}{1-q} - 2(p-q)^2,$$

viewed as a function of q for a fixed p. Then the objective is to show that $f(q) \geq 0$ for all $q \in (0, p]$. Clearly $f(p) = 0$. Moreover,

$$\begin{aligned} f'(q) &= -\frac{p}{q} + \frac{1-p}{1-q} + 4(p-q) \\ &= \frac{-p(1-q) + (1-p)q + 4(p-q)q(1-q)}{q(1-q)} \\ &= -(p-q)\frac{4q^2 - 4q + 1}{q(1-q)} \\ &= -(p-q)\frac{(2q-1)^2}{q(1-q)}. \end{aligned}$$

Since $(2q-1)^2 \geq 0, q(1-q) \geq 0$ always, it follows that $f'(q) \leq 0 \; \forall q < p$. Therefore $f(q) \geq f(p) = 0 \; \forall q < p$. □

The result above is universally known as "Pinsker's inequality," but it would be fairer to credit also Csiszár; see [33]. It turns out that if we seek a bound of the form

$$D(\boldsymbol{\mu}\|\boldsymbol{\nu}) \geq C[\rho(\boldsymbol{\mu},\boldsymbol{\nu})]^2,$$

then $C = 2$ is the best possible constant. More generally, suppose we seek a bound of the form

$$D(\boldsymbol{\mu}\|\boldsymbol{\nu}) \geq \sum_{i=1}^{\infty} C_{2i} V^{2i},$$

where $V = \rho(\boldsymbol{\mu},\boldsymbol{\nu})$. Then it is shown in [123] that

$$D(\boldsymbol{\mu}\|\boldsymbol{\nu}) \geq \frac{1}{2}V^2 + \frac{1}{36}V^4 + \frac{1}{270}V^6 + \frac{221}{340200}V^8.$$

The same paper also gives a general methodology for extending the power series further, if one has the patience.

A couple of final comments. First, since the total variation metric is symmetric while the K-L divergence is not, we can also write (2.36) as

$$\rho(\boldsymbol{\mu},\boldsymbol{\nu}) \leq \min\{[(1/2)D(\boldsymbol{\mu}\|\boldsymbol{\nu})]^{1/2}, [(1/2)D(\boldsymbol{\nu}\|\boldsymbol{\mu})]^{1/2}\}. \tag{2.38}$$

Second, there is no *lower bound* for the total variation metric in terms of the K-L divergence. Indeed, there cannot be. Just choose $\boldsymbol{\nu},\boldsymbol{\mu}$ such that for some index i we have $\nu_i = 0$ but μ_i is very small but positive. Then $\rho(\boldsymbol{\nu},\boldsymbol{\mu})$ is small but $D(\boldsymbol{\mu}\|\boldsymbol{\nu})$ is infinite. So there cannot exist a lower bound analogous to (2.38).

Chapter Three

Nonnegative Matrices

In this chapter, the focus is on nonnegative matrices. They are relevant in the study of Markov processes (see Chapter 4), because the state transition matrix of a Markov process is a special kind of nonnegative matrix, known as a stochastic matrix.[1] However, it turns out that practically all of the useful properties of a stochastic matrix also hold for the more general class of nonnegative matrices. Hence it is desirable to present the theory in the more general setting, and then specialize to Markov processes. The reader may find the more general results useful in some other context.

Nonnegative matrices have two very useful properties. First, through symmetric row and column permutations, every nonnegative matrix can be put into a corresponding "canonical form" that is essentially unique. Second, the eigenvalues of a nonnegative matrix, subject to some additional conditions, have a very special structure. The two sections of the present chapter are devoted respectively to these two properties.

3.1 CANONICAL FORM FOR NONNEGATIVE MATRICES

In this section it is shown that to every nonnegative matrix A there corresponds an essentially unique canonical form C. Moreover, A can be transformed into C via symmetric row and column permutations.

3.1.1 Basic Version of the Canonical Form

Let $\mathbb{R}_+$ denote the set $[0, \infty)$ of nonnegative numbers. Suppose $A \in \mathbb{R}_+^{n \times n}$ is a given nonnegative matrix. It turns out that the canonical form associated with A depends only on which elements of A are positive, but does not otherwise depend on the *size* of the positive elements. Thus, to capture the *location* of the positive elements of A, we define the associated **incidence matrix** T corresponding to A as follows: $T \in \{0,1\}^{n \times n}$, and

$$t_{ij} = \begin{cases} 1 & \text{if} \quad a_{ij} > 0, \\ 0 & \text{if} \quad a_{ij} = 0. \end{cases}$$

Since A has n rows and columns, we can think of $\mathbb{N} := \{1, \ldots, n\}$ as the set of nodes of a directed graph,[2] and place a directed edge from node i to

---

[1] Both of these terms are defined in Chapter 4.

[2] Note that in Chapter 1, the symbol $\mathbb{N}$ is used to denote the set of natural numbers.

node j if and only if $a_{ij} > 0$ (or equivalently, $t_{ij} = 1$). A **path** of length l from node i to node j is a sequence of length $l+1$ of the form $\{i_0, i_1, \ldots, i_l\}$, where $i_0 = i$, $i_l = j$, and in addition $a_{i_s, i_{s+1}} > 0$ for all $s = 0, \ldots, l-1$. Let $a_{ij}^{(l)}$ denote the ij-th element of A^l. From the matrix multiplication formula

$$a_{ij}^{(l)} = \sum_{s_1=1}^{n} \cdots \sum_{s_{l-1}=1}^{n} a_{is_1} a_{s_1 s_2} \cdots a_{s_{l-1} j}$$

it is obvious that $a_{ij}^{(l)} > 0$ if and only if there exists a path of length l from i to j.

At this point the reader may wonder why we don't define a path from i to j as a sequence *of length* $\leq n$ (or equivalently, restrict a path to have length $\leq n-1$). Clearly, in any sequence of nodes of length $\geq n+1$, at least one index $k \in \mathbb{N}$ must occur twice. As a consequence the path must contain at least one cycle (a path whose starting and end nodes are the same). This cycle can be removed without affecting the existence of a path from i to j. In short, there exists a path from i to $j \neq i$ if and only if there exists a path of length $\leq n-1$ from i to j. There exists a path from i back to i if and only if there exists a path from i to itself of length $\leq n$. So why then do we not restrict the length of the path to be $\leq n$?

The answer is found in the sentence at the end of the next to previous paragraph. Given a matrix A, we wish to study the pattern of nonzero (or positive) elements of successive power A^l for *all* values of l, not just when $l \leq n$. Note that the statement "$a_{ij}^{(l)} > 0$ if and only if there exists a path of length l from i to j" is valid for *every* value of l, even if $l \geq n+1$.

We say that a node i **leads to** another node j if there exists a path of *positive* length from i to j. We write $i \to j$ if i leads to j. In particular, $i \to i$ if and only if there is a cycle from node i to itself. Thus it is quite possible for a node *not* to lead to itself.

Definition 3.1 *A node i is said to be* **inessential** *if there exists a node j (of necessity, not equal to i) such that $i \to j$, but $j \not\to i$. Otherwise, i is said to be* **essential**.

Note that if i is essential, $i \to j$ implies $j \to i$. By convention, if a node i does not lead to any other node j, then i is taken to be inessential. For instance, if row i of the matrix A is identically zero, then i is inessential.

With the above definition, we can divide the set of nodes $\mathbb{N} = \{1, \ldots, n\}$ into two disjoint sets: $\mathcal{I}$ denoting the set of inessential nodes, and $\mathcal{E}$ denoting the set of essential nodes.

Now we present a very preliminary canonical form of a nonnegative matrix.

Lemma 3.2 *Renumber the rows and columns of A in such a way that all the nodes in $\mathcal{E}$ come first, followed by all the nodes in $\mathcal{I}$. Let Π denote the*

However, throughout the remainder of the book, the symbol $\mathbb{N}$ is used to denote the set of integers $\{1, \ldots, n\}$.

permutation matrix corresponding to the renumbering. Then

$$\Pi^{-1}A\Pi = \begin{array}{c} \\ \mathcal{E} \\ \mathcal{I} \end{array} \begin{array}{c} \begin{array}{cc} \mathcal{E} & \mathcal{I} \end{array} \\ \left[\begin{array}{cc} P & \mathbf{0} \\ R & Q \end{array} \right] \end{array}. \tag{3.1}$$

Proof. It is claimed that if $i \in \mathcal{E}$ and $i \to j \in \mathbb{N}$, then $j \in \mathcal{E}$. In other words, an essential node can lead only to another essential node. The proof of this claim makes use of the property that the relation $\to$ is transitive, or in other words, $i \to j$ and $j \to k$ implies that $i \to k$. The transitivity of $\to$ is clear from the definition of a path, but we give below an algebraic proof. If $i \to j$, then there is an integer l such that $a_{ij}^{(l)} > 0$. Similarly, if $j \to k$, then there exists another integer r such that $a_{jk}^{(r)} > 0$. Now from the formula for matrix multiplication, it follows that

$$a_{ik}^{(l+r)} = \sum_{s=1}^{n} a_{is}^{(l)} a_{sk}^{(r)} \geq a_{ij}^{(l)} a_{jk}^{(r)} > 0.$$

Thus $i \to k$. Coming back to the main issue, we are given that $i \in \mathcal{E}$ and that $i \to j \in \mathbb{N}$; we wish to show that $j \in \mathcal{E}$. For this purpose, we must show that if $j \to k \in \mathbb{N}$, then $k \to j$. Accordingly, suppose that $j \to k$. Since $i \to j$ and $j \to k$, the transitivity of $\to$ implies that $i \to k$. Next, since $i \in \mathcal{E}$, it follows that $k \to i$. Invoking once again the transivity property, we conclude from $k \to i$ and $i \to j$ that $k \to j$. Since k is arbitrary other than that $j \to k$, this shows that $j \in \mathcal{E}$.

We have just shown that if $i \in \mathcal{E}$ and $i \to j$, then $j \in \mathcal{E}$. As a consequence, if $i \in \mathcal{E}$ and $j \in \mathcal{I}$, then $i \not\to j$. In particular, if $i \in \mathcal{E}$ and $j \in \mathcal{I}$, then $a_{ij} = 0$. Otherwise, if $a_{ij} > 0$, then $i \to j$ (because there is a path of length one), which is a contradiction. Therefore by symmetric permutation of rows and columns, the matrix A can be put into the canonical form (3.1). □

Example 3.1 Suppose a 10×10 matrix A has the following structure, where $\times$ denotes a positive element. This notation is chosen deliberately to highlight the fact that the actual values of the elements of A don't matter, only whether they are positive or zero. Note that the incidence matrix corresponding to A can be obtained simply by changing all occurrences of $\times$ to 1.

$$A = \left[\begin{array}{cccccccccc} 0 & \times & 0 & 0 & \times & 0 & 0 & \times & 0 & 0 \\ 0 & 0 & \times & 0 & 0 & 0 & \times & 0 & \times & 0 \\ 0 & \times & 0 & \times & 0 & \times & \times & 0 & 0 & 0 \\ 0 & 0 & 0 & 0 & 0 & 0 & 0 & \times & 0 & 0 \\ 0 & 0 & 0 & 0 & 0 & 0 & \times & 0 & 0 & 0 \\ 0 & 0 & 0 & \times & 0 & 0 & 0 & 0 & 0 & 0 \\ 0 & 0 & 0 & 0 & \times & 0 & 0 & 0 & 0 & 0 \\ 0 & 0 & 0 & 0 & 0 & \times & 0 & 0 & 0 & 0 \\ 0 & 0 & 0 & 0 & 0 & 0 & 0 & 0 & 0 & \times \\ 0 & 0 & 0 & 0 & 0 & 0 & 0 & 0 & \times & 0 \end{array} \right].$$

With this matrix we can associate a directed graph where there is an edge from node i to another node j if and only if $a_{ij} > 0$.

Let us compute the **reachability matrix** M of this graph, where we write $m_{ij} = +$ if $i \to j$ and $m_{ij} = \cdot$ if $i \not\to j$. Thus

$$M = \begin{bmatrix} \cdot & + & + & + & + & + & + & + & + & + \\ \cdot & + & + & + & + & + & + & + & + & + \\ \cdot & + & + & + & + & + & + & + & + & + \\ \cdot & \cdot & \cdot & + & \cdot & + & \cdot & + & \cdot & \cdot \\ \cdot & \cdot & \cdot & \cdot & + & \cdot & + & \cdot & \cdot & \cdot \\ \cdot & \cdot & \cdot & + & \cdot & + & \cdot & + & \cdot & \cdot \\ \cdot & \cdot & \cdot & \cdot & + & \cdot & + & \cdot & \cdot & \cdot \\ \cdot & \cdot & \cdot & + & \cdot & + & \cdot & + & + & + \\ \cdot & \cdot & \cdot & \cdot & \cdot & \cdot & \cdot & \cdot & + & + \\ \cdot & \cdot & \cdot & \cdot & \cdot & \cdot & \cdot & \cdot & + & + \end{bmatrix}.$$

From the reachability matrix we can see that nodes $1, 2, 3$ are inessential, whereas nodes 4 through 10 are essential. So if we simply make a cyclic permutation and shift 1, 2, 3 to the end, the matrix A now looks like

$$\Pi^{-1}A\Pi = \begin{array}{c|ccccccc|ccc} & 4 & 5 & 6 & 7 & 8 & 9 & 10 & 1 & 2 & 3 \\ 4 & 0 & 0 & 0 & 0 & \times & 0 & 0 & 0 & 0 & 0 \\ 5 & 0 & 0 & 0 & \times & 0 & 0 & 0 & 0 & 0 & 0 \\ 6 & \times & 0 & 0 & 0 & 0 & 0 & 0 & 0 & 0 & 0 \\ 7 & 0 & \times & 0 & 0 & 0 & 0 & 0 & 0 & 0 & 0 \\ 8 & 0 & 0 & \times & 0 & 0 & 0 & 0 & 0 & 0 & 0 \\ 9 & 0 & 0 & 0 & 0 & 0 & 0 & \times & 0 & 0 & 0 \\ 10 & 0 & 0 & 0 & 0 & 0 & \times & 0 & 0 & 0 & 0 \\ \hline 1 & 0 & \times & 0 & 0 & \times & 0 & 0 & 0 & \times & 0 \\ 2 & 0 & 0 & 0 & \times & 0 & \times & 0 & 0 & 0 & \times \\ 3 & \times & 0 & \times & \times & 0 & 0 & 0 & 0 & \times & 0 \end{array}$$

Thus the triangular structure of A is brought out clearly. As we develop the theory further, we shall return to this example and refine further the matrix on the right side.

Example 3.2 The purpose of this example is to demonstrate the possibility that *every* node can be inessential. Suppose

$$A = \begin{bmatrix} 0 & \times \\ 0 & 0 \end{bmatrix}.$$

Then it is easy to see that both nodes are inessential.

The next lemma shows when such a phenomenon can occur.

Lemma 3.3 *Suppose $A \in \mathbb{R}_+^{n \times n}$ and that every row of A contains at least one positive element. Then there exists at least one essential node.*

Proof. Suppose by way of contradiction that every node is inessential. Recall that there are two ways in which a node i can be inessential: (i) i does not lead to any other node, or (ii) there exists a $j \in \mathbb{N}$ such that $i \rightarrow j$ but $j \not\rightarrow i$. By assumption, no row of A is identically zero. Hence every node i leads to at least one other node j (which could be the same as i). Hence the first possibility is ruled out. Now choose a node $i_0 \in \mathbb{N}$ arbitrarily. Since i_0 is inessential, there exists a node $i_1 \in \mathbb{N}$ such that $i_0 \rightarrow i_1$ but $i_1 \not\rightarrow i_0$. Clearly this implies that $i_1 \neq i_0$. Now i_1 is also inessential. Hence there exists an $i_2 \in \mathbb{N}$ such that $i_1 \rightarrow i_2$ but $i_2 \not\rightarrow i_1$. This implies, *inter alia*, that $i_2 \neq i_1$ and that $i_2 \neq i_0$. It is obvious that $i_2 \neq i_1$. If $i_2 = i_0$ then $i_2 = i_0 \rightarrow i_1$, which contradicts the assumption that $i_2 \not\rightarrow i_1$. Now repeat the procedure. At step l, we have $l+1$ nodes $i_0, i_1, \ldots, i_l$ such that

$$i_0 \rightarrow i_1 \rightarrow \cdots \rightarrow i_{l-1} \rightarrow i_l, \text{ but } i_l \not\rightarrow i_{l-1}.$$

We claim that this implies that i_l is not equal to any of $i_0, \ldots, i_{l-1}$; otherwise, the transitivity property of $\rightarrow$ implies that we have $i_l \rightarrow i_{l-1}$, which is a contradiction. Hence all the $l+1$ nodes are distinct. After we repeat the process n times, we will supposedly have $n+1$ distinct nodes $i_0, i_1, \ldots, i_n$. But this is impossible since there are only n nodes. Hence it is not possible for every node to be inessential. □

Lemma 3.4 *Suppose $A \in \mathbb{R}_+^{n \times n}$ and that every row of A contains at least one positive element. Then every inessential node (if any) leads to an essential node.*

Proof. As before let $\mathcal{E}$ denote the set of essential nodes and $\mathcal{I}$ the set of inessential nodes. From Lemma 3.3, we know that $\mathcal{E}$ is nonempty. If $\mathcal{I}$ is nonempty, the lemma states that for each $i \in \mathcal{I}$ there exists a $j \in \mathcal{E}$ such that $i \rightarrow j$. Accordingly, suppose $i \in \mathcal{I}$ is arbitrary. Since every row of A contains at least one positive element, node i leads to at least one other node. Since node i is inessential, there exists a $j \in \mathbb{N}$ such that $i \rightarrow j$ but $j \not\rightarrow i$. If $j \in \mathcal{E}$ the claim is established, so suppose $j \in \mathcal{I}$. So there exists a $k \in \mathbb{N}$ such that $j \rightarrow k$ but $k \not\rightarrow j$. Clearly $k \neq j$ and also $k \neq i$ since $i \rightarrow j$. If $k \in \mathcal{E}$, then $i \rightarrow j \rightarrow k$ and the claim is established, so suppose $k \in \mathcal{I}$. Proceeding as before, we can choose $l \in \mathbb{N}$ such that $k \rightarrow l$ but $l \not\rightarrow k$. Clearly $l \neq i, j, k$. If $l \in \mathcal{E}$ then $i \rightarrow j \rightarrow k \rightarrow l$ implies $i \rightarrow l \in \mathcal{E}$. Proceeding in this fashion, we will construct a sequence of nodes, all of them distinct, and all belonging to $\mathcal{I}$. Since $\mathcal{I}$ is a finite set, sooner or later this process must stop with some node $l \in \mathcal{E}$. Hence $i \rightarrow l \in \mathcal{E}$. □

These two lemmas lead to a slightly refined version of the canonical form.

Lemma 3.5 *Suppose $A \in \mathbb{R}_+^{n \times n}$ and that every row of A contains at least one positive element. Then in the canonical form (3.1), the set $\mathcal{E}$ is nonempty. Moreover, if the set $\mathcal{I}$ is also nonempty, then*

1. *The matrix R contains at least one positive element, i.e., $R \neq \mathbf{0}$.*

2. *Let $m = |\mathcal{I}|$, the number of inessential nodes. Express A^m in the form*

$$\Pi^{-1} A^m \Pi = \begin{array}{c} \\ \mathcal{E} \\ \mathcal{I} \end{array} \begin{array}{c} \begin{array}{cc} \mathcal{E} & \mathcal{I} \end{array} \\ \left[\begin{array}{cc} P^m & \mathbf{0} \\ R^{(m)} & Q^m \end{array} \right] \end{array}. \tag{3.2}$$

Then each row of $R^{(m)}$ contains at least one nonzero element.

Proof. The fact that $\mathcal{E}$ is nonempty follows from Lemma 3.3. If $r_{ij} = 0 \, \forall i \in \mathcal{I}, j \in \mathcal{E}$, then $i \not\to j$ whenever $i \in \mathcal{I}, j \in \mathcal{E}$. We know from Lemma 3.4 that this cannot be. This establishes the first statement.

The proof of the second statement is based on Lemmas 3.2 and 3.4. Note that, since $\Pi^{-1}A\Pi$ is block triangular, the diagonal blocks of $\Pi^{-1}A^m\Pi$ are indeed P^m and Q^m respectively. However, the off-diagonal block is a complicated function of P and Q, so we denote it by $R^{(m)}$. Let $i \in \mathcal{I}$ be arbitrary. Then from Lemma 3.4, there exists a $j \in \mathcal{E}$ such that $i \to j$. It can be assumed without loss of generality that the path from i to j does not contain any other essential nodes. If the path from i to j does indeed pass through another $j' \in \mathcal{E}$, then we can replace j by j'. So suppose $i \to j$, and that the path from i to j does not pass through any other essential node; thus the path must pass through only inessential nodes. Since there are only m inessential nodes, this implies that there exists a path from i to j of length no larger than m that passes through only inessential nodes. Denote this length by $l(i, j)$. Next, since $j \in \mathcal{E}$, it is claimed that there exists a $j' \in \mathcal{E}$ such that $a_{jj'} > 0$. This is because, if $a_{jj'} = 0$ for all $j' \in \mathbb{N}$, then clearly the node j does not lead to any other node, and by definition j would be inessential. Thus there exists a $j' \in \mathbb{N}$ such that $a_{jj'} > 0$. Now, it is obvious that $j \to j'$, and we already know from Lemma 3.2 that an essential node can lead only to another essential node. Thus there exists a $j' \in \mathcal{E}$ (not just $j' \in \mathbb{N}$) such that $a_{jj'} > 0$. Concatenating the path of length $l(i, j)$ from i to j with the path of length one from j to j', we get a path of length $l(i, j) + 1$ from $i \in \mathcal{I}$ to $j' \in \mathcal{E}$. Repeating this process $m - l(i, j)$ times gives a path from $i \in \mathcal{I}$ to some $k \in \mathcal{E}$ that has length exactly equal to m. Thus, in the matrix $R^{(m)}$, the ik-th entry is positive. Since i is arbitrary, this establishes the second statement. □

3.1.2 Irreducible Matrices

In order to proceed beyond the basic canonical form (3.1), we introduce the notion of irreducibility.

Definition 3.6 *A matrix $A \in \mathbb{R}_+^{n \times n}$ is said to be* **reducible** *if there exists a partition of $\mathbb{N}$ into disjoint nonempty sets $\mathcal{I}$ and $\mathcal{J}$ such that, through a symmetric permutation of rows and columns, A can be brought into the form*

$$\Pi^{-1}A\Pi = \begin{array}{c} \\ \mathcal{I} \\ \mathcal{J} \end{array} \begin{array}{c} \begin{array}{cc} \mathcal{I} & \mathcal{J} \end{array} \\ \left[\begin{array}{cc} A_{11} & \mathbf{0} \\ A_{21} & A_{22} \end{array} \right] \end{array}, \tag{3.3}$$

where Π is a permutation matrix. If this is not possible, then A is said to be **irreducible**.

From the above definition, we can give an equivalent characterization of irreducibility: $A \in \mathbb{R}_+^{n \times n}$ is irreducible if and only if, for every partition of $\mathbb{N}$

into disjoint nonempty sets $\mathcal{I}$ and $\mathcal{J}$, there exist $i \in \mathcal{I}$ and $j \in \mathcal{J}$ such that $a_{ij} > 0$.

The next result brings out the importance of irreducibility.

Theorem 3.7 *For a given $A \in \mathbb{R}_+^{n \times n}$, the following statements are equivalent:*

(i) A is irreducible.

(ii) Every $i \in \mathbb{N}$ leads to every other $j \in \mathbb{N}$.

Proof. **(ii) $\Rightarrow$ (i).** Actually we prove the contrapositive, namely: If (i) is false, then (ii) is false. Accordingly, suppose (i) is false and that A is reducible. Put A in the form (3.3). Then for every integer $l \geq 1$, the matrix $\Pi^{-1} A^l \Pi$ has the form

$$\Pi^{-1} A^l \Pi = \begin{array}{c} \\ \mathcal{I} \\ \mathcal{J} \end{array} \begin{array}{c} \begin{array}{cc} \mathcal{I} & \mathcal{J} \end{array} \\ \left[\begin{array}{cc} A_{11}^l & \mathbf{0} \\ A_{21}^{(l)} & A_{22}^l \end{array} \right] \end{array}.$$

Hence $a_{ij}^{(l)} = 0 \; \forall i \in \mathcal{I}, j \in \mathcal{J}, \; \forall l$. In turn this implies that $i \not\to j$ whenever $i \in \mathcal{I}, j \in \mathcal{J}$. Hence (ii) is false.

(i) $\Rightarrow$ (ii). Recall the alternate characterization of irreducibility given after Definition 3.6. Choose $i \in \mathbb{N}$ arbitrarily. It is shown that i leads to every $j \in \mathbb{N}$. To begin the argument, let $l = 1$, $i_1 = i$, and consider the partition $\mathcal{I}_1 = \{i_1\}$, $\mathcal{J}_1 = \mathbb{N} \setminus \mathcal{I}_1$. Since A is irreducible, there exists an $i_2 \in \mathcal{J}_1$ such that $a_{i_1 i_2} > 0$. So $i_1 \to i_2$. Next, let $l = 2$, and consider the partition $\mathcal{I}_2 = \{i_1, i_2\}$ and $\mathcal{J}_2 = \mathbb{N} \setminus \mathcal{I}_2$. Since A is irreducible, there exists an $i_3 \in \mathcal{J}_2$ such that $a_{i_s i_3} > 0$ for either $s = 1$ or $s = 2$. If $s = 1$, then $a_{i_1 i_3} > 0$ implies that $i_1 \to i_3$. If $s = 2$, then $i_2 \to i_3$. Since $i_1 \to i_2$, the transitivity of $\to$ implies that $i_1 \to i_3$. In either case we can conclude that $i_1 \to i_3$. Repeat the argument. At step l we would have identified l *distinct* nodes $i_1, \ldots, i_l$ such that $i_1 \to i_s$ for $s = 2, \ldots, l$. Now consider the partition $\mathcal{I}_l = \{i_1, \ldots, i_l\}$, $\mathcal{J}_l = \mathbb{N} \setminus \mathcal{I}_l$. Since A is irreducible, there exists an $i_{l+1} \in \mathcal{J}_l$ such that $a_{i_s i_{l+1}} > 0$ for some $s = 1, \ldots, l$. Since $i_1 \to i_s$ for $s = 2, \ldots, l$, this implies that $i_1 \to i_{l+1}$. Notice that at each step we pick up yet another *distinct* node. Hence, after $l = n - 1$ steps, we conclude that $i \to j$ for all $j \neq i$, which is (i). □

Corollary 3.8 *For a given $A \in R_+^{n \times n}$, the following statements are equivalent:*

(i) A is irreducible.

(ii) For each $i, j \in \mathbb{N}$, there exists an integer l such that $a_{ij}^{(l)} > 0$.

(iii) Define

$$B := \sum_{l=1}^{n-1} A^l.$$

Then $b_{ij} > 0\ \forall i, j \in \mathbb{N}, i \neq j$. In other words, all off-diagonal elements of B are positive.

Proof. **(i) $\Rightarrow$ (ii).** The only difference between the present Statement (ii) and Statement (ii) of Theorem 3.7 is that there is no restriction here that $i \neq j$. Suppose A is irreducible. Then, as shown in Theorem 3.7, for every $i, j \in \mathbb{N}, i \neq j$, we have that $i \to j$. By the same logic, $j \to i$ also. Thus $i \to i\ \forall i \in \mathbb{N}$.

(ii) $\Rightarrow$ (i). Suppose (ii) is true. Then in particular $i \to j$ whenever $i, j \in \mathbb{N}, i \neq j$. So from Theorem 3.7, A is irreducible.

(iii) $\Leftrightarrow$ (ii). Statement (iii) is equivalent to: For each $i, j \in \mathbb{N}, i \neq j$, there exists an $l \leq n - 1$ such that $a_{ij}^{(l)} > 0$. This is the same as: For each $i, j \in \mathbb{N}, i \neq j$, there exists a path *of length* $\leq n - 1$ from i to j. Clearly (iii) implies (ii). To see that (ii) implies (iii), observe that if there is a path of any length l from i to $j \neq i$, and if $l \geq n$, then the path must include a cycle (a path starting and ending at the same node). This cycle can be removed from the path without affecting the reachability of j from i. Hence it can be assumed that $l \leq n - 1$, which is (iii). □

3.1.3 Final Version of Canonical Form

With the aid of Theorem 3.7, we can refine the basic canonical form introduced in Lemma 3.2.

We begin by reprising some definitions from Subsection 3.1.1. Given a matrix $A \in \mathbb{R}_+^{n \times n}$, we divided the nodes in $\mathbb{N} = \{1, \ldots, n\}$ into two disjoint sets: The set $\mathcal{E}$ of essential nodes, and the set $\mathcal{I}$ of inessential nodes. We also showed that $a_{ij} = 0$ whenever $i \in \mathcal{E}$ and $j \in \mathcal{I}$, leading to the canonical form (3.1). The objective of the present subsection is to refine further the structure of the canonical form.

For this purpose, observe that the binary relation $\to$ ("leads to") is an *equivalence relation* on the set $\mathcal{E}$. In order to be an equivalence relation, $\to$ must satisfy the following three conditions:

(a) $i \to i$ for all $i \in \mathcal{E}$. (Reflexivity)

(b) $i \to j$ implies that $j \to i$. (Symmetry)

(c) $i \to j, j \to k$ together imply that $i \to k$. (Transitivity)

Let us now verify each of these conditions in succession.

Suppose $i \in \mathcal{E}$. Recall the convention that if i does not lead to any node, then i is taken to be inessential. Hence $i \in \mathcal{E}$ implies that there exists some $j \in \mathbb{N}$ (which could be i) such that $i \to j$. By the definition of an essential node, this in turn implies that $j \to i$. Now the transitivity of $\to$ implies that $i \to i$. Hence $\to$ is reflexive. Next, suppose $i \to j \neq i$. Since $i \in \mathcal{E}$, this implies that $j \to i$. Hence $\to$ is symmetric. Finally, the transitivity of $\to$ has already been established in Subsection 3.1.1. Therefore $\to$ is an equivalence relation on $\mathcal{E}$.

Note that $\rightarrow$ need not be an equivalence relation on $\mathbb{N}$, the set of *all* nodes. In particular, only transitivity is guaranteed, and $\rightarrow$ need not be either reflexive or symmetric. If we look at the matrix A of Example 3.1, we see that node 1 is not reachable from itself. Moreover, node 2 can be reached from node 1, but node 1 cannot be reached from node 2.

Since $\rightarrow$ is an equivalence relation on $\mathcal{E}$, we can partition $\mathcal{E}$ into its disjoint equivalence classes under $\rightarrow$. Let s denote the number of these equivalence classes, and let $\mathcal{E}_1, \ldots, \mathcal{E}_s$ denote the equivalence classes. Hence if $i, j \in \mathcal{E}$ belong to disjoint equivalence classes, then it follows that $i \not\rightarrow j$ and $j \not\rightarrow i$. In particular, $a_{ij} = 0$ whenever i, j belong to disjoint equivalence classes. Now let us permute the elements of $\mathcal{E}$ in such a way that all elements of $\mathcal{E}_1$ come first, followed by those of $\mathcal{E}_2$, and ending finally with the elements of $\mathcal{E}_s$. Note that the ordering of the equivalence classes themselves, as well as the ordering of the elements within a particular equivalence class, can both be arbitrary. With this permutation, the matrix P in (3.1) looks like

$$\Pi_P^{-1} P \Pi_P = \begin{array}{c} \\ \mathcal{E}_1 \\ \vdots \\ \mathcal{E}_s \end{array} \begin{array}{c} \begin{array}{ccc} \mathcal{E}_1 & \ldots & \mathcal{E}_s \end{array} \\ \left[\begin{array}{ccc} P_1 & \ldots & \mathbf{0} \\ \vdots & \ddots & \vdots \\ \mathbf{0} & \ldots & P_s \end{array} \right] \end{array}.$$

Moreover, each matrix P_l has the property that every node in $\mathcal{E}_l$ leads to every other node in $\mathcal{E}_l$. Hence, by Theorem 3.7, it follows that each matrix P_l is *irreducible*.

We can also refine the structure of the matrix Q in (3.1). Unfortunately, the results are not so elegant as with P. Let us begin with the set $\mathcal{I}$ of inessential nodes, and partition it into $\mathcal{I}_1, \ldots, \mathcal{I}_r$ such that each $\mathcal{I}_l$ is a **communicating class**, that is, each $\mathcal{I}_l$ has the property that if $i, j \in \mathcal{I}_l$ then $i \rightarrow j$ and also $j \rightarrow i$. From the definition, it is clear that if $i, j \in \mathcal{I}$ belong to disjoint communicating classes, then either $i \rightarrow j$ or $j \rightarrow i$, but not both (or perhaps neither). Hence the communicating classes can be numbered in such a way that if $i \in \mathcal{I}_{r_1}$ and $j \in \mathcal{I}_{r_2}$ and $r_1 < r_2$, then $i \not\rightarrow j$. Note that the ordering of these communicating classes need not be unique. With this renumbering, the matrix Q in (3.1) becomes *block triangular*.

These observations can be captured in the following:

Theorem 3.9 *Given a matrix $A \in \mathbb{R}_+^{n \times n}$, identify the set $\mathcal{E}$ of essential nodes and the set $\mathcal{I}$ of inessential nodes. Let s denote the number of equivalence classes of $\mathcal{E}$ under $\rightarrow$, and let r denote the number of distinct communicating classes of $\mathcal{I}$ under $\rightarrow$. Then there exists a permutation matrix Π over $\mathbb{N}$ such that*

$$\Pi^{-1} A \Pi = \begin{array}{c} \\ \mathcal{E} \\ \mathcal{I} \end{array} \begin{array}{c} \begin{array}{cc} \mathcal{E} & \mathcal{I} \end{array} \\ \left[\begin{array}{cc} P & \mathbf{0} \\ R & Q \end{array} \right] \end{array}. \tag{3.4}$$

Moreover, P and Q have the following special forms:

$$P = \begin{array}{c} \\ \mathcal{E}_1 \\ \vdots \\ \mathcal{E}_s \end{array} \begin{array}{c} \begin{array}{ccc} \mathcal{E}_1 & \ldots & \mathcal{E}_s \end{array} \\ \left[\begin{array}{ccc} P_1 & \ldots & \mathbf{0} \\ \vdots & \ddots & \vdots \\ \mathbf{0} & \ldots & P_s \end{array} \right] \end{array}, \tag{3.5}$$

$$Q = \begin{array}{c} \\ \mathcal{I}_1 \\ \mathcal{I}_2 \\ \vdots \\ \mathcal{I}_r \end{array} \begin{array}{c} \begin{array}{cccc} \mathcal{I}_1 & \mathcal{I}_2 & \ldots & \mathcal{I}_r \end{array} \\ \left[\begin{array}{cccc} Q_{11} & \mathbf{0} & \ldots & \mathbf{0} \\ Q_{21} & Q_{22} & \ldots & \mathbf{0} \\ \vdots & \vdots & \ddots & \vdots \\ Q_{r1} & Q_{r2} & \ldots & Q_{rr} \end{array} \right] \end{array}. \tag{3.6}$$

If every row of A contains a positive element, then the set $\mathcal{E}$ is nonempty. Moreover, if the set $\mathcal{I}$ is also nonempty, then every row of the matrix R contains at least one positive element.

Example 3.3 Let us return to the matrix A of Example 3.1. We have already seen that the essential nodes are $\mathcal{E} = \{4, \ldots, 10\}$ and the inessential nodes are $\mathcal{I} = \{1, 2, 3\}$. To complete the canonical form, we need to do two things. First, we need to identify the equivalence classes of $\mathcal{E}$ under $\rightarrow$. Second, we need to identify the communicating classes of $\mathcal{I}$ under $\rightarrow$. From the reachability matrix, we can see that there are three equivalence classes of $\mathcal{E}$, namely $\mathcal{E}_1 = \{4, 6, 8\}$, $\mathcal{E}_2 = \{5, 7\}$, and $\mathcal{E}_3 = \{9, 10\}$. Note that the ordering of these sets is arbitrary. Next, the reachability matrix also shows that there are two communicating classes in $\mathcal{I}$, namely $\mathcal{I}_1 = \{2, 3\}$ and $\mathcal{I}_2 = \{1\}$. Here the ordering is *not* arbitrary if we wish Q to have a block-triangular structure. Hence the matrix A can be permuted into the following canonical form:

$$\Pi^{-1} A \Pi = \begin{array}{c} \\ 4 \\ 6 \\ 8 \\ 5 \\ 7 \\ 9 \\ 10 \\ 2 \\ 3 \\ 1 \end{array} \begin{array}{c} \begin{array}{ccccccccccc} 4 & 6 & 8 & 5 & 7 & 9 & 10 & & 2 & 3 & 1 \end{array} \\ \left[\begin{array}{ccc|cc|cc|cc|c} 0 & 0 & \times & 0 & 0 & 0 & 0 & 0 & 0 & 0 \\ \times & 0 & 0 & 0 & 0 & 0 & 0 & 0 & 0 & 0 \\ 0 & \times & 0 & 0 & 0 & 0 & 0 & 0 & 0 & 0 \\ \hline 0 & 0 & 0 & 0 & \times & 0 & 0 & 0 & 0 & 0 \\ 0 & 0 & 0 & \times & 0 & 0 & 0 & 0 & 0 & 0 \\ \hline 0 & 0 & 0 & 0 & 0 & 0 & \times & 0 & 0 & 0 \\ 0 & 0 & 0 & 0 & 0 & \times & 0 & 0 & 0 & 0 \\ \hline 0 & 0 & 0 & 0 & \times & \times & 0 & 0 & \times & 0 \\ \times & \times & 0 & 0 & \times & 0 & 0 & \times & 0 & 0 \\ \hline 0 & 0 & \times & \times & 0 & 0 & 0 & \times & 0 & 0 \end{array} \right] \end{array}$$

3.1.4 Irreducibility, Aperiodicity, and Primitivity

In this subsection we go beyond studying merely irreducible matrices, and introduce two more notions, namely: the period of an irreducible matrix, and primitive matrices. An important tool in this theory is the greatest common divisor (g.c.d.) of a set of integers. The g.c.d. of a *finite* set of positive integers is the stuff of high school arithmetic, but here we study the g.c.d. of an *infinite* set of integers, some or all of which could be negative.

In the sequel, $\mathbb{Z}$ denotes the set of integers, and $\mathbb{Z}_+$ the set of nonnegative integers. Throughout we use the notation "$a|b$" to denote the fact that $a \in \mathbb{Z}$ divides $b \in \mathbb{Z}$, i.e., there exists another integer $c \in \mathbb{Z}$ such that $b = ac$. Note that $a|b$ if and only if the nonnegative integer $|a|$ divides $|b|$.

Suppose $b_1, \ldots, b_m \in \mathbb{Z}$. Then their **greatest common divisor (g.c.d.)** is an integer a that satisfies two conditions:

(i) $a|b_i$ for all i.

(ii) If $c|b_i$ for all i, then $c|a$.

The g.c.d. is unique except for its sign. Thus if a is a g.c.d. of $\{b_1, \ldots, b_m\}$, then $-a$ is the only other g.c.d. By convention, here we always take the g.c.d. to be a positive number.

The following elementary fact is well known. Suppose that $b_1, \ldots, b_m \in \mathbb{Z}_+$ and let $a \in \mathbb{Z}_+$ denote their g.c.d. Then there exist $c_1, \ldots, c_m \in \mathbb{Z}$ such that

$$a = \sum_{i=1}^{m} c_i b_i.$$

Note that the c_i need not be nonnegative; they will just be integers, and in general some of them will be negative. Since each b_i is a positive multiple of a, an easy consequence of this is that every *sufficiently large* multiple of a can be expressed as a *nonnegative* "linear combination" of the b_i's. More precisely, there exists an integer l_0 such that, whenever $l \geq l_0$, we can find *nonnegative* constants $c_1, \ldots, c_m \in \mathbb{Z}_+$ such that

$$la = \sum_{i=1}^{m} c_i b_i.$$

Just how large is "large enough"? It is shown in [48] that $l_0 \leq b_0^2$, where $b_0 := \max_i b_i$.

All of this is fine for a *finite* set of integers. What about an *infinite* set of integers? Suppose $1 \leq b_1 < b_2 < \ldots$ is an infinite collection of integers. We can define their g.c.d. in terms of the same two conditions (i) and (ii) above. Notice that the definition of a g.c.d. is unambiguous even when there are infinitely many integers. Define a_i to be the g.c.d. of $\{b_1, \ldots, b_i\}$. Then $1 \leq a_{i+1} \leq a_i$. So the sequence $\{a_i\}$ is bounded below and therefore converges to some value, call it a. Moreover, since each a_i is an *integer*, there exists an integer i_0 such that $a_i = a \ \forall i \geq i_0$. These observations can be summarized as follows: Every set of integers $\{b_1, b_2, \ldots\}$, ordered such that $1 \leq b_1 < b_2 < \ldots$, has a g.c.d., call it a. Moreover, there exists an integer i_0 such that a is the g.c.d. of $\{b_1, \ldots, b_{i_0}\}$. Every sufficiently large multiple of a can be represented as a nonnegative linear combination of $b_1, \ldots, b_{i_0}$.

The next important concept is that of the period of a primitive matrix, given in Definition 3.14. We lead up to this important definition in stages.

Lemma 3.10 *Suppose $A \in \mathbb{R}_+^{n \times n}$ is irreducible. Then for each $i \in \mathbb{N}$, there exists an integer $r > 0$ such that $a_{ii}^{(r)} > 0$.*

Proof. Given $i \in \mathbb{N}$, choose an arbitrary $j \neq i$. Since A is irreducible, Theorem 3.7 implies that $i \to j$ and $j \to i$. Now the transitivity of $\to$ implies that $i \to i$. Hence $a_{ii}^{(r)} > 0$ for some integer $r > 0$. □

Definition 3.11 *Suppose $A \in \mathbb{R}_+^{n \times n}$ is irreducible, and let $i \in \mathbb{N}$. The* **period** *of i is defined as the g.c.d. of all integers r such that $a_{ii}^{(r)} > 0$.*

Example 3.4 Suppose A has the form

$$A = \begin{bmatrix} 0 & \times & 0 \\ \times & 0 & \times \\ \times & 0 & 0 \end{bmatrix}.$$

Then there are two cycles from node 1 to itself, namely: $1 \to 2 \to 1$ (length $= 2$) and $1 \to 2 \to 3 \to 1$ (length $= 3$). Hence $a_{ii}^{(2)} > 0$ and $a_{ii}^{(3)} > 0$. Since the g.c.d. of 2 and 3 is one, the period of node 1 is 1.

Example 3.5 Suppose A has the form

$$A = \begin{bmatrix} 0 & \times & 0 & 0 \\ \times & 0 & 0 & \times \\ \times & 0 & 0 & 0 \\ 0 & 0 & \times & 0 \end{bmatrix}.$$

Then there are cycles from node 1 to itself of length 2 ($1 \to 2 \to 1$) and length 4 ($1 \to 2 \to 4 \to 3 \to 1$). All cycles from node 1 to itself are concatenations of these two cycles. Hence every cycle from node 1 to itself has even length. Therefore the period of node 1 is 2.

Now consider node 4. There is no cycle of length 2 from node 4 to itself, but there are cycles of length 4 ($4 \to 3 \to 1 \to 2 \to 4$) and length 6 ($4 \to 3 \to 1 \to 2 \to 1 \to 2 \to 4$). The g.c.d. of 4 and 6 is also 2, so the period of node 4 is also 2. Thus nodes 1 and 4 have the same period. We shall see shortly that this is not a coincidence, but is a general property of irreducible matrices. Incidentally, this example also shows the rationale behind permitting the cycles to have lengths larger than the size of the matrix.

Theorem 3.12 *Suppose $A \in \mathbb{R}_+^{n \times n}$ is irreducible. Then every node in $\mathbb{N}$ has the same period.*

Proof. For a node i, define

$$S_i := \{r : a_{ii}^{(r)} > 0\}.$$

In words, S_i consists of the lengths of all cycles from node i to itself. Let p_i denote the period of node i. Then by definition p_i is the g.c.d. of all integers in the set S_i. In particular, $p_i | r \; \forall r \in S_i$.

Choose any $j \neq i$. Since A is irreducible, we have from Corollary 3.8 that $i \to j$ and $j \to i$. So there exist integers l and m such that $a_{ij}^{(l)} > 0$ and $a_{ji}^{(m)} > 0$. As a consequence,

$$a_{ii}^{(l+m)} = \sum_{s=1}^{n} a_{is}^{(l)} a_{si}^{(m)} \geq a_{ij}^{(l)} a_{ji}^{(m)} > 0.$$

So $l + m \in S_i$ and as a result $p_i | (l + m)$.

Now let $r \in S_j$ be arbitrary. Then by definition $a_{jj}^{(r)} > 0$. Therefore

$$a_{ii}^{(l+m+r)} = \sum_{s=1}^{n} \sum_{t=1}^{n} a_{is}^{(l)} a_{st}^{(r)} a_{ti}^{(m)} \geq a_{ij}^{(l)} a_{jj}^{(r)} a_{ji}^{(m)} > 0.$$

So $l + m + r \in S_i$, and $p_i | (l + m + r)$. Since it has already been shown that $p_i | (l + m)$, we conclude that $p_i | r$. But since r is an *arbitrary* element of S_j, it follows that p_i divides *every* element of S_j, and therefore $p_i | p_j$ (where p_j is the period of node j).

However, i and j are arbitrary nodes, so their roles can be interchanged throughout, leading to the conclusion that $p_j | p_i$. This shows that $p_i = p_j$ for all i, j. □

Theorem 3.12 shows that we can speak of "the period of an irreducible matrix" without specifying which node we are speaking about.

Definition 3.13 *An irreducible matrix is said to be* **aperiodic** *if its period is one.*

Definition 3.14 *A matrix $A \in \mathbb{R}_+^{n \times n}$ is said to be* **primitive** *if there exists an integer m such that $A^m > \mathbf{0}$.*

The next theorem is one of the key results in the theory of nonnegative matrices.

Theorem 3.15 *Given a matrix $A \in \mathbb{R}_+^{n \times n}$, the following statements are equivalent:*

(i) A is irreducible and aperiodic.

(ii) A is primitive.

Proof. We begin with a simple observation. Consider two statements:

1. There exists an integer m such that $A^m > \mathbf{0}$ (that is, A is primitive as defined in Definition 3.14).

2. There exists an integer l_0 such that $A^l > \mathbf{0} \; \forall l \geq l_0$.

We claim that both statements are equivalent. It is clear that Statement 2 implies Statement 1; just take $m = l_0$. To prove the implication in the opposite direction, suppose Statement 1 is true. Then clearly no row of A can be identically zero. (If a row of A is identically zero, then the same row of A^m would continue to be identically zero for all values of m.) Suppose $a_{ij}^{(m)} > 0 \, \forall i, j \in \mathbb{N}$. Then

$$a_{ij}^{(m+1)} = \sum_{s=1}^{n} a_{is} a_{sj}^{(m)} > 0 \, \forall i, j,$$

since $a_{is} > 0$ for at least one value of s for each given i, and $a_{sj}^{(m)} > 0$ for every s, j. This shows that $A^{m+1} > \mathbf{0}$. Now repeat with m replaced by $m + 1$, and use induction.

(ii) $\Rightarrow$ (i). Suppose A is primitive. Then A has to be irreducible. If A has the form

$$\Pi^{-1} A \Pi = \begin{bmatrix} \times & \mathbf{0} \\ \times & \times \end{bmatrix}$$

after symmetric permutation of rows and columns, then A^m has the form

$$\Pi^{-1} A^m \Pi = \begin{bmatrix} \times & \mathbf{0} \\ \times & \times \end{bmatrix}$$

for every value of m. So a reducible matrix cannot be primitive. Second, suppose A is irreducible but has period $p > 1$. Then, by the definition of the period, it follows that $a_{ii}^{(l)} = 0$ whenever l is not a multiple of p. Hence Statement 2 above is false and A cannot be primitive.

(i) $\Rightarrow$ (ii). Suppose A is irreducible and aperiodic. Fix a value $i \in \mathbb{N}$. Let $d_1, d_2, \ldots$ denote the lengths of all the cycles from i to itself. By the definition of the period, the g.c.d. of all these lengths is one. Hence the g.c.d. of a finite subset of these lengths is also one. Let $d_1, \ldots, d_s$ denote a finite subset of the cycle lengths whose g.c.d. is one. Then, as discussed at the beginning of this subsection, *every* sufficiently large integer r can be expressed as a nonnegative linear combination of the form $r = \sum_{t=1}^{s} \mu_t d_t$. Clearly, if there are cycles of lengths $d_1, \ldots, d_s$ from node i to itself, then there is a cycle of length $\sum_{t=1}^{s} \mu_t d_t$ for every set of nonnegative integers $\mu_1, \ldots, \mu_s$. (Follow the cycle of length d_t μ_t times, and do this for $t = 1, \ldots, s$.) So the conclusion is that there exists an integer r_i such that, for every integer $r \geq r_i$, there is a cycle of length r from node i to itself. Note that the smallest such integer r_i may depend on i. Now define $r^* := \max\{r_1, \ldots, r_n\}$. Then by the manner in which in which r^* has been chosen, it follows that

$$a_{ii}^{(r)} > 0, \ \forall i \in \mathbb{N}, \ \forall r \geq r^*.$$

Now it is shown that $A^{r^*+n-1} > \mathbf{0}$; this is enough to show that A is primitive. To show that $A^{r^*+n-1} > \mathbf{0}$, we must show that for each $i, j \in \mathbb{N}$, there exists a path of length *exactly equal* to $r^* + n$ from node i to node j. It can be assumed that $i \neq j$, since $a_{ii}^{(r^*+n-1)}$ is already known to be positive for all i. Since A is irreducible, we have from Theorem 3.7 that every i leads to every other j. Choose a path of length $\mu(i,j)$ from node i to node j. Without loss of generality, it can be assumed that $\mu(i,j) \leq n-1$. Now by the characterization of the integer r^*, there exists a path of length $r^*+n-1-\mu(i,j)$ from node i to itself. If this is concatenated with a path of length $\mu(i,j)$ from node i to node j, we will get a path of length r^*+n-1 from node i to node j. Hence $A^{r^*+n-1} > \mathbf{0}$ and A is primitive. □

Theorem 3.15 shows that, if A is irreducible and aperiodic, then $A^l > \mathbf{0}$ for all sufficiently large l. Now we answer the question of how large l needs to be.

Theorem 3.16 *Suppose $A \in \mathbb{R}_+^{n\times n}$ is irreducible and aperiodic. Define*

$$m_0(A) := \min\{m : A^m > \mathbf{0}\}.$$

Next, define

$$\mu(n) := \max_{A\in\mathbb{R}_+^{n\times n}} m_0(A),$$

where it is understood that the maximum is taken only over the set of irreducible and aperiodic matrices. Then

$$(n-2)(n-1) \leq \mu(n) \leq 3n^2+n-1. \tag{3.7}$$

Proof. As a preliminary first step, we repeat the trick from the proof of Theorem 3.15. Define

$$l_0(A) := \min\{l^* : A^l > \mathbf{0}\ \forall l \geq l^*\}.$$

Then it is claimed that $l_0(A) = m_0(A)$. It is obvious that $m_0(A) \leq l_0(A)$, since by definition $A^{l_0(A)} > \mathbf{0}$. To prove the converse, observe as before that if A is irreducible, then no row of A can be identically zero. Let $m = m_0(A)$. Then by assumption $a_{ij}^{(m)} > 0\ \forall i,j$. So

$$a_{ij}^{(m+1)} = \sum_{s=1}^{n} a_{is}a_{sj}^{(m)} > 0\ \forall i,j,$$

since $a_{is} > 0$ for some s and $a_{sj}^{(m)} > 0$ for every s. Repeating this argument with m replaced by $m+1, m+2$ etc. shows that $A^l > \mathbf{0}\ \forall l \geq m_0(A)$. Hence $l_0(A) \leq m_0(A)$.

First we establish the lower bound for $\mu(n)$. Note that the bound is trivial if $n=1$ or 2, since the left side equals zero in either case. So suppose $n \geq 3$, and define $A \in \mathbb{R}_+^{n\times n}$ to be a cyclic permutation matrix on $\mathbb{N} = \{1,\ldots,n\}$ with one extra positive element, as follows:

$$a_{1,2}=1, a_{2,3}=1, \ldots, a_{i,i+1}=1 \text{ for } i=1,\ldots,n-1, a_{n,1}=1,$$

and in addition $a_{n-1,1}=1$. All other elements of A are zero. To illustrate, if $n=5$, then

$$A = \begin{bmatrix} 0 & 1 & 0 & 0 & 0 \\ 0 & 0 & 1 & 0 & 0 \\ 0 & 0 & 0 & 1 & 0 \\ 1 & 0 & 0 & 0 & 1 \\ 1 & 0 & 0 & 0 & 0 \end{bmatrix}.$$

Now there are two cycles of length $n-1$ and n respectively from node 1 to itself; these are $1 \to 2 \to \cdots \to n-1 \to 1$ and $1 \to 2 \to \cdots \to n-1 \to n \to 1$. There are no other cycles of length $\leq n-1$. Hence all cycles from node 1 to itself have lengths of the form $\alpha(n-1)+\beta n$ for nonnegative integers α and β. It is easy to verify that $(n-2)(n-1)$ is the smallest integer l_0 with the property that *every* integer $l \geq l_0$ can be expressed in the form $\alpha(n-1)+\beta n$ for nonnegative integers α and β. Hence $l_0(A)=(n-2)(n-1)$, whence $\mu(n) \geq (n-2)(n-1)$.

The proof of the upper bound requires a result that is not proven here; see [1] or [2, Lemma 2.3]. It follows from this lemma that if A is irreducible and aperiodic, then for *every* integer $s \geq 3n^2$, there exits a cycle of length s from node i to itself for every i. Now we can repeat the argument in the proof of Theorem 3.15 to show that for every integer $l \geq 3n^2+n-1$, there exists a path of length l from node i to node j, for every i, j. Hence $l_0(A) \leq 3n^2+n-1$, for every irreducible and aperiodic A. □

3.1.5 Canonical Form for Periodic Irreducible Matrices

Up to now we have been studying *aperiodic* irreducible matrices. In this section, the focus is on irreducible matrices whose period is greater than one.

Lemma 3.17 *Suppose $A \in \mathbb{R}_+^{n\times n}$ is irreducible and has period $p \geq 2$. Then for each $i, j \in \mathbb{N}=\{1,\ldots,n\}$, there exists a unique integer $r(i,j) \in \{0,\ldots,p-1\}$ such that the length of every path from node i to node j equals $r(i,j) \bmod p$.*

Proof. Consider two distinct paths from node i to node j, of lengths l_1 and l_2 respectively. The desired conclusion is that $l_1 = l_2 \bmod p$, or equivalently, that $p|(l_1-l_2)$. Choose some path of length m from node j to node i. Such a path exists because A is irreducible. By concatenating this path with the two paths from node i to node j, we get two cycles from node i to itself, of lengths l_1+m and l_2+m respectively. Now, since p is the period of A, it divides the length of each cycle. Thus $p|(l_1+m)$ and $p|(l_2+m)$. Hence p divides the difference, i.e., $p|[(l_1+m)-(l_2+m)]$, or $p|(l_1-l_2)$. □

Lemma 3.18 *Suppose $A \in \mathbb{R}_+^{n\times n}$ is irreducible and has period $p \geq 2$. Define the integer $r(\cdot,\cdot)$ as in Lemma 3.17. Then for all $i, j, k \in \mathbb{N}$, we have that*

$$r(i,k) = [r(i,j)+r(j,k)] \bmod p.$$

Proof. Choose a path of length l from node i to node j, and of length m from node j to node k. By concatenating the two paths, we get a path of length $l+m$ from node i to node k. Now Lemma 3.17 implies that $r(i,k) = (l+m) \bmod p$, irrespective of how the paths are chosen. But clearly

$$\begin{aligned} r(i,k) &= (l+m) \bmod p = [l \bmod p + m \bmod p] \bmod p \\ &= [r(i,j)+r(j,k)] \bmod p. \end{aligned}$$

This is the desired conclusion. □

The above two lemmas suggest a way of partitioning the nodes in $\mathbb{N}$. Choose an arbitrary node $i \in \mathbb{N}$. Define

$$\mathcal{C}_s := \{j \in \mathbb{N} : r(i,j) = s\}, s = 1, \ldots, p-1.$$

Thus $\mathcal{C}_s$ consists of all nodes reachable from node i by paths of length s mod p. Finally, define

$$\mathcal{C}_0 := \mathbb{N} \setminus \left[\bigcup_{s=1}^{p-1} \mathcal{C}_s \right].$$

Clearly $i \in \mathcal{C}_0$, since every cycle from node i to itself has length 0 mod p, and as a result $i \notin \mathcal{C}_s$ for $s = 1, \ldots, p-1$. But $\mathcal{C}_0$ could contain other nodes as well. (See Example 3.6 below.) In fact, $\mathcal{C}_0$ consists of all nodes that are reachable from node i by paths whose lengths are multiples of p. Now the following observation is a ready consequence of Lemma 3.18.

Lemma 3.19 *Partition $\mathbb{N}$ into disjoint sets $\mathcal{C}_0, \ldots, \mathcal{C}_{p-1}$ as above. Then, for $j_1 \in \mathcal{C}_{s_1}$ and $j_2 \in \mathcal{C}_{s_2}$, we have*

$$r(j_1, j_2) = (s_2 - s_1) \bmod p.$$

Proof. Note that

$$j_l \in \mathcal{C}_{s_l} \Rightarrow r(i, s_l) = s_l \bmod p, \text{ for } l = 1, 2.$$

Hence, from Lemma 3.18,

$$r(j_1, j_2) = [r(i, j_2) - r(i, j_1)] \bmod p = (s_2 - s_1) \bmod p.$$

This is the desired conclusion. □

Now we are in a position to state the main result of this section.

Theorem 3.20 *Suppose $A \in \mathbb{R}_+^{n \times n}$ is irreducible and has period $p \geq 2$. Partition $\mathbb{N}$ into disjoint sets $\mathcal{C}_0, \ldots, \mathcal{C}_{p-1}$ as above. Then, by a symmetric permutation of rows and columns, A can be put in the form*

$$\Pi^{-1} A \Pi = \begin{array}{c} \\ \mathcal{C}_0 \\ \mathcal{C}_1 \\ \mathcal{C}_2 \\ \vdots \\ \mathcal{C}_{p-2} \\ \mathcal{C}_{p-1} \end{array} \begin{array}{c} \begin{array}{cccccc} \mathcal{C}_0 & \mathcal{C}_1 & \mathcal{C}_2 & \ldots & \mathcal{C}_{p-2} & \mathcal{C}_{p-1} \end{array} \\ \left[\begin{array}{cccccc} \mathbf{0} & A_{01} & \mathbf{0} & \ldots & \mathbf{0} & \mathbf{0} \\ \mathbf{0} & \mathbf{0} & A_{12} & \ldots & \mathbf{0} & \mathbf{0} \\ \mathbf{0} & \mathbf{0} & \mathbf{0} & \ldots & \mathbf{0} & \mathbf{0} \\ \vdots & \vdots & \vdots & \ddots & \vdots & \vdots \\ \mathbf{0} & \mathbf{0} & \mathbf{0} & \ldots & \mathbf{0} & A_{p-2,p-1} \\ A_{p-1,0} & \mathbf{0} & \mathbf{0} & \ldots & \mathbf{0} & \mathbf{0} \end{array} \right] \end{array} =: B, \tag{3.8}$$

where the matrix $A_{s,s+1}$ has dimension $|\mathcal{C}_s| \times |\mathcal{C}_{s+1}|$. This canonical form is unique to within (i) a cyclic permutation of the classes $\mathcal{C}_0, \ldots, \mathcal{C}_{p-1}$, and (ii) an arbitrary permutation of the indices within each class. Moreover, each of the p cyclic products

$$M_0 := A_{01} A_{12} \cdots A_{p-1,0},$$

$$M_1 := A_{12} \cdots A_{p-1,0} A_{01}, \ldots$$

$$M_{p-1} := A_{p-1,0} A_{01} \cdots A_{p-2,p-1} \tag{3.9}$$

is primitive.

Proof. It readily follows from Lemma 3.19 that if $i \in \mathcal{C}_s$ and $j \in \mathcal{C}_t$, then $a_{ij} = 0$ unless $t = (s+1) \bmod p$. This is because if $a_{ij} > 0$, then there exists a path of length one from node i to node j, which implies that $r(i,j) = 1$, which in turn implies that $t = (s+1) \bmod p$. This shows that the permuted matrix $\Pi^{-1} A \Pi$ has the block cyclic form shown in (3.8).

It remains only to show that each of the matrices in (3.9) is primitive. Note that the matrix B defined in (3.8) has a nonzero matrix only in block $(i, i+1)$ for $i = 0, \ldots, p-1$. (Here and elsewhere in this proof, any index $i \geq p$ should be replaced by $i \bmod p$.) Hence, for every integer $m \geq 1$, B^m has a nonzero matrix only in blocks $(i, i+m)$ for $i = 0, \ldots, p-1$. In particular, B^p is block diagonal and equals

$$B^p = \begin{array}{c} \\ \mathcal{C}_0 \\ \vdots \\ \mathcal{C}_{p-1} \end{array} \begin{array}{c} \begin{array}{ccc} \mathcal{C}_0 & \ldots & \mathcal{C}_{p-1} \end{array} \\ \left[\begin{array}{ccc} M_0 & \ldots & \mathbf{0} \\ \vdots & \ddots & \vdots \\ \mathbf{0} & \ldots & M_{p-1} \end{array} \right] \end{array}.$$

So for every integer l, B^{pl} equals

$$B^{pl} = (B^p)^l = \begin{array}{c} \\ \mathcal{C}_0 \\ \vdots \\ \mathcal{C}_{p-1} \end{array} \begin{array}{c} \begin{array}{ccc} \mathcal{C}_0 & \ldots & \mathcal{C}_{p-1} \end{array} \\ \left[\begin{array}{ccc} M_0^l & \ldots & \mathbf{0} \\ \vdots & \ddots & \vdots \\ \mathbf{0} & \ldots & M_{p-1}^l \end{array} \right] \end{array}.$$

The assertion that each M_i is primitive is equivalent to the statement that each of the diagonal blocks of B^{pl} is a strictly positive matrix for all sufficiently large values of l. Hence the assertion that each M_i is primitive is equivalent to the following statement: There exists an integer l_0 such that, for all $l \geq l_0$, there exists a path of length lp from node i to node j whenever they both belong to the same class $\mathcal{C}_s$, for some $s \in \{0, \ldots, p-1\}$. Accordingly, suppose i, j belong to the same class $\mathcal{C}_s$. By the definition of these classes, it follows that there is a path from node i to node j whose length is a multiple of p; call it $m(i,j)p$. By the definition of the period, the g.c.d. of the lengths of all cycles from node i to itself is p. Hence there exists an integer $l_0(i)$ such that there exist cycles of length lp from node i to itself for all $l \geq l_0(i)$. Hence there exist paths of length lp from node i to node j for all $l \geq l_0(i) + m(i,j)$. Since there are only finitely many i, j, the quantity

$$l_0 := \max_{i,j} \{l_0(i) + m(i,j)\}$$

is finite. Moreover, for all $l \geq l_0$, there exist paths of length lp from node i to node j whenever i, j belong to the same class $\mathcal{C}_s$. This shows that each matrix M_i is primitive. □

Example 3.6 Suppose A is a 9×9 matrix of the form

$$A = \begin{array}{c} \\ 1 \\ 2 \\ 3 \\ 4 \\ 5 \\ 6 \\ 7 \\ 8 \\ 9 \end{array} \begin{array}{c} \begin{array}{ccccccccc} 1 & 2 & 3 & 4 & 5 & 6 & 7 & 8 & 9 \end{array} \\ \left[\begin{array}{ccccccccc} 0 & \times & 0 & \times & 0 & 0 & \times & 0 & \times \\ 0 & 0 & \times & 0 & 0 & \times & 0 & \times & 0 \\ 0 & 0 & 0 & 0 & \times & 0 & 0 & 0 & 0 \\ 0 & 0 & 0 & 0 & 0 & \times & 0 & 0 & 0 \\ 0 & \times & 0 & 0 & 0 & 0 & \times & 0 & 0 \\ \times & 0 & 0 & 0 & 0 & 0 & 0 & 0 & 0 \\ 0 & 0 & \times & 0 & 0 & 0 & 0 & \times & 0 \\ \times & 0 & 0 & 0 & \times & 0 & 0 & 0 & 0 \\ 0 & 0 & \times & 0 & 0 & 0 & 0 & 0 & 0 \end{array} \right] \end{array}$$

It can be verified that A has period 3. The classes (starting from node 1) are:

$$\mathcal{C}_0 = \{1, 5\}, \mathcal{C}_1 = \{2, 4, 7, 9\}, \mathcal{C}_2 = \{3, 6, 8\}.$$

After permuting rows and columns accordingly, the matrix A can be put in the form

$$\Pi^{-1} A \Pi = \begin{array}{c} \\ 1 \\ 5 \\ 2 \\ 4 \\ 7 \\ 9 \\ 3 \\ 6 \\ 8 \end{array} \begin{array}{c} \begin{array}{ccccccccc} 1 & 5 & 2 & 4 & 7 & 9 & 3 & 6 & 8 \end{array} \\ \left[\begin{array}{cc|cccc|ccc} 0 & 0 & \times & \times & \times & \times & 0 & 0 & 0 \\ 0 & 0 & \times & 0 & \times & 0 & 0 & 0 & 0 \\ \hline 0 & 0 & 0 & 0 & 0 & 0 & \times & \times & 0 \\ 0 & 0 & 0 & 0 & 0 & 0 & 0 & \times & 0 \\ 0 & 0 & 0 & 0 & 0 & 0 & \times & 0 & \times \\ 0 & 0 & 0 & 0 & 0 & 0 & \times & 0 & 0 \\ \hline 0 & \times & 0 & 0 & 0 & 0 & 0 & 0 & 0 \\ \times & 0 & 0 & 0 & 0 & 0 & 0 & 0 & 0 \\ \times & 0 & 0 & 0 & 0 & 0 & 0 & 0 & 0 \end{array} \right] \end{array}$$

3.2 PERRON-FROBENIUS THEORY

In this section, we present various theorems about primitive and irreducible matrices. The first such theorems are due to Perron [103] and Frobenius [57, 58, 59]. Perron's original paper was for *positive* matrices, while Frobenius extended the theory to *nonnegative* matrices. The paper by Wielandt [139] was very influential in that most subsequent expositions of the theory follow his approach. But the theory continues to be known by the names of the two originators. A substantial generalization of the theory results when nonnegative matrices are replaced by matrices that leave a given "cone" invariant. For an exposition of this approach, see [16]. Much of the discussion in this section follows [116, Chapter 1].

Throughout this section, we write $\mathbf{x} \geq \mathbf{0}$ to indicate that every component of a (row or column) vector $\mathbf{x}$ is nonnegative, and $\mathbf{x} > \mathbf{0}$ to indicate that every component of $\mathbf{x}$ is positive. Similarly, $A \geq \mathbf{0}$ ($A > \mathbf{0}$) indicates that every component of the matrix A is nonnegative (positive). Expressions such as $\mathbf{x} \geq \mathbf{y}$ or $B < A$ are self-explanatory. Also, given an arbitrary row vector $x \in \mathbb{C}^n$, the symbol $\mathbf{x}_+$ denotes the vector $[|x_1| \ldots |x_n|] \in \mathbb{R}^n_+$. Note that $\mathbf{x}_+$ always belongs to $\mathbb{R}^n_+$. The notation A_+ is defined analogously. Finally, as in Chapter 1, the symbol $\mathbb{S}_n$ denotes the n-dimensional simplex, i.e.,

$$\mathbb{S}_n = \{\mathbf{x} \in \mathbb{R}^n_+ : \sum_{i=1}^{n} x_i = 1\}.$$

Given a matrix A, the **spectrum** of A consists of all eigenvalues of A and is denoted by spec(A). If spec$(A) = \{\lambda_1, \ldots, \lambda_n\}$, then

$$\rho(A) := \max\{|\lambda_i| : \lambda_i \in \text{spec}(A)\}$$

is called the **spectral radius** of A. Note that spec(A) can contain complex numbers, but $\rho(A)$ is always real and nonnegative.

3.2.1 Perron-Frobenius Theorem for Primitive Matrices

In this section, we state and prove the principal result for primitive matrices. In the next subsection, it is shown that very similar results hold for irreducible matrices even if they are not primitive.

We begin with an upper bound for the spectral radius of an arbitrary matrix.

Lemma 3.21 *Given a matrix $A \in \mathbb{R}^{n \times n}$, define*

$$r(A; \mathbf{x}) := \min_{j \in \mathbb{N}} \frac{(\mathbf{x}A_+)_j}{x_j} \; \forall \mathbf{x} \in \mathbb{S}_n,$$

and

$$r(A) := \max_{\mathbf{x} \in \mathbb{S}_n} r(A; \mathbf{x}). \tag{3.10}$$

Then $\rho(A) \leq r(A)$.

Proof. Note that in the definition of $r(A; \mathbf{x})$, we take the ratio $(\mathbf{x}A_+)_j / x_j$ to be ∞ if $x_j = 0$. However, since $\mathbf{x} \in \mathbb{S}_n$, clearly $x_j \neq 0$ for at least one index $j \in \mathbb{N}$. Hence $r(A; \mathbf{x})$ is finite for all $\mathbf{x} \in \mathbb{S}_n$. An equivalent and alternate definition of $r(A; \mathbf{x})$ is:

$$r(A; \mathbf{x}) := \max\{\lambda \in \mathbb{R}_+ : \mathbf{x}A_+ \geq \lambda \mathbf{x}\}.$$

This definition brings out clearly the fact that, for a fixed matrix A, the map $\mathbf{x} \mapsto r(A; \mathbf{x})$ is **upper semi-continuous**.[3] Suppose $\lambda_i \to \lambda_0, \mathbf{x}_i \to \mathbf{x}_0, A_+\mathbf{x}_i \geq \lambda_i \mathbf{x}_i$ where $\lambda_i, \lambda_0 \in \mathbb{R}_+$ and $\mathbf{x}_i, \mathbf{x}_0 \in \mathbb{S}_n$. Then it is obvious that $A_+\mathbf{x}_0 \geq \lambda_0 \mathbf{x}_0$, whence $r(A; \mathbf{x}_0) \geq \lambda_0$. In particular, if $\mathbf{x}_i \to \mathbf{x}_0$ and $r_0 = \limsup_{i \to \infty} r(A; \mathbf{x}_i)$, then $r(A; \mathbf{x}_0) \geq r_0$. Since $\mathbb{S}_n$ is a compact subset of $\mathbb{R}^n$, $r(A; \cdot)$ attains its maximum on $\mathbb{S}_n$.

[3] The reader is referred to [112] for various concepts and results from real analysis, such as compactness, semi-continuity, etc.

To show that $\rho(A) \leq r(A)$, let $\lambda \in \mathbb{C}$ be an arbitrary eigenvalue of A, and let $\mathbf{z} \in \mathbb{C}^n$ be an associated row eigenvector. Without loss of generality, we can scale $\mathbf{z}$ such that $\mathbf{z}_+ \in \mathbb{S}_n$. Now

$$\lambda z_j = \sum_{i=1}^{n} z_i a_{ij} \ \forall j \in \mathbb{N},$$

$$|\lambda||z_j| = \left| \sum_{i=1}^{n} z_i a_{ij} \right| \leq \sum_{i=1}^{n} |z_i||a_{ij}| \ \forall j \in \mathbb{N},$$

$$|\lambda| \leq \min_{j \in \mathbb{N}} \frac{(\mathbf{z}_+ A_+)_j}{(\mathbf{z}_+)_j} \leq r(A).$$

Hence $\rho(A) \leq r(A)$. □

Now we come to the Perron-Frobenius theorem for primitive matrices.

Theorem 3.22 *Suppose $A \in \mathbb{R}_+^{n \times n}$ is primitive. Then*

(i) $\rho(A)$ is an eigenvalue of A.

(ii) There exists a row eigenvector $\mathbf{v} > \mathbf{0}$ of A corresponding to the eigenvalue $\rho(A)$.

(iii) If $B \in \mathbb{R}^{n \times n}$ satisfies $\mathbf{0} \leq B \leq A$, then $\rho(B) \leq \rho(A)$, with equality if and only if $B = A$.

(iv) Rank$[\rho(A)I - A] = n - 1$, so that the eigenvectors of A associated with the eigenvalue $\rho(A)$ are multiples of each other.

(v) $\rho(A)$ is a simple eigenvalue of A.

(vi) If λ is any other eigenvalue of A, then $|\lambda| < \rho(A)$.

Proof. Since $r(A; \cdot)$ is an upper semi-continuous function on $\mathbb{S}_n$ and $\mathbb{S}_n$ is a compact subset of $\mathbb{R}_+^n$, there exists a vector $\mathbf{v} \in \mathbb{S}_n$ such that $r(A; \mathbf{v}) = r(A)$. For brevity, let r stand for $r(A)$. Also, let l be an integer such that $A^l > \mathbf{0}$. Such an integer l exists because A is primitive.

(i). Since $r(A; \mathbf{v}) = r(A)$, it follows from the definition of $r(\cdot)$ that $\mathbf{v}A - r\mathbf{v} \geq \mathbf{0}$. (Note that $A_+ = A$ since A is a nonnegative matrix.) Let $\mathbf{z}$ denote $\mathbf{v}A - r\mathbf{v}$ and note that $\mathbf{z} \geq \mathbf{0}$. If $\mathbf{z} = \mathbf{0}$ then $\mathbf{v}A = r\mathbf{v}$ and we are through, so suppose by way of contradiction that $\mathbf{z} \neq \mathbf{0}$. Then, since $\mathbf{z} \geq \mathbf{0}, \mathbf{z} \neq \mathbf{0}$, and $A^l > \mathbf{0}$, it follows that

$$\mathbf{0} < \mathbf{z}A^l = \mathbf{v}A^{l+1} - r\mathbf{v}A^l = (\mathbf{v}A^l)A - r(\mathbf{v}A^l).$$

Now $\mathbf{v}A^l > \mathbf{0}$ since $\mathbf{v} \geq \mathbf{0}, \mathbf{v} \neq \mathbf{0}$ and $A^l > \mathbf{0}$. Hence $\mathbf{v}A^l$ can be scaled by a constant μ such that $\mathbf{y} := (1/\mu)\mathbf{v}A^l$ belongs to $\mathbb{S}_n$. Since dividing by the constant μ does not affect the sign of anything, we have

$$\mathbf{0} < \mathbf{y}A - r\mathbf{y}.$$

Hence we can choose a number $\lambda > r$ such that $\mathbf{y}A - \lambda\mathbf{y} \geq \mathbf{0}$. So $r(\mathbf{y})$ is strictly larger than $r = r(A)$. But this contradicts the definition of $r(A)$ in (3.10). Hence it must be the case that $\mathbf{z} = \mathbf{0}$, that is, $\mathbf{v}A = r\mathbf{v}$. Therefore $r(A)$ is an eigenvalue of A. In turn this implies that $\rho(A) \geq r(A)$. But $\rho(A) \leq r(A)$ from Lemma 3.21. We conclude that $\rho(A) = r(A)$ and that $\rho(A)$ is an eigenvalue of A.

(ii). It has already been established that $\mathbf{v}A = r\mathbf{v}$ for some $\mathbf{v} \geq \mathbf{0}$, $\mathbf{v} \neq \mathbf{0}$. Hence $\mathbf{v}A^l = [\rho(A)]^l\mathbf{v}$. Moreover, $\mathbf{v} \geq \mathbf{0}, \mathbf{v} \neq \mathbf{0}$ and $A^l > \mathbf{0}$ together imply that $\mathbf{v}A^l > \mathbf{0}$. Finally $\mathbf{v} = (1/[\rho(A)]^l)\mathbf{v}A^l > \mathbf{0}$. Hence there exists a strictly positive row eigenvector $\mathbf{v}$ of A corresponding to the eigenvalue $\rho(A)$.

(iii). Suppose $\mathbf{0} \leq B \leq A$. Let β be an eigenvalue of B and let $\mathbf{y} \in \mathbb{C}^n$ be an associated *column* eigenvector of B; that is, $B\mathbf{y} = \beta\mathbf{y}$. Since $B \leq A$, it follows that $\mathbf{x}B \leq \mathbf{x}A \;\forall \mathbf{x} \in \mathbb{R}^n_+$. Also, $B\mathbf{y} = \beta\mathbf{y}$ implies that

$$|\beta|\mathbf{y}_+ \leq B\mathbf{y}_+ \leq A\mathbf{y}_+. \tag{3.11}$$

Multiplying both sides by $\mathbf{v}$ gives

$$|\beta|\mathbf{v}\mathbf{y}_+ \leq \mathbf{v}A\mathbf{y}_+ = \rho(A)\mathbf{v}\mathbf{y}_+.$$

Now let $\mathbf{v}$ denote a strictly positive row eigenvector of A corresponding to the eigenvalue $\rho(A)$, as identified in Statement (ii). Now $\mathbf{v}\mathbf{y}_+ > 0$ since $\mathbf{v} > \mathbf{0}$ and $\mathbf{y}_+ \geq \mathbf{0}, \mathbf{y}_+ \neq \mathbf{0}$. So we can divide both sides of the above inequality by $\mathbf{v}\mathbf{y}_+ > 0$, and conclude that $|\beta| \leq \rho(A)$. Since β is an arbitrary eigenvalue of A, it follows that $\rho(B) \leq \rho(A)$.

To prove the second part of the claim, suppose $|\beta| = \rho(A) = r$. Then (3.11) implies that

$$\mathbf{z} := A\mathbf{y}_+ - r\mathbf{y}_+ \geq \mathbf{0}.$$

As in the proof of Statement (i), if $\mathbf{z} \neq \mathbf{0}$, then we get a contradiction to the definition of $r(A)$. Hence $\mathbf{z} = \mathbf{0}$, or $A\mathbf{y}_+ = r\mathbf{y}_+$. Now we can multiply both sides by $A^l > \mathbf{0}$ to get $A(A^l)\mathbf{y}_+ = rA^l\mathbf{y}_+$ and in addition, $A^l\mathbf{y}_+ > \mathbf{0}$. Also, since $A\mathbf{y}_+ = r\mathbf{y}_+$, we have that $A^l\mathbf{y}_+ = r^l\mathbf{y}_+ > \mathbf{0}$, which means that $\mathbf{y}_+ > \mathbf{0}$. Now (3.11) also implies that

$$r\mathbf{y}_+ = B\mathbf{y}_+ = A\mathbf{y}_+, \text{ or } (B - A)\mathbf{y}_+ = \mathbf{0},$$

since the two extreme inequalities are in fact equalities. Since $B - A \leq \mathbf{0}$ and $\mathbf{y}_+ > \mathbf{0}$, the only way for $(B - A)\mathbf{y}_+$ to equal $\mathbf{0}$ is to have $B = A$. Hence Statement (iii) is proved.

(iv). Suppose $r\mathbf{y} = \mathbf{y}A$, so that $\mathbf{y}$ is also a row eigenvector of A corresponding to the eigenvalue $r = \rho(A)$. Then, as in the proof of Lemma 3.19 and Statement (i) above, it follows that $\mathbf{y}_+$ also satisfies $r\mathbf{y}_+ = \mathbf{y}_+A$. Also, by noting that $\mathbf{y}_+A^l = r^l\mathbf{y}_+ > \mathbf{0}$, and that $\mathbf{y}_+ \geq \mathbf{0}$ and $\mathbf{y}_+ \neq \mathbf{0}$, it follows that $\mathbf{y}_+A^l > \mathbf{0}$ and hence $\mathbf{y}_+ > \mathbf{0}$. Hence *every* row eigenvector of A corresponding to the eigenvalue $r = \rho(A)$ must have all nonzero components. Now let $\mathbf{v}$ denote the vector identified in the proof of Statement (i). Then $r(\mathbf{v} - c\mathbf{y}) = (\mathbf{v} - c\mathbf{y})A$ for all complex constants c. Suppose $\mathbf{y}$ is not a multiple of $\mathbf{v}$. Then it is possible to choose the constant c in such a way that $\mathbf{v} - c\mathbf{y} \neq \mathbf{0}$, but at least one component of $\mathbf{v} - c\mathbf{y}$ equals zero. Since $\mathbf{v} - c\mathbf{y} \neq \mathbf{0}$, $\mathbf{v} - c\mathbf{y}$ is a row eigenvector of A corresponding to the eigenvalue $r = \rho(A)$, and it has at least one component equal to zero, which is a contradiction. Hence $\mathbf{y}$ is a multiple of $\mathbf{v}$. Thus it has been shown that the equation $\mathbf{y}(rI - A) = \mathbf{0}$ has only one independent solution, i.e., that $\text{Rank}(rI - A) = n - 1$.

(v). Let $\phi(\lambda) := \det(\lambda I - A)$ denote the characteristic polynomial of A, and let $\text{Adj}(\lambda I - A)$ denote the adjoint matrix of $\lambda I - A$. We have already seen that $\rho(A) = r$ is an eigenvalue of A; hence $\phi(r) = 0$. Now it is shown that $\phi'(r) \neq 0$, which is enough to show that r is a *simple* zero of the polynomial $\phi(\cdot)$ and hence a simple eigenvalue of A.

For brevity let $M(\lambda)$ denote $\text{Adj}(\lambda I - A)$. Then it is easy to see that each element of $M(\lambda)$ is a polynomial of degree $\leq n - 1$ in λ. For every value of λ, we have

$$(\lambda I_n - A)M(\lambda) = \phi(\lambda)I_n.$$

Differentiating both sides with respect to λ leads to

$$M(\lambda) + (\lambda I_n - A)M'(\lambda) = \phi'(\lambda)I.$$

Let $\mathbf{v} > \mathbf{0}$ denote the eigenvector found in the proof of Statement (ii). Then $r\mathbf{v} = \mathbf{v}A$, or $\mathbf{v}(rI_n - A) = \mathbf{0}$. So if we right-multiply both sides of the above equation by $\mathbf{v}$ and substitute $\lambda = r$, we get

$$\phi'(r)\mathbf{v} = \mathbf{v}M(r) + \mathbf{v}(rI_n - A)M'(\lambda) = \mathbf{v}M(r). \tag{3.12}$$

Thus the proof is complete if it can be shown that $\mathbf{v}M(r) > \mathbf{0}$, because that is enough to show that $\phi'(r) > 0$.

We begin by establishing that no row of $M(r)$ is identically zero. Specifically, it is shown that the diagonal elements of $M(r)$ are all strictly positive. For each index $i \in \mathbb{N}$, the ii-th element of $M(r)$ is the principal minor given by

$$m_{ii}(r) = \det(rI_{n-1} - \bar{A}_i),$$

where $\bar{A}_i \in \mathbb{R}_+^{(n-1)\times(n-1)}$ is obtained from A by deleting the i-th row and i-th column. Now it is claimed that $\rho(\bar{A}_i) < \rho(A) = r$. To see this, suppose first (for notational convenience only) that $i = 1$. Then the two matrices

$$\bar{A}_1 = \begin{bmatrix} a_{22} & \dots & a_{2n} \\ \vdots & \ddots & \vdots \\ a_{n2} & \dots & a_{nn} \end{bmatrix}, \text{ and } A_1^* = \begin{bmatrix} 0 & 0 & \dots & 0 \\ 0 & a_{22} & \dots & a_{2n} \\ \vdots & \vdots & \ddots & \vdots \\ 0 & a_{n2} & \dots & a_{nn} \end{bmatrix}$$

have exactly the same spectrum, except that A_1^* has an extra eigenvalue at zero. For other values of i, A_i^* equals A with the i-th row and i-th column set equal to zero, while $\bar{A}_i$ is obtained from A by deleting the i-th row and i-th column. Hence, for all i, the matrices A_i^* and $\bar{A}_i$ have the same spectrum, except that A_i^* has an extra eigenvalue at zero. Now $\mathbf{0} \leq A_i^* \leq A$. Moreover, $A_i^* \neq A$ since A is irreducible and thus cannot have an identically zero row or zero column. Hence by Statement (iii) above, $\rho(A_i^*) < \rho(A) = r$. In particular, r is not an eigenvalue of $\bar{A}_i$, and as a consequence,

$$m_{ii}(r) = \det(rI_{n-1} - \bar{A}_i) \neq 0 \ \forall i \in \mathbb{N}.$$

Actually, we can conclude that

$$m_{ii}(r) = \det(rI_{n-1} - \bar{A}_i) > 0 \ \forall i \in \mathbb{N}. \tag{3.13}$$

To see this, define the characteristic polynomial $\phi_i(\lambda) := \det(\lambda I - \bar{A}_i)$. Then $\phi_i(\lambda) > 0$ when l is positive and sufficiently large. So if $\phi_i(r) < 0$, then $\phi_i(\cdot)$ would have a real zero $> r$, which contradicts the fact that $\rho(A_i^*) < r$. Hence (3.13) is established.

We have thus far established that every row of $M(r)$ contains a positive element. Now

$$M(r)(rI_n - A) = \phi(r)I_n = \mathbf{0}.$$

Hence every nonzero row of $M(r)$ is a row eigenvector of A corresponding to the eigenvalue r. Since $m_{ii}(r) > 0$ for each i, it follows that every row of $M(r)$ is a positive multiple of $\mathbf{v}$. In other words, $M(r)$ has the form

$$M(r) = \mathbf{wv}, \mathbf{w} \in \mathbb{R}_+^n.$$

Now let us return to (3.12). Substituting for $M(r)$ and "cancelling" $\mathbf{v}$ (which is permissible since $\mathbf{v} > \mathbf{0}$) shows that

$$\phi'(r) = \mathbf{wv} > 0.$$

Hence r is a simple eigenvalue of A.

(vi). The claim is that if λ is any eigenvalue of A such that $|\lambda| = \rho(A) = r$, then of necessity $\lambda = \rho(A)$. Suppose $|\lambda| = \rho(A) = r$ with corresponding row eigenvector $\mathbf{z} \in \mathbb{C}^n$. Then

$$\lambda z_j = \sum_{i=1}^{n} z_i a_{ij}, \ \forall j \in \mathbb{N}, \text{ and } \lambda^l z_j = \sum_{i=1}^{n} z_i a_{ij}^{(l)}, \ \forall j \in \mathbb{N}.$$

As before, let $\mathbf{z}_+ \in \mathbb{R}_+^n$ denote the vector $[|z_1| \ldots |z_n|]$. Then, just as in the proof of Lemma 3.19, $\lambda \mathbf{z} = \mathbf{z}A$ implies that $|\lambda|\mathbf{z}_+ \leq \mathbf{z}_+ A$. However, $|\lambda| = r$ by assumption. Hence, as in the proof of Statement (i), $r\mathbf{z}_+ \leq \mathbf{z}_+ A$ implies that in fact $r\mathbf{z}_+$ *equals* $\mathbf{z}_+ A$, and as a consequence $r^l \mathbf{z}_+ = \mathbf{z}_+ A^l$. In other words,

$$|\lambda^l z_j| = \left| \sum_{i=1}^{n} z_i a_{ij}^{(l)} \right| = \sum_{i=1}^{n} |z_i| a_{ij}^{(l)}, \ \forall j \in \mathbb{N}.$$

Since the z_i are complex numbers in general, and since $a_{ij}^{(l)} > 0$ for all i, j, the magnitude of the sum of $z_i a_{ij}^{(l)}$ equals the sum of the magnitudes of $z_i a_{ij}^{(l)}$ if and only if all these complex numbers are *aligned*, i.e., there is a *common* number θ such that

$$z_i a_{ij}^{(l)} = |z_i| a_{ij}^{(l)} \exp(\mathbf{i}\theta) \ \forall i, j \in \mathbb{N},$$

where $\mathbf{i} = \sqrt{-1}$. Dividing through by the positive number a_{ij} shows that

$$z_i = |z_i| \exp(\mathbf{i}\theta) \; \forall i \in \mathbb{N}.$$

Let us return to $r\mathbf{z}_+ = \mathbf{z}_+ A$. By Statement (iv), this implies that $\mathbf{z}_+ = c\mathbf{v}$ for some positive constant c, or

$$\mathbf{z} = c \exp(\mathbf{i}\theta)\mathbf{v}.$$

Therefore

$$\mathbf{z}A = c\exp(\mathbf{i}\theta)\mathbf{v}A = c\exp(\mathbf{i}\theta)r\mathbf{v} = r\mathbf{z}.$$

Since $\mathbf{z}A = \lambda z$, this shows that $r\mathbf{z} = \lambda\mathbf{z}$, and since at least one component of $\mathbf{z}$ is nonzero, this shows that $\lambda = r$. □

3.2.2 Perron-Frobenius Theorem for Irreducible Matrices

In this section, we study irreducible matrices that have a period $p \geq 2$. From Theorem 3.15, we know that an aperiodic irreducible matrix is primitive. So Theorem 3.22 applies to such matrices.

There are two distinct ways to approach the study of such matrices. First, we can examine the proof of Theorem 3.22 and see how much of it can be salvaged for periodic matrices. Second, we can use Theorem 3.20 in conjunction with Theorem 3.22. It turns out that each approach leads to its own distinctive set of insights. We begin with a preliminary result that is useful in its own right.

Lemma 3.23 *Suppose $A \in \mathbb{R}^{n\times m}$ and $B \in \mathbb{R}^{m\times n}$. Then every nonzero eigenvalue of AB is also an eigenvalue of BA.*

Proof. Suppose $\lambda \neq 0$ is an eigenvalue of AB, and choose $\mathbf{x} \neq \mathbf{0}$ such that $AB\mathbf{x} = \lambda\mathbf{x}$. Clearly $B\mathbf{x} \neq \mathbf{0}$. Now

$$BAB\mathbf{x} = \lambda B\mathbf{x}.$$

Since $B\mathbf{x} \neq \mathbf{0}$, this shows that λ is also an eigenvalue of BA with corresponding eigenvector $B\mathbf{x}$. The converse follows by interchanging A and B throughout. □

Note that the above reasoning breaks down if $\lambda = 0$. Indeed, if $n \neq m$ (say $n > m$ to be precise), then the larger matrix (AB if $n > m$) must necessarily be rank deficient, and thus must have at least $n - m$ eigenvalues at zero. But if BA is nonsingular, then it will not have any eigenvalues at zero. Therefore Lemma 3.23 can be stated as follows: Given two matrices A, B of complementary dimensions (so that both AB and BA are well-defined), both AB and BA have the same spectrum, except for the possibility that one of them may have some extra eigenvalues at zero. In particular, $\rho(AB) = \rho(BA)$.

Theorem 3.24 *Suppose $A \in \mathbb{R}_+^{n\times n}$ is irreducible and has period $p \geq 2$. Then*

(i) The eigenvalues of A exhibit cyclic symmetry with period p. Specifically, if λ is an eigenvalue of A, so are

$$\lambda \exp(\mathbf{i}2\pi k/p), k = 1, \ldots, p-1.$$

(ii) In particular, $r = \rho(A)$ is an eigenvalue of A, and so are

$$r \exp(\mathbf{i}2\pi k/p), k = 1, \ldots, p-1.$$

Proof. **(i).** Suppose $\lambda \in \mathbb{C}$ is an eigenvalue of A, and note that if $\lambda = 0$, then $\lambda \exp(\mathbf{i}2\pi k/p) = 0$ for all k. So we need to study only the case where $\lambda \neq 0$. Without loss of generality, we can assume that A is in the canonical form (3.8), because permuting the rows and columns of a matrix does not change its spectrum. So we use the symbol A (instead of B) for the canonical form. Suppose $\lambda \in \mathbb{C}$ is an eigenvalue of A, and choose $\mathbf{x} \neq \mathbf{0}$ such that $\lambda \mathbf{x} = \mathbf{x}A$. Partition $\mathbf{x}$ commensurately with the canonical form. Then $\lambda \mathbf{x} = \mathbf{x}A$ can be also be partitioned, and implies that

$$\mathbf{x}_j A_{j,j+1} = \lambda \mathbf{x}_{j+1}, j = 0, \ldots, p-1, \tag{3.14}$$

where if $j = p-1$ we take $j+1 = 0$ since $p = 0 \bmod p$. Now let $\alpha := \exp(\mathbf{i}2\pi/p)$ denote the p-th root of 1. Fix an integer k between 1 and $p-1$, and define the vector $\mathbf{y}^{(k)} \in \mathbb{C}^n$ by

$$\mathbf{y}_j^{(k)} = \alpha^{-jk} \mathbf{x}_j = \exp(-\mathbf{i}2\pi jk/p)\mathbf{x}_j, j = 0, \ldots, p-1.$$

Now it follows from (3.14) that

$$\begin{aligned} \mathbf{y}_j^{(k)} A_{j,j+1} &= \alpha^{-jk} \mathbf{x}_j A_{j,j+1} \\ &= \alpha^{-jk} \lambda \mathbf{x}_{j+1} = \lambda \alpha^k \alpha^{-(j+1)k} \mathbf{x}_{j+1} \\ &= \lambda \alpha^k \mathbf{y}_{j+1}. \end{aligned}$$

Hence $\mathbf{y}^{(k)}$ is an eigenvector of A corresponding to the eigenvalue $\lambda\alpha^k$. This argument can be repeated for each integer k between 1 and $p-1$. This establishes Statement (i).

(ii). Again, assume without loss of generality that A is in the canonical form (3.8), and define matrices $M_0, \ldots, M_{p-1}$ as in (3.9). Thus, as in the proof of Theorem 3.20, it follows that

$$A^p = \text{Block Diag}\{M_0, \ldots, M_{p-1}\}.$$

Hence the spectrum of A^p is the union of the spectra of $M_0, \ldots, M_{p-1}$. Moreover, it is clear from (3.9) that these p matrices are just cyclic products of A_{01} through $A_{p-1,0}$. Hence repeated application of Lemma 3.23 shows that each of these matrices has the same set of nonzero eigenvalues. (Since these matrices may have different dimensions, they will in general have different numbers of eigenvalues at zero.) As a consequence

$$\rho(M_0) = \rho(M_1) = \cdots = \rho(M_{p-1}) =: c, \text{ say}.$$

We also know from Theorem 3.20 that each of the p matrices $M_0, \ldots, M_{p-1}$ is primitive. Hence $\rho(M_i) = c$ is an eigenvalue of M_i for each i. Therefore

c is a p-fold eigenvalue of A^p. Since the spectrum of A^p consists of just the p-th powers of the spectrum of A, we see that some p-th roots of c are also eigenvalues of A. However, it now follows from Statement (i) that in fact *every* p-th root of c is an eigenvalue of A. This is precisely Statement (ii). □

Theorem 3.25 *Suppose $A \in \mathbb{R}_+^{n \times n}$ is irreducible. Then*

(i) $\rho(A)$ is an eigenvalue of A.

(ii) There exists a row eigenvector $\mathbf{v} > \mathbf{0}$ of A corresponding to the eigenvalue $\rho(A)$.

(iii) If $B \in \mathbb{R}^{n \times n}$ satisfies $\mathbf{0} \leq B \leq A$, then $\rho(B) \leq \rho(A)$, with equality if and only if $B = A$.

(iv) Rank$[\rho(A)I - A] = n - 1$, so that the eigenvectors of A associated with the eigenvalue $\rho(A)$ are multiples of each other.

(v) $\rho(A)$ is a simple eigenvalue of A.

(vi) If A is aperiodic, then every other eigenvalue λ of A satisfies $|\lambda| < \rho(A)$. Otherwise, if A has period $p \geq 2$, then each of the p numbers $\rho(A)\exp(\mathbf{i}2\pi k/p), k = 0, \ldots, p-1$ is an eigenvalue of A. All other eigenvalues λ of A satisfy $|\lambda| < \rho(A)$.

Remarks: Comparing Theorem 3.22 and 3.25, we see that only Statement (vi) is different in the two theorems.

Proof. The proof consists of mimicking the proof of Theorem 3.22 with minor changes. Hence we give the proof in a very sketchy form.

Since A is irreducible, it follows from Corollary 3.8 that the matrix

$$M := I + B = \sum_{i=0}^{n-1} A^i$$

is strictly positive. Moreover, if $\mathbf{z}A = \lambda\mathbf{z}$, then

$$\mathbf{z}M = \left(\sum_{i=0}^{n-1} \lambda^i \right) \mathbf{z}.$$

(i). This is already established in Theorem 3.24, but we give an independent proof paralleling that of Theorem 3.22. Define $r(A)$ as in Lemma 3.21, and choose $\mathbf{v} \in \mathbb{S}_n$ such that $r(\mathbf{v}) = r(A)$. Define $\mathbf{z} = \mathbf{v}A - r\mathbf{v}$ where $r = r(A)$. Then $\mathbf{z} \geq \mathbf{0}$. If $\mathbf{z} = \mathbf{0}$ then we are through, so suppose by way of contradiction that $\mathbf{z} \neq \mathbf{0}$. Then $\mathbf{z}M > \mathbf{0}$ because $M > \mathbf{0}$. Thus $\mathbf{0} < \mathbf{z}M = \mathbf{v}AM - r\mathbf{v}M = \mathbf{v}MA - r\mathbf{v}M$ because A and M commute. Moreover $\mathbf{v}M > \mathbf{0}$ since $\mathbf{v} \in \mathbb{S}_n$ and $M > \mathbf{0}$. So if we define $\mathbf{y} = (1/\mu)\mathbf{v}M$, then $\mathbf{y} \in \mathbb{S}_n$ for a suitable scaling constant μ, and $\mathbf{y}A - r\mathbf{y} > \mathbf{0}$. This contradicts the definition of $r(A)$. Hence $\mathbf{z} = \mathbf{0}$ and r is an eigenvalue of A.

(ii). We know that $\mathbf{v}A = r\mathbf{v}$. So $\mathbf{v}AM = \mathbf{v}MA = r\mathbf{v}M$, and $\mathbf{v}M > \mathbf{0}$. So there exists a strictly positive eigenvector of A corresponding to r.

(iii). The proof is identical to the proof of Statement (iii) in Theorem 3.22, except that instead of multiplying by A^l we multiply by M.

(iv). Ditto.

(v). The proof is identical to that of Statement (v) in Theorem 3.22.

(vi). It has already been shown in Statement (ii) of Theorem 3.24 that each of the complex numbers $\rho(A) \exp \mathbf{i} 2\pi k/p, k = 0, \ldots, p-1$ is an eigenvalue of A. So it remains only to show that there are no other eigenvalues of A with magnitude $\rho(A)$. Suppose λ is an eigenvalue of A. Then λ^p is an eigenvalue of the matrix A^p. Since A has the canonical form (3.8), it follows that λ^p is an eigenvalue of one of the matrices M_i. Each of these matrices is primitive, so every eigenvalue other than $[\rho(A)]^p$ has magnitude strictly less than $[\rho(A)]^p$. Hence we either have $\lambda^p = [\rho(A)]^p$, or else $|\lambda|^p < [\rho(A)]^p$. This is precisely Statement (vi). $\square$

PART 2
Hidden Markov Processes

Chapter Four

Markov Processes

In this chapter, we begin our study of Markov processes, which in turn lead to "hidden" Markov processes, the core topic of the book. We define the "Markov property," and show that all the relevant information about a Markov process assuming values in a finite set of cardinality n can be captured by a nonnegative $n \times n$ matrix called the "state transition matrix," and an n-dimensional probability distribution of the initial state. Then we invoke the results of Chapter 3 on nonnegative matrices to analyze the temporal evolution of Markov processes. Then we proceed to a discussion of more advanced topics such as hitting times and ergodicity.

4.1 BASIC DEFINITIONS

4.1.1 The Markov Property and the State Transition Matrix

For the purposes of the simplified setting studied here, we define a "stochastic process" to be a sequence of random variables $\{X_0, X_1, X_2, \ldots\}$ or $\{X_t\}_{t=0}^{\infty}$ for short, where each X_t assumes values in a common set of finite cardinality, known as the "alphabet" of the process. To make it clear that the alphabet could consist of just abstract symbols, we could denote the alphabet as $\mathbb{A} = \{a_1, \ldots, a_n\}$. However, to facilitate the use of matrix notation, we will use the symbol $\mathbb{N} = \{1, \ldots, n\}$ to represent the alphabet. This leads to some potential abuse wherein the elements of $\mathbb{N}$ are sometimes viewed as abstract labels and at other times as integers. But this is a mild abuse and it will be easy for the reader to determine the appropriate interpretation. Note that in Chapter 1, the symbol $\mathbb{N}$ is used to denote the set of natural numbers. Henceforth however, the symbols $\mathbb{N}$ and $\mathbb{M}$ will be used solely to denote finite sets of cardinality n and m respectively. Though the index t could in principle stand for just about anything, it is most common to think of t as representing "time." If we think of the parameter t as representing "time," then notions such as "past" and "future" make sense. Thus if $t < t'$, then X_t is a "past" variable for $X_{t'}$, while $X_{t'}$ is a "future" variable for X_t. However, the index need not always represent time. For instance, when the stochastic process corresponds to the genome sequence of an organism, the set $\mathbb{N}$ is the four symbol nucleotide alphabet $\{A, C, G, T\}$, and the ordering is spatial rather than temporal. Similarly when we think of the primary structure of a protein, the set $\mathbb{N}$ is the 20-symbol set of amino acids. In the case of both DNA and proteins, the sequences have a definite spatial direction and therefore cannot be "read backwards." This spatial orientation allows us to replace the "past" and "future" by "earlier" and "later."

For each integer T, the random variables $(X_0, X_1, \ldots, X_T)$ have a *joint distribution*, which is a probability distribution over the *finite* set $\mathbb{N}^{T+1}$. Since in this book we try to "avoid the infinite" to the extent possible, we will not speak about the "joint law" of all the infinitely many random variables taken together. However, it is essential to emphasize that our exposition is somewhat constricted due to this self-imposed restriction, and is definitely somewhat impoverished as a consequence. The ability to "cope with the infinite" in a mathematically meaningful and consistent manner is one of the substantial achievements of axiomatic probability theory.

Now we introduce a very fundamental property called the Markov property.

Definition 4.1 *The process* $\{X_t\}_{t=0}^{\infty}$ *is said to* **possess the Markov property**, *or to be a* **Markov process**, *if for every* $t \geq 1$ *and every sequence* $u_0 \ldots u_{t-1} u_t \in \mathbb{N}^{t+1}$, *it is true that*

$$\Pr\{X_t = u_t | X_0 = u_0, \ldots X_{t-1} = u_{t-1}\} = \Pr\{X_t = u_t | X_{t-1} = u_{t-1}\}. \quad (4.1)$$

Recall that a conditional probability of X_t, irrespective of on what measurements it is conditioned, is a probability distribution on the set $\mathbb{N}$. The Markov property states therefore that the conditional probability distribution of the "current state" X_t depends only on the "immediate past" state X_{t-1}, and not on any of the previous states. Thus adding some more measurements prior to time $t-1$ does not in any way alter the conditional probability distribution of X_t.

For convenience, we introduce the notation X_j^k to denote the entity $(X_i, j \leq i \leq k$. Alternatively, $X_j^k = (X_j, X_{j+1}, \ldots, X_{k-1}, X_k)$. Clearly this notation makes sense only if $j \leq k$. Similarly, given an infinite sequence $\{u_t\}$ where each $u_t \in \mathbb{N}$, we define $\mathbf{u}_j^k = (u_j, \ldots, u_k) \in \mathbb{N}^{k-j+1}$. With this notation, we can rephrase Definition 4.1 as follows: The stochastic process $\{X_t\}$ is a Markov process if, for every $\mathbf{u}_0^t \in \mathbb{N}^{t+1}$, it is true that

$$\Pr\{X_t = u_t | X_0^{t-1} = \mathbf{u}_0^{t-1}\} = \Pr\{X_t = u_t | X_{t-1} = u_{t-1}\}.$$

For *any* stochastic process $\{X_t\}$ and *any* sequence $\mathbf{u}_0^t \in \mathbb{N}^{t+1}$, we can apply the law of iterated conditioning and write

$$\Pr\{X_0^t = \mathbf{u}_0^t\} = \Pr\{X_0 = u_0\} \cdot \prod_{i=0}^{t-1} \Pr\{X_{i+1} = u_{i+1} | X_0^i = \mathbf{u}_0^i\}.$$

However, if the process under study is a Markov process, then the above formula can be simplified to

$$\Pr\{X_0^t = \mathbf{u}_0^t\} = \Pr\{X_0 = u_0\} \cdot \prod_{i=0}^{t-1} \Pr\{X_{i+1} = u_{i+1} | X_i = u_i\}. \quad (4.2)$$

Thus, with a Markov process, the length of the "tail" on which the conditioning is carried out is always one.

In the probability literature, one often uses the name "Markov chain" to denote a Markov process $\{X_t\}$ where the underlying parameter t assumes only discrete values (as opposed to taking values in a continuum, such as $\mathbb{R}_+$ for example). In the present book, attention is restricted only to the case where t assumes values in $\mathbb{Z}_+$, the set of nonnegative integers. Accordingly, we use the expressions "Markov process" and "Markov chain" interchangeably.

The formula (4.2) demonstrates the importance of the quantity

$$\Pr\{X_{t+1} = j | X_t = i\},$$

viewed as a function of three entities: The "current" state $i \in \mathbb{N}$, the "next state" $j \in \mathbb{N}$, and the "current time" $t \in \mathbb{Z}_+$. Recall that $\mathbb{N}$ is a finite set. Let us fix the time t for the time being, and define the quantity

$$a_{ij}(t) := \Pr\{X_{t+1} = j | X_t = i\}, i, j \in \mathbb{N} = \{1, \ldots, n\}, t \in \mathbb{Z}_+. \tag{4.3}$$

Thus $a_{ij}(t)$ is the probability of making a transition from the current state i at time t to the next state j at time $t+1$.

Definition 4.2 *The $n \times n$ matrix $A(t) = [a_{ij}(t)]$ is called the* **state transition matrix** *of the Markov process at time t. The Markov chain is said to be* **homogenous** *if $A(t)$ is a constant matrix for all $t \in \mathbb{Z}_+$, and* **inhomogenous** *otherwise.*

Lemma 4.3 *Suppose $\{X_t\}$ is a Markov process assuming values in a finite set $\mathbb{N}$ of cardinality n, and let $A(t)$ denote its state transition matrix at time t. Then $A(t)$ is a* **stochastic matrix** *for all t. That is:*

$$a_{ij}(t) \in [0, 1] \; \forall i, j \in \mathbb{N}, t \in \mathbb{Z}_+.$$

$$\sum_{j=1}^{n} a_{ij}(t) = 1 \; \forall i \in \mathbb{N}, t \in \mathbb{Z}_+.$$

Proof. Both properties are obvious from the definition. A conditional probability always lies between 0 and 1. Moreover, the sum of the conditional probabilities over all possible outcomes equals one. □

Lemma 4.4 *Suppose $\{X_t\}$ is a Markov process assuming values in a finite set $\mathbb{N}$ of cardinality n, and let $A(t)$ denote its state transition matrix at time t. Suppose the initial state X_0 is distributed according to $\mathbf{c}_0 \in \mathbb{S}_n$. That is,*

$$\Pr\{X_0 = x_i\} = c_i \; \forall i \in \mathbb{N}.$$

Then for all $t \geq 0$, the state X_t is distributed according to

$$\mathbf{c}_t = \mathbf{c}_0 A(0) A(1) \ldots A(t-1). \tag{4.4}$$

Proof. Pick an arbitrary tuple $\mathbf{u}_0^t \in \mathbb{N}^{t+1}$. Then it follows from (4.2) that

$$\begin{aligned} \Pr\{X_t = u_t\} &= \sum_{\mathbf{u}_0^{t-1} \in \mathbb{N}^t} \Pr\{X_0 = u_0\} \cdot \prod_{i=0}^{t-1} \Pr\{X_{i+1} = u_{i+1} | X_i = u_i\} \\ &= \sum_{\mathbf{u}_0^{t-1} \in \mathbb{N}^t} c_{u_0} a_{u_0,u_1}(0) \ldots a_{u_{t-1},u_t}(t-1). \end{aligned}$$

But the right side is just the u_t-th component of $\mathbf{c}_t = \mathbf{c}_0 A(0) A(1) \ldots A(t-1)$ written out in expanded form. □

Now we introduce the notion of a multistep Markov process, which is widely used in biological applications and elsewhere.

Definition 4.5 *A stochastic process $\{X_t\}$ over a finite alphabet $\mathbb{N}$ is said to be an s-***step Markov process** *if, for every $\mathbf{u}_0^t \in \mathbb{N}^{t+1}$ with $t \geq s$, it is true that*

$$\Pr\{X_t = u_t | X_0^{t-1} = \mathbf{u}_0^{t-1}\} = \Pr\{X_t = u_t | X_{t-s}^{t-1} = \mathbf{u}_{t-s}^{t-1}\}. \tag{4.5}$$

By comparing Definitions 4.1 and 4.5, we see that (4.5) reduces to (4.1) if $s = 1$. Hence some authors refer to a stochastic process that satisfies (4.1) as a "one-step" Markov process. In a "one-step" Markov process, the conditional distribution of the present state depends only on the immediately preceding state, whereas in an s-step Markov process the conditional distribution of the present state depends only on the s immediately preceding states.

Actually, an s-step Markov process can be interpreted as a one-step Markov process over the state space $\mathbb{N}^s$. To see this, suppose the original process $\{X_t\}$ is an s-step Markov process, and define the associated process $\{Z_t\}$ by $Z_t = X_{t-s+1}^t$. In other words, each Z_t consists of a "block" of s states of the original process $\{X_t\}$. Note that $Z_t = (X_{t-s+1}, \ldots X_t)$ and thus belongs to $\mathbb{N}^s$. Of course if $s = 1$ then $Z_t = X_t$. It is now established that $\{Z_t\}$ is a Markov process over the set $\mathbb{N}^s$, and that its transition matrix has a very special form.

Let us first examine the conditional probability $\Pr\{Z_t = \mathbf{u} | Z_{t-1} = \mathbf{v}\}$ when $\mathbf{u}, \mathbf{v} \in \mathbb{N}^s$ are arbitrary. Since

$$Z_t = (X_{t-s+1}, \ldots X_t), Z_{t-1} = (X_{t-s}, \ldots, X_{t-1}),$$

it is obvious that if the value of Z_{t-1} is "given," then $X_{t-s+1}, \ldots X_t$ are already specified. Therefore, unless the first $s-1$ components of $\mathbf{u}$ are equal to the last $s-1$ components of $\mathbf{v}$, the conditional probability is zero. Now suppose it is indeed the case that $\mathbf{u}, \mathbf{v}$ have the form $\mathbf{u} = (w_1, \ldots, w_{s-1}, u_s)$, $\mathbf{v} = (v_0, w_1, \ldots, w_{s-1})$. Suppose $Z_{t-1} = \mathbf{v}$. Then $X_{t-s+1}, \ldots, X_{t-1}$ are already known, and only X_t is random. Therefore

$$\Pr\{Z_t = \mathbf{u} | Z_{t-1} = \mathbf{v}\} = \Pr\{X_t = u_s | X_{t-s}^{t-1} = \mathbf{v}\}.$$

Conditioning Z_t on values of Z_τ for $\tau < t-1$ does not change the conditional probability, because $X_{t-s}, \ldots, X_{t-1}$ are in any case fixed by specifying Z_{t-1}, and the above conditional probability does not change by adding further conditions (because the original process $\{X_t\}$ is s-step Markov.) This shows that the process $\{Z_t\}$ is a Markov process over the set $\mathbb{N}^s$. Now, since X_t is just the *first component* of the vector Z_t, strictly speaking we should not say that $\{X_t\}$ itself is a Markov process over the set $\mathbb{N}^s$. Rather, we should say that X_t is a "function of a Markov process over the set $\mathbb{N}^s$. This is an example of a "hidden Markov process," a concept introduced in Chapter 6.

Next we examine the structure of the $n^s \times n^s$ transition matrix B associated with the Markov process $\{Z_t\}$. Suppose $\mathbf{v} = (v_0, w_1, \ldots, w_{s-1}) \in \mathbb{N}^s$. Then, as shown above, a transition from $\mathbf{v}$ to another state $\mathbf{u} \in \mathbb{N}^s$ is possible only if $\mathbf{u}$ has the form $\mathbf{u} = (w_1, \ldots, w_{s-1}, u_s)$. Therefore, in the row of B corresponding to the entry $\mathbf{v} = (v_0, w_1, \ldots, w_{s-1})$, only the entries in the column corresponding to $\mathbf{u} = (w_1, \ldots, w_{s-1}, u_s)$ can be nonzero; the remaining entries are perforce equal to zero. In other words, even though the transition matrix B has n^s columns, in each row at most n entries can be nonzero, and the rest are zero.

Example 4.1 Suppose $\{X_t\}$ is a third-order Markov chain over the two-symbol alphabet $\mathbb{A} = \{a, b\}$. Then the associated process of triplets $\{Z_t\}_{t \geq 0}$ where $Z_t = X_{t-2}^t$ is a one-step Markov process over the alphabet $\mathbb{A}^3$. The transition matrix has the form shown below, where $\times$ indicates an entry that *could be nonzero*, while 0 indicates an element that *must be zero*.

$A =$

	aaa	*aab*	*aba*	*abb*	*baa*	*bab*	*bba*	*bbb*
aaa	$\times$	$\times$	0	0	0	0	0	0
aab	0	0	$\times$	$\times$	0	0	0	0
aba	0	0	0	0	$\times$	$\times$	0	0
abb	0	0	0	0	0	0	$\times$	$\times$
baa	$\times$	$\times$	0	0	0	0	0	0
bab	0	0	$\times$	$\times$	0	0	0	0
bba	0	0	0	0	$\times$	$\times$	0	0
bbb	0	0	0	0	0	0	$\times$	$\times$

Example 4.2 This example illustrates a one-step Markov chain, and is a variation on the card game "blackjack," in which the objective is to keep drawing cards until the total value of the cards drawn equals or exceeds twenty one. In the present simplified version, the 13 cards are replaced by a four-sided die. (It may be mentioned that in many ancient cultures, dice were made from animal bones, and had only four sides since they were oblong.) The four sides are labeled as 0, 1, 2, and 3 (as opposed to the more conventional 1, 2, 3, and 4), and are equally likely to appear on any one throw. A player rolls the die again and again, and X_t denotes the accumulated score after t throws. If the total *exactly* equals nine, the player wins; otherwise the player loses. It is assumed that the outcome of each throw is independent of all the previous throws.

It is now shown that $\{X_t\}$ is a Markov process assuming values in the set $\mathbb{N} := \{0, 1, \ldots, 8, W, L\}$ of cardinality eleven. Let Y_t denote the outcome of the roll of the die at time t. Then

$$\Pr\{Y_t = 0\} = \Pr\{Y_t = 1\} = \Pr\{Y_t = 2\} = \Pr\{Y_t = 3\} = 1/4.$$

Now let us examine the distribution of X_t. We know that $X_t = X_{t-1} + Y_t$, except that if $X_{t-1} + Y_t = 9$ we take $X_t = W$ (win), and if $X_{t-1} + Y_t > 9$ we take $X_t = L$ (loss). If $X_{t-1} = W$ or L, then the game is effectively over, and we take $X_t = X_{t-1}$. These observations can be summarized in the following rules: If $X_{t-1} \leq 5$, then

$$\Pr\{X_t = X_{t-1}\} = \Pr\{X_t = X_{t-1} + 1\} =$$
$$\Pr\{X_t = X_{t-1} + 2\} = \Pr\{X_t = X_{t-1} + 3\} = 1/4.$$

If $X_{t-1} = 6$, then

$$\Pr\{X_t = 6\} = \Pr\{X_t = 7\} = \Pr\{X_t = 8\} = \Pr\{X_t = W\} = 1/4.$$

If $X_{t-1} = 7$, then

$$\Pr\{X_t = 7\} = \Pr\{X_t = 8\} = \Pr\{X_t = W\} = \Pr\{X_t = L\} = 1/4.$$

If $X_{t-1} = 8$, then

$$\Pr\{X_t = 8\} = \Pr\{X_t = W\} = 1/4, \Pr\{X_t = L\} = 2/4.$$

Finally, if $X_{t-1} = W$ or L, then $\Pr\{X_t = X_{t-1}\} = 1$.

The process $\{X_t\}$ is a Markov process because the probability distribution of X_t depends *only* on the value of X_{t-1}, and does not at all depend on *how* the score happened to reach X_{t-1}. In other words, when it comes to determining the probability distribution of X_t, only the value of X_{t-1} is relevant, and all past values of $X_i, i \leq t-2$ are irrelevant.

The state transition matrix of the Markov process is an 11×11 matrix given by

$$A = \begin{bmatrix} 1/4 & 1/4 & 1/4 & 1/4 & 0 & 0 & 0 & 0 & 0 & 0 & 0 \\ 0 & 1/4 & 1/4 & 1/4 & 1/4 & 0 & 0 & 0 & 0 & 0 & 0 \\ 0 & 0 & 1/4 & 1/4 & 1/4 & 1/4 & 0 & 0 & 0 & 0 & 0 \\ 0 & 0 & 0 & 1/4 & 1/4 & 1/4 & 1/4 & 0 & 0 & 0 & 0 \\ 0 & 0 & 0 & 0 & 1/4 & 1/4 & 1/4 & 1/4 & 0 & 0 & 0 \\ 0 & 0 & 0 & 0 & 0 & 1/4 & 1/4 & 1/4 & 1/4 & 0 & 0 \\ 0 & 0 & 0 & 0 & 0 & 0 & 1/4 & 1/4 & 1/4 & 1/4 & 0 \\ 0 & 0 & 0 & 0 & 0 & 0 & 0 & 1/4 & 1/4 & 1/4 & 1/4 \\ 0 & 0 & 0 & 0 & 0 & 0 & 0 & 0 & 1/4 & 1/4 & 2/4 \\ 0 & 0 & 0 & 0 & 0 & 0 & 0 & 0 & 0 & 1 & 0 \\ 0 & 0 & 0 & 0 & 0 & 0 & 0 & 0 & 0 & 0 & 1 \end{bmatrix},$$

where the rows and columns are labeled as $\{0, \ldots, 8, W, L\}$. Since the transition matrix does not explicitly depend on t, the Markov chain is homogeneous.

It is natural that the game begins with the initial score equal to zero. Thus the "random variable" X_0 equals zero, or in other words, the initial distribution $\mathbf{c} \in \mathbb{R}^{1 \times 11}$ has a 1 in the first component and zeros elsewhere.

Now repeated application of the formula (4.4) gives the distributions of the random variables X_1, X_2, etc. If $\mathbf{c}_t$ denotes the distribution of X_t, then we have

$$\mathbf{c}_0 = \mathbf{c} = [1 \ 0 \ldots 0],$$

$$\mathbf{c}_1 = \mathbf{c}_0 A = [1/4 \ 1/4 \ 1/4 \ 1/4 \ 0 \ldots 0],$$

$$\mathbf{c}_2 = [1/16 \ 2/16 \ 3/16 \ 4/16 \ 3/16 \ 2/16 \ 1/16 \ 0 \ 0 \ 0 \ 0],$$

and so on. One noteworthy feature of this Markov process is that it is *nonstationary*. Note that each X_t has a different distribution. The precise definition of a stationary process is that the joint distribution (or law) of all the infinitely many random variables $\{X_0, X_1, X_2, \ldots\}$ is the same as the joint law of the variables $\{X_1, X_2, X_3, \ldots\}$. A necessary, but not sufficient, condition for stationarity is that each individual random variable must have the same distribution. This condition does not hold in the present case. So it is important to note that a homogeneous Markov chain can still be nonstationary—it depends on what the initial distribution is.

Another noteworthy point about this Markov chain is that if we examine the distribution $\mathbf{c}_t$, then $\Pr\{X_t = i\}$ approaches zero as $t \to \infty$ for every $i \in \{0, \ldots, 8\}$. In plain English, all games will "eventually" wind up in either a win or a loss. All other states are "transient." This idea is made precise in the next section.

4.1.2 Estimating the State Transition Matrix

In this section we derive a simple closed-form formula for the maximum likelihood estimate for the state transition matrix of a Markov chain, based on a sample path.

Suppose we know that a Markov process $\{X_t\}$ assumes values in a finite alphabet (or state space) $\mathbb{N}$, but we do not know its state transition matrix A. Suppose further that we observe a sample path of the Markov process, in the form of a sequence $\mathbf{x}_1^l := x_1 \ldots x_l$. Using this observation, we would like to find the value of the state transition matrix that makes this observed sequence most likely. Or in other words, we would like to determine the maximum likelihood estimate for A.

This actually turns out to be quite easy. Let $L(\mathbf{x}_1^l|A)$ and $LL(\mathbf{x}_1^l|A)$ denote respectively the likelihood and the log likelihood of the observed sequence $\mathbf{x}_1^l$ if A is the state transition matrix of the underlying Markov process. Then

$$L(\mathbf{x}_1^l|A) = \Pr\{x_1\} \cdot \prod_{t=2}^{l} \Pr\{x_t|x_{t-1}\} = \Pr\{x_1\} \cdot \prod_{t=2}^{l} a_{x_{t-1}x_t}.$$

This formula becomes simpler if we compute the log likelihood.

$$LL(\mathbf{x}_1^l|A) = \log \Pr\{x_1\} + \sum_{t=2}^{l} \log a_{x_{t-1}x_t}.$$

A further simplification is possible. For each pair $(i, j) \in \mathbb{N}^2$, let ν_{ij} denote the number of times that the state sequence ij appears in the sample path $\mathbf{x}_1^l$. In symbols

$$\nu_{ij} = \sum_{t=2}^{l} I_{\{x_{t-1}x_t = ij\}}.$$

Next, define

$$\bar{\nu}_i = \sum_{j=1}^{n} \nu_{ij}.$$

Then it is easy to see that, instead of summing over the sample path, we can also sum over the indices i and j. Then $a_{x_{t-1}x_t}$ equals a_{ij} precisely ν_{ij} times. Therefore

$$LL(\mathbf{x}_1^l | A) = \log \Pr\{x_1\} + \sum_{i=1}^{n} \sum_{j=1}^{n} \nu_{ij} \log a_{ij}.$$

We can ignore the first term as it does not depend on A. Also, the A matrix needs to satisfy the stochasticity constraint that

$$\sum_{j=1}^{n} a_{ij} = 1, \ \forall i \in \mathbb{N}.$$

Therefore we form the Lagrangian

$$J = \sum_{i=1}^{n} \sum_{j=1}^{n} \nu_{ij} \log a_{ij} + \sum_{i=1}^{n} \lambda_i \left(1 - \sum_{j=1}^{n} a_{ij} \right)$$

and minimize that with respect to the various a_{ij}. Next, observe that

$$\frac{\partial J}{\partial a_{ij}} = \frac{\nu_{ij}}{a_{ij}} - \lambda_i.$$

Setting the partial derivatives equal to zero gives

$$\lambda_i = \frac{\nu_{ij}}{a_{ij}}, \text{ or } a_{ij} = \frac{\nu_{ij}}{\lambda_i}.$$

The value of the Lagrange multiplier λ_i is readily obtained from the constraint on A. Specifically, we have

$$\sum_{j=1}^{n} a_{ij} = \sum_{j=1}^{n} \frac{\nu_{ij}}{\lambda_i} = 1 \Rightarrow \lambda_i = \bar{\nu}_i.$$

Therefore the maximum likelihood estimate for the matrix A is given by

$$a_{ij} = \frac{\nu_{ij}}{\bar{\nu}_i}. \tag{4.6}$$

The above technique can be extended quite readily to multistep Markov processes. Suppose we know that $\{X_t\}$ is an s-step Markov process assuming values in a finite alphabet $\mathbb{N}$ of cardinality n, where the value of s is known.

Suppose further that we observe a sample path $\mathbf{x}_1^l = x_1 \ldots x_l$. Then, as we have already seen, $\{X_t\}$ can be viewed as a function of the Markov process $\{Z_t\}$ where $Z_t = X_{t-s+1}^t$. Associated with the Markov process $\{Z_t\}$ is an $n^s \times n^s$ transition matrix B, where in each row at most n elements can be nonzero. To determine the maximum likelihood estimate of the entries of this matrix, we proceed exactly as above. The final answer is as follows: For each $\mathbf{v} = (v_1, v_2, \ldots, v_s) \in \mathbb{N}^s$ and each $u \in \mathbb{N}$, let $\nu_{\mathbf{v}u}$ denote the number of times that the string $\mathbf{v}u$ occurs in the sample path. In symbols,

$$\nu_{\mathbf{v}u} = \sum_{t=s+1}^{l} I_{\{x_{t-s} \ldots x_{t-1} x_t = \mathbf{v}u\}}.$$

Next, define

$$\bar{\nu}_{\mathbf{v}} = \sum_{u \in \mathbb{N}} \nu_{\mathbf{v}u}.$$

Then the maximum likelihood estimate of the transition matrix B is given by

$$b_{\mathbf{v},\mathbf{u}} = \begin{cases} 0 & \text{if } \mathbf{v}_2^s \neq \mathbf{u}_1^{s-1}, \\ \frac{\nu_{\mathbf{v}u_s}}{\bar{\nu}_{\mathbf{v}}} & \text{if } \mathbf{v}_2^s = \mathbf{u}_1^{s-1} \end{cases} \tag{4.7}$$

In practical situations, the above maximum likelihood estimate is often adjusted to avoid zero entries in the estimated transition matrix. This adjustment is often referred to as the **Laplacian correction**.

Example 4.3 Suppose $\{X_t\}$ is a Markov process on the finite alphabet $\mathbb{N} = \{a, b, c\}$ and we observe the sample path *abacabacabc* of length 11. The set $\mathbb{N}^2$ consists of 9 elements, namely $aa, \ldots, cc$. It can be readily verified that

$$\nu_{ab} = 3, \nu_{ac} = \nu_{ba} = \nu_{ca} = 2, \nu_{bc} = 1,$$

and $\nu_{ij} = 0$ for the remaining four pairs ij. In turn this implies that

$$\bar{\nu}_a = 5, \bar{\nu}_b = 3, \bar{\nu}_c = 2.$$

Therefore the maximum likelihood estimate of the transition matrix is given by

$$A = \begin{bmatrix} 0 & 3/5 & 2/5 \\ 2/3 & 0 & 1/3 \\ 1 & 0 & 0 \end{bmatrix}.$$

The above example shows a major weakness of the maximum likelihood estimate. Though the estimate is mathematically correct, it often contains a lot of zero entries. Specifically, if a pair ij does not occur in the sample path, then the associated element a_{ij} is set equal to zero. This can lead to anomalous situations. With the A matrix in Example 4.3 for example, any sample path that contains anyone of the pairs aa, bb, cb, cc has a likelihood of zero, or a log likelihood of minus infinity! Moreover, if the sample path length l is of the same order as n^2, the number of pairs, then it is often the case that several pairs do not occur in the sample path.

To avoid this anomaly, it is common to use an "add one correction," which is also called a "Laplacian correction" even though it is not clear whether Laplace actually advocated it; we shall adopt the same nomenclature in this book as well. The Laplacian correction consists of just artificially adding 1 to all integers ν_{ij}, so as to ensure that there are no zero entries. With the Laplacian correction, the integers become

$$\nu_{ab} = 4, \nu_{ac} = \nu_{ba} = \nu_{ca} = 3, \nu_{bc} = 2,$$

and $\nu_{ij} = 1$ for the remaining four pairs ij. The resulting estimate of A is given by

$$A = \begin{bmatrix} 1/8 & 4/8 & 3/8 \\ 3/6 & 1/6 & 2/6 \\ 2/4 & 1/4 & 1/4 \end{bmatrix}.$$

Example 4.4 The purpose of this example is to illustrate that, when the Laplacian correction is applied to estimate the transition matrix of a multistep Markov process, any element that is "structurally equal to zero" must be left unchanged.

Suppose $\{X_t\}$ is a second-order Markov process over the two-symbol alphabet $\{a, b\}$, and suppose we observe a sample path

$$abaababbababaaabba$$

of length 18. Since $\{X_t\}$ is a second-order Markov process, the associated process $\{Z_t = (X_{t-1}, X_t)\}$ is a Markov process over the set $\{a, b\}^2$. In accordance with (4.7), we compute the integers ν_{ijk} for all triplets $ijk \in \{a, b\}^3$. The "raw" or unadjusted counts are

String	aaa	aab	aba	abb	baa	bab	bba	bbb
Count	1	2	4	2	2	3	2	0

Therefore the Laplacian correction consists of adding one to each of these raw counts, resulting in

String	aaa	aab	aba	abb	baa	bab	bba	bbb
Count	2	3	5	3	3	4	3	1

With these augmented counts, the associated transition matrix is given by

$$A = \left[\begin{array}{c|cccc} & aa & ab & ba & bb \\ \hline aa & 2/5 & 3/5 & 0 & 0 \\ ab & 0 & 0 & 5/8 & 3/8 \\ ba & 3/7 & 4/7 & 0 & 0 \\ bb & 0 & 0 & 3/4 & 1/4 \end{array} \right].$$

Note again that the entries of the matrix A that are zero for "structural reasons" remain at zero.

4.2 DYNAMICS OF STATIONARY MARKOV CHAINS

In this section, we study the dynamics of Markov chains where the state transition matrix is constant with time. By applying the general results on nonnegative matrices from Chapter 3, we show that it is possible to partition the state space into "recurrent" and "transient" states. Then we analyze the dynamics in greater detail, and derive explicit formulas for the probability that a trajectory will hit a specified subset of the state space, as well as the average time needed to do so.

4.2.1 Recurrent and Transient States

In this section, we specialize the contents of Chapter 3 to stochastic matrices, which are a special kind of nonnegative matrix, and thereby draw some very useful conclusions about the temporal evolution of stationary Markov chains.

Definition 4.6 *Suppose A is a stochastic matrix of dimension $n \times n$; that is, $A \in [0,1]^{n \times n}$, and $\sum_{j=1}^{n} a_{ij} = 1$ for all $i \in \mathbb{N}$. Then a vector $\boldsymbol{\pi} \in \mathbb{S}_n$ is said to be a* **stationary distribution** *of A if $\boldsymbol{\pi} A = \boldsymbol{\pi}$.*

The significance of a stationary distribution is obvious. Suppose $\{X_t\}$ is a homogeneous Markov chain with the state transition matrix A. We have seen (as in Example 4.2) that, depending on the initial distribution, the resulting process $\{X_t\}$ may have different distributions at different times. However, suppose A has a stationary distribution $\boldsymbol{\pi}$ (and at the moment we don't know whether a given matrix *has* a stationary distribution). Then $\boldsymbol{\pi} A^t = \boldsymbol{\pi}$ for all t. So if X_0 has the distribution $\boldsymbol{\pi}$, then it follows from (4.4) that X_t also has the same distribution $\boldsymbol{\pi}$ for all values of t. To put it in words, if a Markov chain is started off in a stationary distribution (assuming there is one), then all future states also have the same stationary distribution. It is therefore worthwhile to ascertain whether a given stochastic matrix A does indeed have a stationary distribution, and if so, to determine the set of *all* stationary distributions of A.

Theorem 4.7 *Suppose A is a stochastic matrix. Then*

1. *$\rho(A) = 1$ where $\rho(\cdot)$ is the spectral radius.*

2. *By a symmetric permutation of rows and columns, A can be put into the canonical form*

$$A = \begin{array}{c} \\ \mathcal{E} \\ \mathcal{I} \end{array} \begin{array}{c} \begin{array}{cc} \mathcal{E} & \mathcal{I} \end{array} \\ \left[\begin{array}{cc} P & \mathbf{0} \\ R & Q \end{array} \right] \end{array}, \tag{4.8}$$

where $\mathcal{E}$ is the set of essential states and $\mathcal{I}$ is the set of inessential states.

3. *If the set $\mathcal{I}$ is nonempty, then R contains at least one positive element.*

4. *By further row and column permutations, P can be put in the form*

$$P = \begin{array}{c} \\ \mathcal{E}_1 \\ \vdots \\ \mathcal{E}_s \end{array} \begin{array}{c} \begin{array}{ccc} \mathcal{E}_1 & \ldots & \mathcal{E}_s \end{array} \\ \left[\begin{array}{ccc} P_1 & \ldots & \mathbf{0} \\ \vdots & \ddots & \vdots \\ \mathbf{0} & \ldots & P_s \end{array} \right] \end{array}, \tag{4.9}$$

where each $\mathcal{E}_i$ is a communicating class and each P_i is irreducible.

5. $\rho(P_i) = 1$ *for* $i = 1, \ldots, s$, *and each* P_i *has a unique invariant distribution* $\mathbf{v}_i \in \mathbb{S}_{n_i}$, *where* $n_i = |\mathcal{E}_i|$.

6. *If* $\mathcal{I}$ *is nonempty, then* $\rho(Q) < 1$.

7. *The matrix A has at least one stationary distribution. The set of all stationary distributions of A is given by*

$$\{[\lambda_1 \mathbf{v}_1 \ldots \lambda_s \mathbf{v}_s \;\; \mathbf{0}], \lambda_i \geq 0 \, \forall i, \sum_{i=1}^{s} \lambda_i = 1\}. \tag{4.10}$$

8. *Let* $\mathbf{c} \in \mathbb{S}_n$ *be an arbitrary initial distribution. If* $\mathcal{I}$ *is nonempty, permute the components of* $\mathbf{c}$ *to be compatible with (4.8), and write* $\mathbf{c} = [\mathbf{c}_{\mathcal{E}} \;\; \mathbf{c}_{\mathcal{I}}]$. *Partition* $\mathbf{c}_t = \mathbf{c}A^t$ *as* $\mathbf{c}_t = [\mathbf{c}_{\mathcal{E}}^{(t)} \;\; \mathbf{c}_{\mathcal{I}}^{(t)}]$. *Then* $\mathbf{c}_{\mathcal{I}}^{(t)} \to \mathbf{0}$ *as* $t \to \infty$, *irrespective of* $\mathbf{c}$.

As a prelude to the proof, we present a lemma that may be of some interest in its own right.

Lemma 4.8 *Suppose* $M \in \mathbb{R}_+^{n \times n}$, *and define*

$$\mu(M) := \max_{i \in \mathbb{N}} \sum_{j=1}^{n} m_{ij}.$$

Then

$$\rho(M) = r(M) \leq \mu(M), \tag{4.11}$$

where $r(\cdot)$ *is defined in Lemma 3.4 and (3.9).*

Proof. The equality of $\rho(M)$ and $r(M)$ follows from Statement (i) of Theorem 3.25. Select an arbitrary vector $\mathbf{x} \in \mathbb{S}_n$. It is shown that $r(\mathbf{x}) \leq \mu(M)$; then (4.11) follows from (3.9), the definition of $r(M)$. So suppose $\mathbf{x} \in \mathbb{S}_n$ is arbitrary, and choose an index $j^* \in \mathbb{N}$ such that

$$x_{j^*} = \max_{j \in \mathbb{N}} x_j.$$

The index j^* need not be unique, but this does not matter. We can choose any one index j^* such that the above holds. Then

$$(\mathbf{x}M)_{j^*} = \sum_{i=1}^{n} x_i m_{ij^*} \leq \sum_{i=1}^{n} x_{j^*} m_{ij^*} \leq \mu(M) x_{j^*}.$$

So

$$r(\mathbf{x}) = \min_{j \in \mathbb{N}} \frac{(\mathbf{x}M)_j}{x_j} \leq \frac{(\mathbf{x}M)_{j^*}}{x_{j^*}} = \mu(M).$$

The desired conclusion (4.11) follows. □

Proof. (Of Theorem 4.7): Everything here follows as almost a routine consequence of the results that have already been proved in Chapter 3.

(1) Let $\mathbf{e}_n$ denote the column vector consisting of n ones. Then the fact that A is stochastic can be expressed as

$$A\mathbf{e}_n = \mathbf{e}_n.$$

This shows that $\lambda = 1$ is an eigenvalue of A, with corresponding column eigenvector $\mathbf{e}_n$. So $\rho(A) \geq 1$. On the other hand, since every row of A sums to one, it follows from Lemma 4.8 that $\mu(A) = 1$, whence $\rho(A) \leq r(A) \leq \mu(A) = 1$. Combining these two observations shows that indeed $\rho(A) = 1$.

(2) The canonical form (4.8) is a ready consequence of Theorem 3.9. In fact (4.8) is the same as (3.4).

(3) Since A is stochastic, no row of A can be identically zero. So the statement follows from Theorem 3.9.

(4) This statement also follows from Theorem 3.9.

(5) Since A is stochastic, it is obvious from (4.8) and (4.9) that each P_i is also stochastic. Hence by Statement 1 it follows that $\rho(P_i) = 1$ for all i. But now we have the additional information that each P_i is irreducible. Hence we conclude from Statement (iv) of Theorem 3.25 that there is a positive eigenvector $\phi_i \in \mathbb{R}_+^{n_i}$ such that $\phi_i P_i = \phi_i$, and that all row eigenvectors of P_i corresponding to the eigenvalue one are multiples of this ϕ_i. Choose $\mathbf{v}_i \in \mathbb{S}_{n_i}$ to be a multiple of ϕ_i. Obviously $\mathbf{v}_i$ is unique.

(6) Suppose $\mathcal{I}$ is nonempty, and let $m = |\mathcal{I}|$ denote the cardinality of $\mathcal{I}$. Partition A^m as

$$A^m = \begin{array}{c} \\ \mathcal{E} \\ \mathcal{I} \end{array} \begin{array}{c} \begin{array}{cc} \mathcal{E} & \mathcal{I} \end{array} \\ \left[\begin{array}{cc} P^m & \mathbf{0} \\ R^{(m)} & Q^m \end{array} \right] \end{array}.$$

Then, from the second statement of Lemma 3.5, it follows that each row of $R^{(m)}$ contains a nonzero element. Thus each row sum of Q^m is strictly less than one. Now it follows from Lemma 4.8 that $\rho(Q^m) < 1$. Since $\rho(Q^m) = [\rho(Q)]^m$, it follows that $\rho(Q) < 1$.

(7) Put A in the form (5.5) and look at the equation $\boldsymbol{\pi} A = \boldsymbol{\pi}$. If $\mathcal{I}$ is nonempty, partition $\boldsymbol{\pi}$ as $[\boldsymbol{\pi}_{\mathcal{E}} \ \boldsymbol{\pi}_{\mathcal{I}}]$. Then $\boldsymbol{\pi} A = \boldsymbol{\pi}$ becomes

$$[\boldsymbol{\pi}_{\mathcal{E}} \ \boldsymbol{\pi}_{\mathcal{I}}] \left[\begin{array}{cc} P & \mathbf{0} \\ R & Q \end{array} \right] = [\boldsymbol{\pi}_{\mathcal{E}} \ \boldsymbol{\pi}_{\mathcal{I}}].$$

Expanding this leads to

$$\boldsymbol{\pi}_{\mathcal{E}} P + \boldsymbol{\pi}_{\mathcal{I}} R = \boldsymbol{\pi}_{\mathcal{E}}, \text{ and } \boldsymbol{\pi}_{\mathcal{I}} Q = \boldsymbol{\pi}_{\mathcal{I}}.$$

But since $\rho(Q) < 1$, the matrix $Q - I$ is nonsingular, and so $\boldsymbol{\pi}_{\mathcal{I}} = \mathbf{0}$. This shows that all stationary distributions have zeros in all components corresponding to $\mathcal{I}$. So we seek all solutions to $\boldsymbol{\pi}_{\mathcal{E}} P = \boldsymbol{\pi}_{\mathcal{E}}$. If we now put P in the form (4.9) and partition $\boldsymbol{\pi}$ as $[\boldsymbol{\pi}_1 \ldots \boldsymbol{\pi}_s]$, then *each* $\boldsymbol{\pi}_i$ must satisfy $\boldsymbol{\pi}_i P_i = \boldsymbol{\pi}_i$. Each of these equations has a unique solution $\mathbf{v}_i \in \mathbb{S}_{n_i}$, by Statement 5. Hence the set of all stationary distributions is given by (4.10).

(8) If A is put in the form (4.8) and $\mathcal{I}$ is nonempty, then A^t has the form

$$A^t = \begin{array}{c} \\ \mathcal{E} \\ \mathcal{I} \end{array} \begin{array}{c} \begin{array}{cc} \mathcal{E} & \mathcal{I} \end{array} \\ \left[\begin{array}{cc} P^t & \mathbf{0} \\ R^{(t)} & Q^t \end{array} \right] \end{array} .$$

Now, since $\rho(Q) < 1$, it follows that $Q^t \to \mathbf{0}$ as $t \to \infty$. Hence $\mathbf{c}_t = \mathbf{c}A^t$ implies that

$$\mathbf{c}_{\mathcal{I}}^{(t)} = \mathbf{c}_{\mathcal{I}} Q^t \to \mathbf{0} \text{ as } t \to \infty,$$

irrespective of what $\mathbf{c}_{\mathcal{I}}$ is. This is the desired conclusion. □

In the Markov chain parlance, the states in $\mathcal{I}$ are referred to as **transient states**, and those in $\mathcal{E}$ are referred to as **recurrent states**. Statement 8 of Theorem 4.7 gives the rationale for this nomenclature. Irrespective of the initial distribution of the Markov chain, we have that

$$\Pr\{X_t \in \mathcal{I}\} \to 0 \text{ as } t \to \infty,$$

while

$$\Pr\{X_t \in \mathcal{E}\} \to 1 \text{ as } t \to \infty.$$

Within the set $\mathcal{E}$ of recurrent states, the disjoint equivalence classes $\mathcal{E}_1, \ldots, \mathcal{E}_s$ are referred to as the **communicating classes**.

Example 4.5 Let us return to the modified blackjack game of Example 4.2. From the state transition matrix, it is obvious that each of the states 0 through 8 leads to W, but W does not lead to any of these states. (The same statement is also true with W replaced by L.) Thus all of these states are inessential and therefore transient. In contrast, both W and L are essential states, because they do not lead to any other states, and thus vacuously satisfy the condition for being essential. Hence it follows from Theorem 4.7 that

$$\Pr\{X_t \in \{W, L\}\} \to 1 \text{ as } t \to \infty.$$

Thus all games end with the player either winning or losing; all intermediate states are transient. Within the set of essential states, since W does not lead to L and vice versa, both $\{W\}$ and $\{L\}$ are communicating classes (consisting of singleton sets).

4.2.2 Hitting Probabilities and Mean Hitting Times

Until now we have introduced the notions of recurrent states and transient states, and have shown that eventually all trajectories enter the set of recurrent states. In this subsection we analyze the dynamics of a Markov chain in somewhat greater detail and study the probability that a trajectory will hit a specified subset of the state space, as well as the average time needed to do so.

To motivate these concepts, let us re-examine the modified blackjack game of Example 4.2. It is shown in Example 4.5 that all trajectories will hit either W (win) or L (loss) with probability one. That is a very coarse analysis of the trajectories. It would be desirable to know the *probability of winning (or losing)* from a given starting position. The hitting probability formalizes this notion. Further, it would be desirable to know *the expected number of moves* that would result in a win or loss. The mean hitting time formalizes this notion.

Suppose $\{X_t\}_{t\geq 0}$ is a Markov process assuming values in a finite state space $\mathbb{N} = \{1, \ldots, n\}$. A subset $S \subseteq \mathbb{N}$ is said to be **absorbing** if $X_t \in S \Rightarrow X_{t+1} \in S$ (and by extension, that $X_{t+k} \in S$ for all $k \geq 1$). Clearly S is absorbing if and only if $a_{ij} = 0$ for all $i \in S, j \notin S$.

Next, suppose $S \subseteq \mathbb{N}$, not necessarily an absorbing set. Define

$$h(S; i, t) := \Pr\{X_t \in S | X_0 = i\}.$$

Thus $h(S; i, t)$ is the probability that a trajectory of the Markov process starting at time 0 in state i "hits" the set S at time t. In the same spirit, define

$$\bar{h}(S; i, t) := \Pr\{\exists l, 0 \leq l \leq t, \text{ s.t. } X_l \in S | X_0 = i\}. \tag{4.12}$$

Thus $\bar{h}(S; i, t)$ is the probability that a trajectory of the Markov process starting at time 0 in state i "hits" the set S *at or before* time t. Note that if S is an absorbing set, then $h(S; i, t) = \bar{h}(S; i, t)$. However, if S is not an absorbing set, then the two quantities need not be the same. In particular, $\bar{h}(S; i, t)$ is a nondecreasing function of t, whether or not S is an absorbing set. The same cannot be said of $h(S; i, t)$. Now let us define

$$g(S; i, t) := \Pr\{X_s \notin S \text{ for } 0 \leq s \leq t-1 \wedge X_t \in S | X_0 = i\}, \tag{4.13}$$

where, as before, the wedge symbol $\wedge$ denotes "and." Then $g(S; i, t)$ is the probability that a trajectory of the Markov process starting at time 0 in state i "hits" the set S *for the first time* at time t. From this definition, it is easy to see that

$$\bar{h}(S; i, t) = \bar{h}(S; i, t-1) + g(S; i, t),$$

and as a result

$$\bar{h}(S; i, l) = \sum_{l=0}^{t} g(S; i, l).$$

Now, since the sequence $\{\bar{h}(S; i, t)\}$ is increasing and bounded above by 1 (since every $\bar{h}(S; i, t)$ is a probability), the sequence converges to some limit as $t \to \infty$. This limit is denoted by $\bar{h}(S; i)$ and called the **hitting probability** of the set S given the initial state i. Incidentally, the same argument also shows that $g(S; i, t) \to 0$ as $t \to \infty$. In words, the probability of hitting a set for the first time at time t approaches zero as $t \to \infty$.

Next, let us define an integer-valued random variable denoted by $\lambda(S;i)$, that assumes values in the set $\mathcal{N} \cup \{0\} \cup \{\infty\}$, where $\mathcal{N}$ denotes the set of natural numbers $\{1, 2, \ldots\}$.[1] Thus $\lambda(S;i)$ assumes values in the set of nonnegative integers, plus infinity. We assign

$$\Pr\{\lambda(S;i) = t\} = g(S;i,t), \Pr\{\lambda(S;i) = \infty\} = 1 - \sum_{t=0}^{\infty} g(S;i,t).$$

We refer to $\lambda(S;i)$ as the "hitting time" of set S when starting from the initial state i.

The **mean hitting time** $\tau(S;i)$ is defined as the expected value of the random hitting time $\lambda(S;i)$; thus

$$\tau(S;i) := \sum_{t=0}^{\infty} t g(S;i,t) + \infty \cdot \left[1 - \sum_{t=0}^{\infty} g(S;i,t) \right]. \tag{4.14}$$

If the quantity inside the square brackets is positive, then the mean hitting time is taken as infinity, by convention. Even if this quantity is zero, the mean hitting time could still be infinite if the hitting time is a heavy-tailed random variable.

In spite of the apparent complexity of the above definitions, there are very simple explicit characterizations of both the hitting probability and mean hitting time. The derivation of these characterizations is the objective of this subsection.

Theorem 4.9 *The vector* $\bar{\mathbf{h}}(S) := [\bar{h}(S;1) \ldots \bar{h}(S;n)]$ *is the minimal nonnegative solution of the set of equations*

$$v_i = 1 \text{ if } i \in S, v_i = \sum_{j \in \mathbb{N}} a_{ij} v_j \text{ if } i \notin S. \tag{4.15}$$

Proof. Here, by a "minimal nonnegative solution," we mean that (i) $\bar{\mathbf{h}}(S)$ satisfies (4.15), and (ii) if $\mathbf{v}$ is any other nonnegative vector that satisfies the same equations, then $\mathbf{v} \geq \bar{\mathbf{h}}(S)$. To prove (i), observe that if $i \in S$, then $h(S;i,0) = 1$, whence $\bar{h}(S;i,t) = \bar{h}(S;i,0) = 1$ for all t. Hence $\bar{h}(S;i)$ being the limit of this sequence also equals one. On the other hand, if $i \notin S$, then

$$h(S;i,t+1) = \sum_{j \in \mathbb{N}} a_{ij} h(S;j,t).$$

This equation states that the probability of hitting S at time $t+1$ starting from the initial state i at time 0 equals the probability of hitting S at time t starting from the initial state j at time 0, weighted by the probability of making the transition from i to j. In writing this equation, we have used both the Markovian nature of the process as well as its stationarity. Now the same reasoning shows that

$$\bar{h}(S;i,t+1) = \sum_{j \in \mathbb{N}} a_{ij} \bar{h}(S;j,t).$$

[1] This is a one-time departure from earlier notation, where $\mathbb{N}$ is used for the set of natural numbers. However, since this set occurs so infrequently, it is preferred to use $\mathbb{N}$ for the state space of a Markov process.

Letting $t \to \infty$ in both sides of the above equation shows that

$$\bar{h}(S;i) = \sum_{j \in \mathbb{N}} a_{ij} \bar{h}(S;j).$$

Thus the vector of hitting probabilities satisfies (4.15).

To establish the second statement, suppose $\mathbf{v} \in \mathbb{R}_+^n$ is some solution of (4.15). Then $v_i = 1$ for all $i \in S$. For $i \notin S$, we get from (4.15),

$$\begin{aligned} v_i &= \sum_{j \in \mathbb{N}} a_{ij} v_j = \sum_{j \in S} a_{ij} v_j + \sum_{j \notin S} a_{ij} v_j \\ &= \sum_{j \in S} a_{ij} + \sum_{j \notin S} a_{ij} v_j. \end{aligned}$$

Now we can substitute a second time for v_j to get

$$\begin{aligned} v_i &= \sum_{j \in S} a_{ij} + \sum_{j \notin S} a_{ij} v_j \\ &= \sum_{j \in S} a_{ij} + \sum_{j \notin S} \sum_{k \in S} a_{ij} a_{jk} + \sum_{j \notin S} \sum_{k \notin S} a_{ij} a_{jk} v_k. \end{aligned}$$

This process can be repeated. If we do this l times, we get

$$\begin{aligned} v_i = & \sum_{j_1 \in S} a_{ij_1} \\ & + \sum_{j_1 \notin S} \sum_{j_2 \in S} a_{ij_1} a_{j_1 j_2} + \ldots \\ & + \sum_{j_1 \notin S, \ldots, j_{l-1} \notin S} \sum_{j_l \in S} a_{ij_1} a_{j_1 j_2} \ldots a_{j_{l-1} j_l} \\ & + \sum_{j_1 \notin S, \ldots, j_{l-1} \notin S} \sum_{j_l \notin S} a_{ij_1} a_{j_1 j_2} \ldots a_{j_{l-1} j_l} v_{j_l}. \end{aligned} \tag{4.16}$$

Next, observe that for each index l, we have

$$\sum_{j_1 \notin S, \ldots, j_{l-1} \notin S} \sum_{j_l \in S} a_{ij_1} a_{j_1 j_2} \ldots a_{j_{l-1} j_l} = g(S; i, l),$$

because the left side is precisely the probability that, starting in state i at time 0, the trajectory hits S for the first time at time l. Moreover, since $i \notin S$, it is clear that $g(S; i, 0) = 0$. Finally, from (4.16), it is clear that the last term in the summation is nonnegative, because $\mathbf{v}$ is a nonnegative vector. Thus it follows from (4.16) that

$$v_i \geq \sum_{s=1}^{l} g(S; i, l) = \sum_{s=0}^{l} g(S; i, s) = \bar{h}(S; i, l).$$

Since this is true for each l, letting $l \to \infty$ shows that $v_i \geq \bar{h}_i(S; i)$, as desired. □

i	$\bar{h}(W;i)$	$\bar{h}(L;i)$
8	1/3	2/3
7	4/9	5/9
6	16/27	11/27
5	37/81	44/81
4	121/243	122/243
3	376/729	353/729
2	1072/2187	1118/2187
1	3289/6561	3272/6561
0	9889/19683	9794/19683

Table 4.1 Hitting Probabilities from Various Initial States

In applying (4.15), it is often convenient to rewrite it as

$$\bar{h}(S;i) = 1 \text{ if } i \in S, \bar{h}(S;i) = \frac{1}{1-a_{ii}} \sum_{j \neq i} a_{ij} \bar{h}(S;j) \text{ if } i \notin S. \tag{4.17}$$

If $a_{ii} < 1$, then the fraction above makes sense. If $a_{ii} = 1$, then a trajectory starting in state i just stays there, so clearly $\bar{h}(S;i) = 0$. Note that the right side of (4.17) is a convex combination of the quantities $\bar{h}(S;j)$ for $j \neq i$, because $\sum_{j \neq i} a_{ij} = 1 - a_{ii}$.

Example 4.6 Let us return to the simplified "blackjack" example of Examples 4.2 and 4.5. We have already seen that every trajectory approaches W or L with probability 1. Now let us compute the probability of winning or losing from a given initial state. To simplify notation, let us write $\bar{h}(W;i)$ to denote $\bar{h}(\{W\};i)$, and define $\bar{h}(L;i)$ in an analogous fashion.

From (4.17), it is obvious that $\bar{h}(W;W) = 1$ and $\bar{h}(L;L) = 1$, because both are absorbing states. For the same reason, we have $\bar{h}(W;L) = 0$ and $\bar{h}(L;W) = 0$. Working backwards, we have from (4.15) that

$$\bar{h}(W;8) = \frac{1}{3} a_{8W} \bar{h}(W;W) + \frac{2}{3} a_{8L} \bar{h}(W;L) = \frac{1}{3} a_{WW} = \frac{1}{3}.$$

Similarly

$$\bar{h}(L;8) = \frac{1}{3} a_{8W} \bar{h}(L;W) + \frac{2}{3} a_{8L} \bar{h}(L;L) = \frac{2}{3} a_{LL} = \frac{2}{3}.$$

It is hardly surprising that $\bar{h}(W;8) + \bar{h}(L;8) = 1$, because these are the only two absorbing states. In fact it is true that $\bar{h}(W;i) + \bar{h}(L;i) = 1$ for *every* initial state i. Next,

$$\bar{h}(W;7) = \frac{1}{3}\bar{h}(W;8) + \frac{1}{3}\bar{h}(W;W) + \frac{1}{3}\bar{h}(W;L) = \frac{4}{9}.$$

Proceeding in this manner, we get the results shown in Table 4.1.

Thus we see that the probability of winning from a particular starting state goes up and down like a yo-yo. The states 4 and 1 are closest to being "neutral" in that the odds of winning and losing are roughly equal, while the state 6 offers the best prospects for winning. This is not surprising, because from the initial state 6 one cannot possibly lose in one time step, but winning in one time step is possible.

Thus far we have analyzed the hitting probability $\bar{h}(S;i)$. Next we analyze the mean hitting $\tau(S;i)$ defined in (4.14). It turns out that $\tau(S;i)$ also satisfies a simple linear recursion analogous to (4.15).

Theorem 4.10 *Suppose that the mean hitting time $\tau(S;i)$ is finite for all $i \in \mathbb{N}$. Then the vector $\boldsymbol{\tau}(S)$ of mean hitting times $[\tau(S;1)\ldots\tau(S;n)]$ is the minimal nonnegative solution of the equations*

$$\alpha_i = 0 \text{ if } i \in S, \alpha_i = 1 + \sum_{j \notin S} a_{ij}\alpha_j \text{ if } i \notin S. \tag{4.18}$$

Proof. As in Theorem 4.9, we need to establish two statements: First, the vector of mean hitting times satisfies (4.18). Second, if $\boldsymbol{\alpha}$ is any other solution of (4.18), then $\alpha_i \geq \tau(S;i)$ for all i.

To prove the first statement, suppose first that $i \in S$. Then clearly $\tau(S;i) = 0$. Next, suppose $i \notin S$, and consider all possible next states j; the transition from i to j occurs with probability a_{ij}. With $X_0 = i$ and $X_1 = j$, we have

$$\begin{aligned}
&\Pr\{X_l \notin S \text{ for } 0 \leq l \leq t-1 \wedge X_t \in S | X_0 = i \wedge X_1 = j\} \\
&\quad = \Pr\{X_l \notin S \text{ for } 1 \leq l \leq t-1 \wedge X_t \in S | X_1 = j\} \\
&\quad = \Pr\{X_l \notin S \text{ for } 0 \leq l \leq t-2 \wedge X_t \in S | X_0 = j\}.
\end{aligned}$$

Here the first equation follows from the Markovian nature of the process $\{X_t\}$, because the behavior of X_l for $l \geq 2$ depends only on X_1 and is independent of X_0. The second equation follows from the stationarity of the process. Hence, if $X_0 = i \notin S$, we can distinguish between two possibilities: If $X_1 = j \in S$, then the mean hitting time from then onwards is zero, and the mean hitting time from the start is 1. If $X_1 = j \notin S$, then the mean hitting time from then onwards is $\tau(S;j)$, and the mean hitting time from the beginning is $1 + \tau(S;j)$. We can average over all of these events to get

$$\begin{aligned}
\tau(S;i) &= \sum_{j \in S} a_{ij} + \sum_{j \notin S} a_{ij}[1 + \tau(S;j)] \\
&= 1 + \sum_{j \notin S} a_{ij}\tau(S;j),
\end{aligned}$$

because

$$\sum_{j \in S} a_{ij} + \sum_{j \notin S} a_{ij} = \sum_{j \in \mathbb{N}} a_{ij} = 1.$$

Therefore the vector $\boldsymbol{\tau}(S)$ of mean hitting times satisfies (4.18).

To prove the second statement, suppose $\boldsymbol{\alpha}$ is any nonnegative solution of (4.18). Then $\alpha_i = 0$ for $i \in S$. For $i \notin S$, it follows from (4.18) that

$$\begin{aligned} \alpha_i &= 1 + \sum_{j \notin S} a_{ij} \alpha_j \\ &= 1 + \sum_{j \notin S} a_{ij} \left[1 + \sum_{k \notin S} a_{jk} \alpha_k \right] \\ &= 1 + \sum_{j \notin S} a_{ij} + \sum_{j \notin S} \sum_{k \notin S} a_{ij} a_{jk} \alpha_k . \end{aligned} \tag{4.19}$$

This process can be repeated by substituting for α_k. If we do this l times, we get

$$\begin{aligned} \alpha_i &= 1 + \sum_{j_1 \notin S} a_{ij_1} + \sum_{j_1, j_2 \notin S} a_{ij_1} a_{j_1 j_2} + \dots \\ &+ \sum_{j_1, \dots, j_{l-1} \notin S} a_{ij_1} a_{j_1 j_2} \cdots a_{j_{l-2} j_{l-1}} \\ &+ \sum_{j_1, \dots, j_l \notin S} a_{ij_1} a_{j_1 j_2} \cdots a_{j_{l-1} j_l} \alpha_{j_l} . \end{aligned} \tag{4.20}$$

Now the last term is nonnegative since $\boldsymbol{\alpha} \geq \mathbf{0}$. As for the remaining terms, since $i \notin S$ it is clear that the hitting time $\lambda(S; i) \geq 1$ with probability one. Thus

$$\Pr\{\lambda(S; i) \geq 1\} = 1.$$

More generally, it is easy to see that

$$\begin{aligned} \sum_{j_1, \dots, j_{l-1} \notin S} a_{ij_1} a_{j_1 j_2} \cdots a_{j_{l-2} j_{l-1}} &= \Pr\{X_s \notin S \text{ for } 0 \leq s \leq l-1 | X_0 = i\} \\ &= \Pr\{\lambda(S; i) \geq l\}, \end{aligned}$$

because the left side is the probability that the trajectory starting at $i \notin S$ at time 0 does not hit the set S during the first $l-1$ transitions. Hence, after neglecting the last term on the right side of (4.20) because it is nonnegative, we get

$$\alpha_i \geq \sum_{k=1}^{l} \Pr\{\lambda(S; i) \geq k\}. \tag{4.21}$$

To conclude the proof, we observe that, as a consequence of (4.13), we have

$$\Pr\{\lambda(S; i) \geq k\} = \sum_{t=k}^{\infty} \Pr\{\lambda(S; i) = t\} = \sum_{t=k}^{\infty} g(S; i, t). \tag{4.22}$$

We need not worry about the convergence of the infinite summation because $\Pr\{\lambda(S; i) \geq k\} \leq 1$ and so the sum converges. Substituting from (4.22) into

(4.21) shows that

$$\alpha_i \geq \sum_{k=1}^{l} \sum_{t=k}^{\infty} g(S;i,t) \geq \sum_{k=1}^{l} \sum_{t=k}^{l} g(S;i,t).$$

Now a simple counting argument shows that

$$\sum_{k=1}^{l} \sum_{t=k}^{l} g(S;i,t) = \sum_{t=1}^{l} t g(S;i,t). \tag{4.23}$$

An easy way to see this is to consider the triangular matrix below, where g_t is shorthand for $g(S;i,t)$.

$$\begin{bmatrix} g_1 & g_2 & \dots & g_{l-1} & g_l \\ 0 & g_2 & \dots & g_{l-1} & g_l \\ \vdots & \vdots & \vdots & \vdots & \vdots \\ 0 & 0 & \dots & g_{l-1} & g_l \\ 0 & 0 & \dots & 0 & g_l \end{bmatrix}$$

The left side of (4.23) is obtained by first summing each row and then adding the row sums; the right side of (4.23) is obtained by first summing each column and then adding the column sums. Clearly both procedures give the same answer, which is what (4.23) says. Let us substitute from (4.22) and (4.23) into (4.21). This gives

$$\alpha_i \geq \sum_{k=1}^{l} t g(S;i,t), \ \forall l.$$

Now letting $l \to \infty$ shows that $\alpha_i \geq \tau(S;i) \ \forall i$, which is the desired statement. □

Example 4.7 Let us again return to the simplified blackjack game and compute the expected duration of the game starting from each initial state. Let us define $E = \{W, L\}$, so that $X_t \in E$ suggests that the game has ended. We can apply (4.18) to compute the mean hitting times. It is again convenient to rewrite (4.18) as

$$\tau(S;i) = \frac{1}{1 - a_{ii}} \left[1 + \sum_{j \notin S, j \neq i} a_{ij} \tau(S;j) \right].$$

It is clear that $\tau(E;W) = \tau(E;L) = 0$, as both W and L belong to E. If $X_0 = 8$, then

$$\tau(E;8) = \frac{4}{3},$$

since the summation on the right side is empty and is thus taken as zero. Next,

$$\tau(E;7) = \frac{4}{3}(1 + a_{78}\tau(E;8)) = \frac{4}{3}(1 + 1/3) = \frac{16}{9}.$$

Proceeding in this manner, we get the table below.

i	$\tau(E;i)$	$\approx$
8	4/3	1.333
7	16/9	1.778
6	64/27	2.370
5	256/81	3.160
4	916/243	3.700
3	3232/729	4.433
2	11200/2187	5.121
1	37888/6561	5.775
0	126820/19683	6.443

4.3 ERGODICITY OF MARKOV CHAINS

In this section we study a very important property of Markov chains, known as "ergodicity." The reader is cautioned that the words "ergodic" and "ergodicity" are used by different authors to refer to different properties. This problem seems to exist only in the Markov process literature. Ergodicity is a property that can be defined for *arbitrary* stationary stochastic processes assuming values in uncountably infinite sets, not just homogeneous Markov chains assuming values in a finite set, as studied here. In this very general setting, there is only one universally accepted definition: A process is defined to be ergodic if, in some precisely defined sense, it explores all parts of the state space. The interested reader is referred to advanced texts such as [61, 62] for a precise statement. The definition of ergodicity used in the present book is consistent with the more general definition. That is, a Markov process over a finite state space is ergodic in the sense defined here if and only if it is ergodic in the general sense used in the stochastic process literature. Unfortunately in the Markov process literature, the adjective "ergodic" is often used to describe a much stronger property. Indeed there appear to be two different, and inconsistent, definitions even within the Markov process literature. Thus the reader should verify the definition used in a particular source before comparing across sources.

Suppose X is a random variable assuming values in a finite set $\mathbb{N} := \{x_1, \ldots, x_n\}$ with probability distribution $\boldsymbol{\pi}$, and that $f : \mathbb{N} \to \mathbb{R}$. Then $f(X)$ is a *real-valued* random variable that assumes values in a finite subset $\{f(x_1), \ldots, f(x_n)\} \subseteq \mathbb{R}$. Moreover the expected value of the random variable $f(X)$ is given, as before, by

$$E[f(X), P_{\boldsymbol{\pi}}] = \sum_{i=1}^{n} f(x_i)\pi_i.$$

Therefore, in order to compute the expected value, one has to enumerate $f(x_i)$ at all values of the argument x_i, multiply by π_i, and then sum.

However, there are situations in which the above definition is difficult or impossible to apply directly. For instance, it may be that the cardinality of the underlying set $\mathbb{N}$ is too large to do the actual summation in a reasonable amount of time. In such a case, Monte Carlo simulation as discussed in Section 1.3 is a popular method for estimating the expected value. In this approach, one generates independent samples $x_1, \ldots, x_l$ of the random variable X, and computes the empirical average

$$\hat{E}(f; \mathbf{x}_1^l) := \frac{1}{l} \sum_{i=1}^{l} f(x_i). \tag{4.24}$$

Then it is possible to give a quantitative estimate of the probability that the estimated value $\hat{E}(f; \mathbf{x}_1^l)$ differs from the true value $E[f(X)]$ by a prespecified tolerance ϵ; see Section 1.3 for details.

The so-called **Markov Chain Monte Carlo (MCMC)** simulation consists of choosing the successive samples x_i as the successive states of a Markov chain whose stationary distribution is $\boldsymbol{\pi}$. Suppose $\{X_t\}$ is a Markov process over the finite set $\mathbb{N}$ with state transition matrix A and stationary distribution $\boldsymbol{\pi}$, and suppose $f : \mathbb{N} \to \mathbb{R}$. Let $\mathcal{F} := \{f(x_1, \ldots, f(x_n)\}$ denote the image of f. Suppose the Markov process is started off with the initial distribution $\mathbf{c}_0$ on the set $\mathbb{N}$. Then X_t has the probability distribution $\mathbf{c}_0 A^t$ over the set $\mathbb{N}$, for all $t \geq 0$. Therefore the function $f(X_t)$ also has the probability distribution $\mathbf{c}_0 A^t$, but over the set $\mathcal{F}$. Now suppose that we define the empirical average $\hat{f}_T$ as

$$\hat{f}_T = \frac{1}{T} \sum_{t=0}^{T-1} f(X_t).$$

Then $\hat{f}_T$ is a real-valued random variable, and its expected value is the average of the expected values of $f(X_t), t = 0, \ldots, T-1$. Since $f(X_t)$ assumes values in $\mathcal{F}$ with the distribution $\mathbf{c}_t = \mathbf{c}_0 A^t$, it follows that the expected value of $\hat{f}_T$ equals $E[f, P_{\bar{\mathbf{c}}_T}]$, where

$$\bar{\mathbf{c}}_T = \frac{1}{T} \sum_{t=0}^{T-1} \mathbf{c}_0 A^t.$$

Therefore, if $\bar{\mathbf{c}}_T \to \boldsymbol{\pi}$ as $T \to \infty$, then the expected value of the average $\hat{f}_T$ would approach the quantity that we wish to compute, namely $E[f, \boldsymbol{\pi}]$. This suggests that the Markov process used to generate the successive samples should have the property that

$$\frac{1}{T} \sum_{t=0}^{T-1} \mathbf{c}_0 A^t \to \boldsymbol{\pi} \text{ as } T \to \infty, \ \forall \mathbf{c}_0 \in \mathbb{S}_n.$$

Actually, as stated in Definition 4.14, it is enough if the above property holds for all $\mathbf{c}_0 \in \mathbb{S}_n$ that are dominated by $\boldsymbol{\pi}$. Moreover, in practical implementations of the MCMC algorithm, one does not start averaging at time zero. Instead, one lets the Markov chain run for some τ instants of time, and then defines

$$\hat{f}_T = \frac{1}{T} \sum_{t=\tau+1}^{\tau+T} f(X_t).$$

The modifications of the theory below that are required to handle this change are simple and left to the reader.

The discussion above leads naturally to the concepts introduced next. Suppose $A \in \mathbb{R}_+^{n \times n}$ is a stochastic matrix, and we define

$$\bar{A}(T) := \frac{1}{T} \sum_{t=0}^{T-1} A^t. \tag{4.25}$$

We refer to $\bar{A}(T)$ as the "ergodic average" of A over T time instants, and the limit of $\bar{A}(T)$ as $T \to \infty$, if it exists, as the **ergodic limit** of A. The question of interest here is: Does A have an ergodic limit, and if so, what does the limit look like? It is shown in the sequel that the limit *does* exist for *every* stochastic matrix, and an explicit formula for the limit is derived.

Theorem 4.11 *Suppose $A \in \mathbb{R}_+^{n \times n}$ is a stochastic matrix, and assume without loss of generality that A is in the canonical form (4.8). Define the eigenvectors $\mathbf{v}_i \in \mathbb{S}_{n_i}$ as in Statement 5 of Theorem 4.7, and let $\mathbf{e}_{n_i}$ denote the column vector consisting of n_i ones. Then the ergodic average $\bar{A}(T)$ defined in (4.25) has a limit $\bar{A}$ as $T \to \infty$. Moreover, this limit is given by*

$$\bar{A} = \begin{array}{c} \\ \mathcal{E} \\ \mathcal{I} \end{array} \begin{array}{c} \begin{array}{cc} \mathcal{E} & \mathcal{I} \end{array} \\ \left[\begin{array}{cc} \bar{P} & \mathbf{0} \\ \bar{R} & \mathbf{0} \end{array} \right] \end{array}, \tag{4.26}$$

where

$$\bar{P} = \begin{array}{c} \\ \mathcal{E}_1 \\ \vdots \\ \mathcal{E}_s \end{array} \begin{array}{c} \begin{array}{ccc} \mathcal{E}_1 & \ldots & \mathcal{E}_s \end{array} \\ \left[\begin{array}{ccc} \mathbf{e}_{n_1}\mathbf{v}_1 & \ldots & \mathbf{0} \\ \vdots & \ddots & \vdots \\ \mathbf{0} & \ldots & \mathbf{e}_{n_s}\mathbf{v}_s \end{array} \right] \end{array}, \tag{4.27}$$

$$\bar{R} = (I - Q)^{-1} R \bar{P}. \tag{4.28}$$

The proof of the theorem proceeds via a series of lemmas. But first we give a few explanatory notes.

1. The matrix $\bar{P}$ in (4.27) is block diagonal, and the number of blocks equals the number of communicating classes. Each of the diagonal blocks is a rank one matrix (namely $\mathbf{e}_{n_i}\mathbf{v}_{n_i}$).

2. It is easy to construct examples where A^t by itself does not converge to anything. For instance, let

$$A = \begin{bmatrix} 0 & 1 \\ 1 & 0 \end{bmatrix}.$$

Then $A^2 = I$, and as a result $A^{2k} = I$ and $A^{2k+1} = A$. So A^t does not have a limit as $t \to \infty$. On the other hand,

$$\bar{A}_{2k} = \frac{1}{2k}(A + I), \bar{A}_{2k+1} = \frac{k}{2k+1}(A + I) + \frac{1}{2k+1} A.$$

So the ergodic limit exists and equals

$$\bar{A} = \frac{1}{2}(A+I) = \begin{bmatrix} 0.5 & 0.5 \\ 0.5 & 0.5 \end{bmatrix} = \begin{bmatrix} 1 \\ 1 \end{bmatrix} [0.5 \ \ 0.5].$$

This is consistent with (4.27), since A is irreducible ($s = 1$) and its unique stationary distribution is $[0.5 \ \ 0.5]$.

Lemma 4.12 *Suppose $A \in \mathbb{R}_+^{n \times n}$ is irreducible and stochastic, and let $\mathbf{v} \in \mathbb{S}_n$ denote its stationary distribution. Then the ergodic limit $\bar{A}$ exists and equals $\mathbf{e}_n \mathbf{v}$.*

Proof. Let p denote the period of the matrix A. Then we know from Theorem 3.25 that A has exactly p eigenvalues on the unit circle, namely

$$\lambda_0 = 1, \lambda_k = \exp(\mathbf{i}\omega\pi k/p), k = 1, \ldots, p-1,$$

and the remaining $n-p$ eigenvalues of A all have magnitude strictly less than one. Let $\mathbf{v}_0 = \mathbf{v}, \mathbf{v}_1, \ldots, \mathbf{v}_p$ denote row eigenvectors of A corresponding to the eigenvalues $\lambda_0, \lambda_1, \ldots, \lambda_p$, and choose other row vectors $\mathbf{v}_{p+1}, \ldots, \mathbf{v}_n$ in such a way that

$$VAV^{-1} = \begin{array}{c} \\ p \\ n-p \end{array} \begin{array}{c} \begin{array}{cc} p & n-p \end{array} \\ \begin{bmatrix} \Lambda & \mathbf{0} \\ \mathbf{0} & S \end{bmatrix} \end{array}, \tag{4.29}$$

where $\Lambda = \text{Diag}\{\lambda_0, \ldots, \lambda_{p-1}\}$, and $\rho(S) < 1$. Now note that $\lambda_1, \ldots, \lambda_p$ are all roots of one, and therefore $\lambda_k^p - 1 = 0$ for $k = 1, \ldots, p-1$. Moreover, since

$$\lambda^p - 1 = (\lambda - 1)\left(\sum_{i=0}^{p-1} \lambda^i\right),$$

it follows that

$$\sum_{i=0}^{p-1} \lambda_k^i = 0, k = 1, \ldots, p-1.$$

So for every integer l, we have that

$$\sum_{t=lp}^{(l+1)p-1} \Lambda^t = \begin{array}{c} \\ 1 \\ p-1 \end{array} \begin{array}{c} \begin{array}{cc} 1 & p-1 \end{array} \\ \begin{bmatrix} l & \mathbf{0} \\ \mathbf{0} & \mathbf{0} \end{bmatrix} \end{array}.$$

Moreover, since $\rho(S) < 1$, it follows that $S^t \to \mathbf{0}$ as $t \to \infty$. So

$$\bar{A}_T = \frac{1}{T}\sum_{t=0}^{T-1} A^t = V^{-1}\bar{Z}_T V + O(1/T),$$

where

$$\bar{Z}_T = \begin{array}{c} \\ 1 \\ p-1 \\ n-p \end{array} \begin{array}{c} \begin{array}{ccc} 1 & p-1 & n-p \end{array} \\ \begin{bmatrix} 1 & \mathbf{0} & \mathbf{0} \\ \mathbf{0} & \mathbf{0} & \mathbf{0} \\ \mathbf{0} & \mathbf{0} & \mathbf{0} \end{bmatrix} \end{array}.$$

So the ergodic limit exists and equals $\mathbf{w}_1\mathbf{v}$, where $\mathbf{w}_1$ is the first column of V^{-1}. Hence the proof is complete once it is shown that $\mathbf{w}_1 = \mathbf{e}_n$. For this purpose, note that since A is stochastic, its rows all add up to one; that is, $A\mathbf{e}_n = \mathbf{e}_n$. Now if we let U denote the matrix on the right side of (4.29), define $W = V^{-1}$, and rewrite (4.29) as $AW = WU$, then the first column of $AW = WU$ says that $A\mathbf{w}_1 = \mathbf{w}_1$. Since $\lambda_0 = 1$ is a simple eigenvalue of A, it follows that $\mathbf{w}_1 = \alpha\mathbf{e}_n$ for some proportionality constant α. To determine this constant, observe that $VW = I_n$ implies that $\mathbf{v}\mathbf{w}_1 = 1$. But since $\mathbf{v}\mathbf{e}_n = 1$ (because $\mathbf{v} \in \mathbb{S}_n$), the constant α equals one, and $\mathbf{w}_1 = \mathbf{e}_n$. □

Corollary 4.13 *Suppose $A \in \mathbb{R}_+^{n\times n}$ is primitive and stochastic, and let $\mathbf{v} \in \mathbb{S}_n$ denote its stationary distribution. Then*

$$A^t \to \mathbf{e}_n\mathbf{v} \text{ as } t \to \infty. \tag{4.30}$$

Proof. Since A is primitive, it is irreducible and also aperiodic. So from Theorem 3.22, we know that all eigenvalues of A have magnitude less than one except for $\lambda_0 = 1$. So we can choose

$$V = \begin{bmatrix} \mathbf{v} \\ V_2 \end{bmatrix}, W = V^{-1} = [\mathbf{e}_n | W_2]$$

such that

$$VAV^{-1} = \begin{array}{c} \\ 1 \\ n-1 \end{array} \begin{array}{c} \begin{array}{cc} 1 & n-1 \end{array} \\ \begin{bmatrix} 1 & \mathbf{0} \\ \mathbf{0} & S \end{bmatrix} \end{array} =: U, \text{ say},$$

where $\rho(S) < 1$. So

$$A^t = V^{-1}U^tV = [\mathbf{e}_n|W_2] \begin{bmatrix} 1 & \mathbf{0} \\ \mathbf{0} & S^t \end{bmatrix} \begin{bmatrix} \mathbf{v} \\ V_2 \end{bmatrix} \to \mathbf{e}_n\mathbf{v} \text{ as } t \to \infty,$$

after noting that $S^t \to 0$ as $t \to \infty$ because $\rho(S) < 1$. □

Proof. (Of Theorem 4.11): Suppose A is in the canonical form (4.8) and (4.9). Then

$$A^t = \begin{array}{c} \\ \mathcal{E} \\ \mathcal{I} \end{array} \begin{array}{c} \begin{array}{cc} \mathcal{E} & \mathcal{I} \end{array} \\ \begin{bmatrix} P^t & \mathbf{0} \\ R^{(t)} & Q^t \end{bmatrix} \end{array},$$

where $R^{(t)}$ can be computed recursively from partitioning $A^t = AA^{t-1}$, as follows:

$$R^{(1)} = R, R^{(t)} = RP^{t-1} + QR^{(t-1)}, \text{ for } t \geq 2. \tag{4.31}$$

So the ergodic average $\bar{A}_T$ is given by

$$\bar{A}_T = \frac{1}{T}\sum_{t=0}^{T-1} A^t = \begin{array}{c} \\ \mathcal{E} \\ \mathcal{I} \end{array} \begin{array}{c} \begin{array}{cc} \mathcal{E} & \mathcal{I} \end{array} \\ \begin{bmatrix} \bar{P}_T & \mathbf{0} \\ \bar{R}_T & \bar{Q}_T \end{bmatrix} \end{array},$$

where $\bar{P}_T, \bar{R}_T, \bar{Q}_T$ are defined in the obvious fashion. The ergodic limit of each matrix is now analyzed separately.

First, since we have from (4.9) that

$$P = \text{Block Diag } \{P_1, \ldots, P_s\},$$

it is clear that

$$\bar{P}_T = \text{Block Diag } \{(P_1)_T, \ldots, (P_s)_T\}.$$

Next, observe that each P_i is irreducible. So it follows from Lemma 4.8 that each P_i has an ergodic limit, which equals $\mathbf{e}_{n_i}\mathbf{v}_i$. So the ergodic limit $\bar{P}$ of P is given by (4.27).

Next, note that $\rho(Q) < 1$ from Statement 6 of Theorem 4.7. So $Q^t \to \mathbf{0}$ as $t \to \infty$, whence $\bar{Q} = \mathbf{0}$.

Now it remains only to compute $\bar{R}$. Since we take $A^0 = I_n$, and the lower triangular block of I_n is $\mathbf{0}$, we can take $R^{(0)} = \mathbf{0}$. With this convention, the formula (4.31) is consistent even for $t = 1$. So

$$\begin{aligned}
\bar{R}_T &= \frac{1}{T}\sum_{t=0}^{T-1} R(t) \\
&= \frac{1}{T}\sum_{t=1}^{T-1} R(t) \text{ since } R^{(0)} = \mathbf{0} \\
&= \frac{1}{T}\sum_{t=1}^{T-1} \{RP^{t-1} + QR^{(t-1)}\} \\
&= \frac{1}{T}R\sum_{t=0}^{T-2} P^t + \frac{1}{T}Q\sum_{t=0}^{T-2} R(t) \\
&= \frac{T-1}{T}R\bar{P}_{T-1} + \frac{T-1}{T}Q\bar{R}_{T-1}.
\end{aligned}$$

So $\bar{R}_T$ satisfies this time-varying recursive equation. However, its limit behaviour is easy to analyze. As $T \to \infty$, the constant $(T-1)/T \to 1$, the matrix $\bar{P}_{T-1}$ approaches the ergodic limit $\bar{P}$, and both $\bar{R}_T$ and $\bar{R}_{T-1}$ approach a constant matrix $\bar{R}$. Thus taking the limit as $T \to \infty$ in the preceding equation shows that $\bar{R}$ satisfies

$$\bar{R} = R\bar{P} + Q\bar{R}.$$

Note that $\rho(Q) < 1$ so that $I - Q$ is nonsingular. So the solution of the above equation for $\bar{R}$ is given by (4.28). □

Suppose A is a stochastic matrix and $\boldsymbol{\pi}$ is a stationary distribution of A. Since we make no assumptions about A other than stochasticity, there is no reason to suppose that $\boldsymbol{\pi}$ is unique; the only assumption is that $\boldsymbol{\pi} = \boldsymbol{\pi}A$, or that $\boldsymbol{\pi}$ is *a* stationary distribution of A. We refer to $(\boldsymbol{\pi}, A)$ as a **Markovian pair**. Recall also the notation $\ll$ introduced in Chapter 2, wherein $\mathbf{c} \ll \boldsymbol{\pi}$ if $\pi_i = 0$ implies that $c_i = 0$.

Definition 4.14 *Suppose $(\boldsymbol{\pi}, A)$ is a Markovian pair. Then $(\boldsymbol{\pi}, A)$ is said to be* **ergodic** *if*

$$\mathbf{c}\bar{A} = \boldsymbol{\pi} \text{ whenever } \mathbf{c} \in \mathbb{S}_n \text{ and } \mathbf{c} \ll \boldsymbol{\pi}. \tag{4.32}$$

Note that (4.32) can be expressed equivalently as

$$\lim_{T\to\infty} \frac{1}{T} \sum_{t=0}^{T-1} \mathbf{c}A^t = \boldsymbol{\pi} \ \forall \mathbf{c} \in \mathbb{S}_n \text{ such that } \mathbf{c} \ll \boldsymbol{\pi}. \tag{4.33}$$

Theorem 4.15 *Suppose $(\boldsymbol{\pi}, A)$ is a Markovian pair, and let $\mathcal{C}_1, \ldots, \mathcal{C}_s$ denote the communicating classes of the matrix A. Then $(\boldsymbol{\pi}, A)$ is ergodic if and only if $\boldsymbol{\pi}$ is concentrated on exactly one of the communicating classes.*

Proof. From Theorem 4.7 and (4.10) we know that $\boldsymbol{\pi}$ must have the form

$$\boldsymbol{\pi} = [\lambda_1 \mathbf{v}_1 \ldots \lambda_s \mathbf{v}_s \ \ \mathbf{0}], \text{ where } [\lambda_1 \ldots \lambda_s] \in \mathbb{S}_s.$$

The theorem says that the pair $(\boldsymbol{\pi}, A)$ is ergodic if and only if all the λ_i's are zero except for one.

"If": Suppose only one of the λ_i's is nonzero and the rest are zero. Renumber the communicating classes such that $\lambda_1 = 1$ and $\lambda_i = 0$ for $i \geq 2$. Thus

$$\boldsymbol{\pi} = [\mathbf{v}_1 \ \ \mathbf{0} \ldots \mathbf{0} \ \ \mathbf{0}] = [\mathbf{v}_1 \ \ \mathbf{0}],$$

where we have aggregated all the zero vectors into one. Suppose now $\mathbf{c} \in \mathbb{S}_n$ is arbitrary except that $\mathbf{c} \ll \boldsymbol{\pi}$. Then $\mathbf{c}$ is also concentrated only on the class $\mathcal{C}_1$. Thus $\mathbf{c}$ has the form

$$\mathbf{c} = [\mathbf{c}_1 \ \ \mathbf{0}],$$

where $\mathbf{c}_1 \in \mathbb{S}_{n_1}$ and n_1 is the number of states in the communicating class $\mathcal{C}_1$. Now, from (4.26) and (4.27), we get

$$\mathbf{c}\bar{A} = [\mathbf{c}_1 \mathbf{e}_{n_1} \mathbf{v}_1 \ \ \mathbf{0}] = [\mathbf{v}_1 \ \ \mathbf{0}] = \boldsymbol{\pi},$$

since $\mathbf{c}_1 \mathbf{e}_{n_1} = 1$ by virtue of the fact that $\mathbf{c}_1 \in \mathbb{S}_{n_1}$.

"Only If": Suppose $\boldsymbol{\pi}$ has the form (4.10) and that at least two of the λ_i's are nonzero. Renumber the communicating classes such that $\lambda_1 \neq 0$ and $\lambda_2 \neq 0$. Then $\boldsymbol{\pi}$ has the form

$$\boldsymbol{\pi} = [\lambda_1 \mathbf{v}_1 \ \ \boldsymbol{\pi}_2 \ \ \mathbf{0}],$$

where $\boldsymbol{\pi}_2 \neq \mathbf{0}$ because it contains the nonzero subvector $\lambda_2 \mathbf{v}_2$. Now choose

$$\mathbf{c} = [\mathbf{v}_1 \ \ \mathbf{0} \ \ \mathbf{0}].$$

Then $\mathbf{c} \ll \boldsymbol{\pi}$. However

$$\mathbf{c}\bar{A} = [\mathbf{v}_1 \ \ \mathbf{0} \ \ \mathbf{0}] \neq \boldsymbol{\pi}.$$

Hence it is not true that $\mathbf{c}\bar{A} = \boldsymbol{\pi}$ whenever $\mathbf{c} \ll \boldsymbol{\pi}$. □

Chapter Five

Introduction to Large Deviation Theory

5.1 PROBLEM FORMULATION

In this chapter, we take some baby steps in a very important part of probability theory, known as large deviation theory.[1] We begin by describing briefly the motivation for the problem under study. Suppose $\mathbb{A} = \{a_1, \ldots, a_n\}$ is a finite set. Let $\mathcal{M}(\mathbb{A})$ denote the set of all probability distributions on the set $\mathbb{A}$. Clearly one can identify $\mathcal{M}(\mathbb{A})$ with the n-simplex $\mathbb{S}_n$. Suppose $\boldsymbol{\mu} \in \mathcal{M}(\mathbb{A})$ is a fixed but possibly unknown probability distribution, and X is a random variable assuming values in $\mathbb{A}$ with the distribution $\boldsymbol{\mu}$. In order to estimate $\boldsymbol{\mu}$, we generate independent samples $x_1, \ldots, x_l, \ldots$, where each x_i belongs to $\mathbb{A}$, is distributed according to $\boldsymbol{\mu}$, and is independent of x_j for $j \neq i$. The symbol $\mathbf{x}_1^l := x_1 \ldots x_l \in \mathbb{A}^l$ denotes the multisample that represents the outcome of the first l experiments. Based on this multisample, we can construct an **empirical distribution** $\hat{\boldsymbol{\mu}}(\mathbf{x}_1^l)$ as follows:

$$(\hat{\boldsymbol{\mu}}(\mathbf{x}_1^l))_i := \frac{1}{l} \sum_{j=1}^{l} I_{\{x_j = a_i\}}, \tag{5.1}$$

where I denotes the indicator function. Thus

$$I_{\{x_j = a_i\}} = \begin{cases} 1 & \text{if} \quad x_j = a_i, \\ 0 & \text{if} \quad x_j \neq a_i. \end{cases}$$

In words, (5.1) simply states that $\hat{\boldsymbol{\mu}}_i(\mathbf{x})$ equals the fraction of the samples $x_1, \ldots, x_l$ that equal the symbol a_i. Since every sample x_i has to equal one of the a_i's, it is easy to see that $\hat{\boldsymbol{\mu}}(\mathbf{x}_1^l)$ is also a probability distribution on $\mathbb{A}$. Moreover $\hat{\boldsymbol{\mu}}(\mathbf{x}_1^l)$ is a "random" element of $\mathcal{M}(\mathbb{A})$ since it is based on the random multisample $\mathbf{x}$. Thus we can think of $\{\hat{\boldsymbol{\mu}}(\mathbf{x}_1^l)\}$ as a stochastic process that assumes values in $\mathcal{M}(\mathbb{A})$ and ask: As $l \to \infty$, does this process converge to the true but possibly unknown measure $\boldsymbol{\mu}$ that is generating the samples, and if so, at what rate?

To address this question, the first thing we do is to convert the question from one of studying a stochastic process into one of studying a sequence of real numbers. Suppose $\Gamma \subseteq \mathcal{M}(\mathbb{A})$ is some set of probability distributions. Then $\Pr\{\hat{\boldsymbol{\mu}}(\mathbf{x}_1^l) \in \Gamma\}_{l \geq 1}$ is a sequence of real numbers. So it makes sense to study the behavior of this sequence as $l \to \infty$. What is the interpretation of "Pr" in this context? Clearly the empirical distribution $\hat{\boldsymbol{\mu}}(\mathbf{x}_1^l)$ depends only on the first l samples $\mathbf{x}_1^l$. So $\Pr\{\hat{\boldsymbol{\mu}}(\mathbf{x}_1^l) \in \Gamma\} = P_\mu^l\{\hat{\boldsymbol{\mu}}(\mathbf{x}_1^l) \in \Gamma\}$.

[1] I firmly resisted the temptation to say "some small steps in large deviation theory."

Suppose now that $\boldsymbol{\mu} \notin \bar{\Gamma}$, where $\bar{\Gamma}$ denotes the closure of Γ in the total variation metric; thus the true measure $\boldsymbol{\mu}$ that is generating the random samples does not belong to the closure of the set Γ. Now it is known from Theorem 1.32 that the empirical measure $\hat{\boldsymbol{\mu}}(\mathbf{x}_1^l)$ converges to the true measure $\boldsymbol{\mu}$ as the number of samples approaches infinity. Specifically, (1.51) states that

$$\Pr\{\|\hat{\boldsymbol{\mu}}(\mathbf{x}_1^l) - \boldsymbol{\mu}\|_\infty > \epsilon\} \leq 2n \exp(-2l\epsilon^2).$$

Hence, if $\boldsymbol{\mu} \notin \bar{\Gamma}$, then the sequence of real numbers $\Pr\{\hat{\boldsymbol{\mu}}(\mathbf{x}_1^l) \in \Gamma\}$ converges to zero. Large deviation theory is concerned with the *rate* at which this sequence converges to zero, and how the rate depends on the set Γ and the true distribution $\boldsymbol{\mu}$.

Specifically, suppose this sequence converges to zero at an exponential rate; that is

$$\Pr\{\hat{\boldsymbol{\mu}}(\mathbf{x}_1^l) \in \Gamma\} \sim c_1 \exp(-lc_2).$$

Then the constant c_2 is the rate of convergence (which will in general depend on both Γ and $\boldsymbol{\mu}$). How can we "get at" this rate? We can compute the quantity

$$\frac{1}{l} \log \Pr\{\hat{\boldsymbol{\mu}}(\mathbf{x}_1^l) \in \Gamma\} \sim \frac{\log c_1}{l} - c_2,$$

and observe that as $l \to \infty$, the negative of this quantity approaches c_2. Motivated by this observation, we define something called the "rate function." Since we will modify the definition almost at once, let us call this a "trial definition."

Let us call a function $I : \mathcal{M}(\mathbb{A}) \to \mathbb{R}_+$ a "rate function" if it has the following properties: (i) Whenever $\Gamma \subseteq \mathcal{M}(\mathbb{A})$ is an *open set*, we have

$$-\inf_{\boldsymbol{\nu} \in \Gamma} I(\boldsymbol{\nu}) \leq \liminf_{l \to \infty} \frac{1}{l} \log \Pr\{\hat{\boldsymbol{\mu}}(\mathbf{x}_1^l) \in \Gamma\}. \tag{5.2}$$

(ii) Whenever $\Gamma \subseteq \mathcal{M}(\mathbb{A})$ is a *closed set*, we have

$$\limsup_{l \to \infty} \frac{1}{l} \log \Pr\{\hat{\boldsymbol{\mu}}(\mathbf{x}_1^l) \in \Gamma\} \leq -\inf_{\boldsymbol{\nu} \in \Gamma} I(\boldsymbol{\nu}). \tag{5.3}$$

However, the above definition leaves a few issues unresolved. First, there are no specifications about the nature of the function I, so it could be quite erratic. Second, there is no requirement that the rate function be unique. Third, there are two separate conditions, one about what happens when Γ is an open set and another about what happens when Γ is a closed set, but nothing about what happens for "in-between" sets Γ.

To overcome these issues, we introduce the notion of a lower semi-continuous function, and then that of a lower semi-continuous relaxation. Given a number $\epsilon > 0$, let us define $\mathcal{B}(\boldsymbol{\nu}, \epsilon)$ to be the open ball of radius centered at $\boldsymbol{\nu}$, where the radius is measured in the total variation metric ρ. Thus

$$\mathcal{B}(\boldsymbol{\nu}, \epsilon) := \{\boldsymbol{\phi} \in \mathbb{S}_n : \rho(\boldsymbol{\nu}, \boldsymbol{\phi}) < \epsilon\}.$$

Similarly, we define

$$\bar{\mathcal{B}}(\boldsymbol{\nu},\epsilon) := \{\boldsymbol{\phi} \in \mathbb{S}_n : \rho(\boldsymbol{\nu},\boldsymbol{\phi}) \leq \epsilon\}.$$

A function $f : \mathcal{M}(\mathbb{A}) \to \mathbb{R}$ is said to be **lower semi-continuous** if

$$\boldsymbol{\nu}_i \to \boldsymbol{\nu}^* \;\Rightarrow\; f(\boldsymbol{\nu}^*) \leq \liminf_i f(\boldsymbol{\nu}_i).$$

Contrast this with the definition of continuity, which is

$$\boldsymbol{\nu}_i \to \boldsymbol{\nu}^* \;\Rightarrow\; f(\boldsymbol{\nu}^*) = \lim_i f(\boldsymbol{\nu}_i).$$

Now, given any function $I : \mathcal{M}(\mathbb{A}) \to \mathbb{R}_+$, we define its **lower semi-continuous relaxation** I_* by

$$I_*(\boldsymbol{\nu}) := \lim_{\epsilon \to 0} \inf_{\boldsymbol{\phi} \in \mathcal{B}(\boldsymbol{\nu},\epsilon)} I(\boldsymbol{\phi}).$$

It is left as a problem to verify that $I_*(\cdot)$ is indeed lower semi-continuous.

Now we quote without proof some results that can be found in [54, p. 41]. Suppose that a rate function $I(\cdot)$ satisfies (5.2) and (5.3). Then so does its lower semi-continuous relaxation $I_*(\cdot)$. Moreover, if the rate function is lower semi-continuous, then it is unique. Putting it another way, it is conceivable that two distinct functions can satisfy (5.2) and (5.3), but if so, they will both have exactly the same lower semi-continuous relaxation. Finally, if the rate function $I(\cdot)$ is lower semi-continuous (we drop the subscript for convenience), then the two equations (5.2) and (5.3) can be combined into the single equation

$$\begin{aligned} -\inf_{\boldsymbol{\nu} \in \Gamma^o} I(\boldsymbol{\nu}) &\leq \liminf_{l \to \infty} \frac{1}{l} \log \Pr\{\hat{\boldsymbol{\mu}}(\mathbf{x}_1^l) \in \Gamma\} \\ &\leq \limsup_{l \to \infty} \frac{1}{l} \log \Pr\{\hat{\boldsymbol{\mu}}(\mathbf{x}_1^l) \in \Gamma\} \leq -\inf_{\boldsymbol{\nu} \in \bar{\Gamma}} I(\boldsymbol{\nu}), \end{aligned} \tag{5.4}$$

where Γ^o denotes the interior of the set Γ.

Now we are ready to state the definition of the rate function as it is used here. Note however that in some books and papers the rate function is also defined in terms of (5.2) and (5.3). As pointed out above, both definitions are equivalent if it is required that the rate function is lower semi-continuous, which can be imposed without loss of generality. Also, strictly speaking, (5.4) makes sense only for "Borel" sets $\Gamma \subseteq \mathcal{M}(\mathbb{A})$. In the interests of simplicity, these subtleties are ignored here.

Definition 5.1 *Let $\hat{\boldsymbol{\mu}}(\mathbf{x}_1^l)$ be defined as in (5.1). Then a function $I : \mathcal{M}(\mathbb{A}) \to \mathbb{R}_+$ is said to be a (the)* **rate function** *of the stochastic process $\{\hat{\boldsymbol{\mu}}(\mathbf{x}_1^l)\}$ if*

1. *I is lower semi-continuous.*

2. *For every set $\Gamma \subseteq \mathcal{M}(\mathbb{A})$, the relationships in (5.4) hold.*

In such a case the process $\{\hat{\boldsymbol{\mu}}(\mathbf{x}_1^l)\}$ is said to satisfy the **large deviation property (LDP)** *with rate function $I(\cdot)$.*

In the above definition, the stochastic process under study is $\{\hat{\boldsymbol{\mu}}(\mathbf{x}_1^l)\}$, which assumes values in the *compact set* $\mathbb{S}_n$. We can think of a more general situation (not encountered in this book) where we study a stochastic process $\{X_l\}$ that assumes values in $\mathbb{R}^n$. See for example [41, 54] for statements of the large deviation property in the general case. In such a case the rate function $I(\cdot)$ would have to be defined over all of $\mathbb{R}^n$, wherever the random variables X_l have their range. The rate function $I(\cdot)$ is said to be a **good rate function** if the "level sets"

$$L_\alpha := \{\mathbf{x} \in \mathbb{R}^n : I(\mathbf{x}) \leq \alpha\}$$

are compact for all α. However, we need not worry about "good" rate functions in the present context since the domain of the rate function is anyhow a compact set; so the above condition is automatically satisfied.

Note that the assumption that I is lower semi-continuous makes the rate function unique if it exists. Suppose that the set Γ does not have any isolated points, i.e., that $\Gamma \subseteq \overline{\Gamma^o}$. Suppose also that the function $I(\cdot)$ is continuous, and not merely lower semi-continuous. Then the two extreme infima in (5.4) coincide. As a result the lim inf and lim sup are equal, which means that both equal the limit of the sequence. This means that

$$\lim_{l\to\infty} \frac{1}{l} \log \Pr\{\hat{\boldsymbol{\mu}}(\mathbf{x}_1^l) \in \Gamma\} = -\inf_{\boldsymbol{\nu}\in\Gamma} I(\boldsymbol{\nu}). \tag{5.5}$$

One of the attractions of the large deviation property is this precise estimate of the rate at which $\Pr\{\hat{\boldsymbol{\mu}}(\mathbf{x}_1^l) \in \Gamma\}$ approaches zero. Let us define α to be the infimum on the right side of (5.5). Then (5.5) can be rewritten as

$$\frac{1}{l} \log \Pr\{\hat{\boldsymbol{\mu}}(\mathbf{x}_1^l) \in \Gamma\} \to -\alpha \text{ as } l \to \infty.$$

This means that, very roughly speaking,

$$\Pr\{\hat{\boldsymbol{\mu}}(\mathbf{x}_1^l) \in \Gamma\} \sim c_1(l) \exp(c_2(l)),$$

where $c_1(l)$ is subexponential in the sense that $(\log c_1(l))/l$ approaches zero as $l \to \infty$, while $c_2 \to -\alpha$ as $l \to \infty$.

In the next section, it is shown that for the specific problem discussed here (namely, i.i.d. processes assuming values in a finite set $\mathbb{A}$ with distribution $\boldsymbol{\mu}$), the rate function is actually $I(\boldsymbol{\nu}) = D(\boldsymbol{\nu}\|\boldsymbol{\mu})$, the Kullback-Leibler divergence between $\boldsymbol{\nu}$ and the true distribution $\boldsymbol{\mu}$. But in anticipation of that, let us discuss the implications of the definition. Equation (5.5) means that the rate at which $\Pr\{\hat{\boldsymbol{\mu}}(\mathbf{x}_1^l) \in \Gamma\}$ approaches zero depends only on the *infimum value* of the rate function $I(\cdot)$ over Γ. To understand the implications of this, consider the situation in Figure 5.1, which shows a "large" set Γ_1, and a much smaller set Γ_2 shown in the figure. Suppose the rate function $I(\cdot)$ is continuous and assumes its minimum over Γ_1 at exactly one choice of $\boldsymbol{\nu}_*$, indicated by the dot in the figure. Moreover, assume that $I(\boldsymbol{\nu}_*) > 0$; this is the case if the true probability distribution $\boldsymbol{\mu}$ does not belong to $\bar{\Gamma}$, because in this case $D(\boldsymbol{\nu}_*\|\boldsymbol{\mu}) > 0$. Now, since $\boldsymbol{\nu}_* \in \Gamma_2$, it is clear that

$$\inf_{\boldsymbol{\nu}\in\Gamma_1} I(\boldsymbol{\nu}) = \inf_{\boldsymbol{\nu}\in\Gamma_2} I(\boldsymbol{\nu}).$$

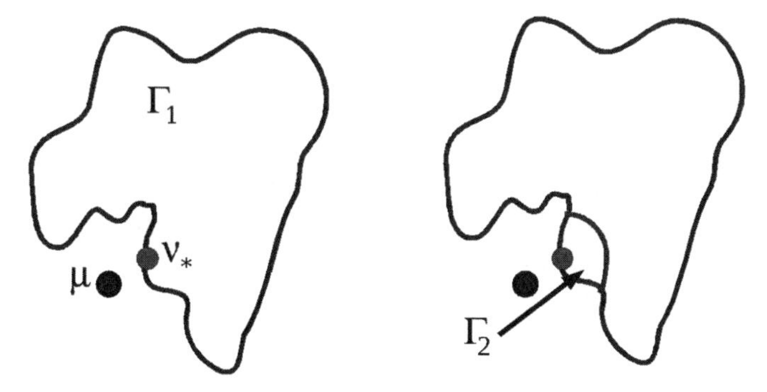

Figure 5.1 Implications of (5.1)

Moreover, if we let $\Gamma_1 \setminus \Gamma_2$ denote the complement of Γ_2 in Γ_1, that is,

$$\Gamma_1 \setminus \Gamma_2 = \{\boldsymbol{\nu} : \boldsymbol{\nu} \in \Gamma_1, \boldsymbol{\nu} \notin \Gamma_2\},$$

then the continuity of the rate function implies that

$$\inf_{\boldsymbol{\nu} \in \Gamma_1 \setminus \Gamma_2} > \inf_{\boldsymbol{\nu} \in \Gamma_1} I(\boldsymbol{\nu}). \tag{5.6}$$

Now it is clear that, for every l, we have

$$\Pr\{\hat{\boldsymbol{\mu}}(\mathbf{x}_1^l) \in \Gamma_1\} = \Pr\{\hat{\boldsymbol{\mu}}(\mathbf{x}_1^l) \in \Gamma_2\} + \Pr\{\hat{\boldsymbol{\mu}}(\mathbf{x}_1^l) \in \Gamma_1 \setminus \Gamma_2\}.$$

Therefore (5.6) implies that the second quantity on the right side approaches zero at a faster exponential rate compared to the first quantity. So we can use this fact to deduce what precisely happens as $l \to \infty$. Since $\Pr\{\hat{\boldsymbol{\mu}}(\mathbf{x}_1^l) \in \Gamma_1\}$ approaches zero at an exponential rate, we can think of $\hat{\boldsymbol{\mu}}(\mathbf{x}_1^l) \in \Gamma_1$ as a "rare" event, which becomes more and more rare as $l \to \infty$. However, *if we now condition on this rare event occurring*, and ask where precisely within the large set Γ_1 the empirical distribution $\hat{\boldsymbol{\mu}}(\mathbf{x}_1^l)$ is likely to lie, we see from the above discussion that

$$\Pr\{\hat{\boldsymbol{\mu}}(\mathbf{x}_1^l) \in \Gamma_2 | \hat{\boldsymbol{\mu}}(\mathbf{x}_1^l) \in \Gamma_1\} = \frac{\Pr\{\hat{\boldsymbol{\mu}}(\mathbf{x}_1^l) \in \Gamma_2\}}{\Pr\{\hat{\boldsymbol{\mu}}(\mathbf{x}_1^l) \in \Gamma_1\}} \to 1 \text{ as } l \to \infty.$$

Since the above argument can be repeated for *every* set Γ_2 such that (5.6) is true, we see that the conditional distribution of $\hat{\boldsymbol{\mu}}(\mathbf{x}_1^l)$, conditioned on the event that $\hat{\boldsymbol{\mu}}(\mathbf{x}_1^l) \in \Gamma_1$, becomes more and more peaked around $\boldsymbol{\nu}_*$ as $l \to \infty$. The above argument does not really depend on the fact that the rate function $I(\cdot)$ assumes its minimum value at exactly one point in the set Γ_1. If the minimum of the rate function occurs at finitely many points in Γ_1, then as $l \to \infty$, the conditional distribution above becomes more and more peaked around these minima.

5.2 LARGE DEVIATION PROPERTY FOR I.I.D. SAMPLES: SANOV'S THEOREM

In this section we derive the exact rate function for the process of empirical measures $\{\hat{\boldsymbol{\mu}}(x_1^l)\}$ defined in (5.1) in the important case where the samples $x_1, x_2, \ldots$ are generated independently using a common distribution $\boldsymbol{\mu}$. Specifically, it is shown that the rate function is $I(\boldsymbol{\nu}) = D(\boldsymbol{\nu}\|\boldsymbol{\mu})$, where $\boldsymbol{\mu}$ is the actual distribution generating the samples. This is known as Sanov's theorem.[2]

Let us begin by restating the problem under study. Suppose $\mathbb{A} = \{a_1, \ldots, a_n\}$ is a finite set. Suppose $\boldsymbol{\mu} \in \mathcal{M}(\mathbb{A})$ is a fixed but possibly unknown probability distribution, and X is a random variable assuming values in $\mathbb{A}$ with the distribution $\boldsymbol{\mu}$. In order to estimate $\boldsymbol{\mu}$, we define an i.i.d. sequence $\{X_t\}$ where each X_t has the distribution $\boldsymbol{\mu}$, and let $\mathbf{x} = x_1, \mathbf{x}_2, \ldots$ denote a sample path (or realization) of this i.i.d. process. Based on the first l samples of this realization, we can construct an **empirical distribution** $\hat{\boldsymbol{\mu}}(\mathbf{x}_1^l)$ as follows:

$$(\hat{\boldsymbol{\mu}}(\mathbf{x}_1^l))_i := \frac{1}{l}\sum_{j=1}^{l} I_{\{x_j = a_i\}}. \tag{5.7}$$

The objective is to make precise estimates of the probability $P_{\mu}^l\{\hat{\boldsymbol{\mu}}(\mathbf{x}_1^l) \in \Gamma\}$ where $\Gamma \subseteq \mathcal{M}(\mathbb{A})$.

For this purpose, we introduce the "method of types." For the moment let us fix the integer l denoting the length of the multisample. Then it is easy to see that the empirical distribution $\hat{\boldsymbol{\mu}}(\mathbf{x}_1^l)$ can take only a finite set of values. For one thing, it is clear from (5.7) that every element of $\hat{\boldsymbol{\mu}}(\mathbf{x}_1^l)$ is a rational number with denominator equal to l. Let $\mathcal{E}(l, n)$ denote the set of all possible empirical distributions that can result from a multisample of length l.[3] We will denote elements of $\mathcal{E}(l, n)$ by symbols such as $\boldsymbol{\nu}, \boldsymbol{\phi}$ etc. Suppose $\mathbf{x}_1^l, \mathbf{y}_1^l \in \mathbb{A}^l$ are two multisamples. We define these two multisamples to be **equivalent** if they generate the same empirical estimate, i.e., if $\hat{\boldsymbol{\mu}}(\mathbf{x}_1^l) = \hat{\boldsymbol{\mu}}(\mathbf{y}_1^l)$. It is easy to see that this is indeed an equivalence relation. For each distribution $\boldsymbol{\nu} \in \mathcal{E}(l, n)$, the set of all multisamples $\mathbf{x}_1^l \in \mathbb{A}$ such that $\hat{\boldsymbol{\mu}}(\mathbf{x}_1^l) = \boldsymbol{\nu}$ is called the **type class** of $\boldsymbol{\nu}$, and is denoted by $T(\boldsymbol{\nu}, l)$.

Example 5.1 Suppose the alphabet $\mathbb{A}$ has cardinality 2. For simplicity we can write $\mathbb{A} = \{1, 2\}$. Suppose $l = 5$. Then there are only six possible empirical distributions, namely

$$\mathcal{E}(5, 2) = \{[0/5 \;\; 5/5], [1/5 \;\; 4/5], [2/5 \;\; 3/5], [3/5 \;\; 2/5], [4/5 \;\; 1/5], [5/5 \;\; 0/5]\}.$$

Next, suppose $\boldsymbol{\nu} = [2/5 \;\; 3/5]$. Then the type class $T(\boldsymbol{\nu}, 5)$ consists of all elements of $\mathbb{A}^5$ that contain precisely two 1's and three 2's. These can be

[2] However, it must be pointed out that the original Sanov's theorem is not exactly what is stated here. A precise statement and proof of what Sanov proved can be found in [30].

[3] It is clear that the possible empirical distributions depend only on n, the cardinality of the alphabet $\mathbb{A}$, and not on the nature of the elements of $\mathbb{A}$.

written out explicitly, as

$$T([2/5 \;\; 3/5], 5) = \begin{array}{ll} \{ & [11222], [12122], [12212], [12221], [21122], \\ & [21212], [21221], [22112], [22121], [22211] \quad \}. \end{array}$$

Note that it is necessary to identify the type class not only with the empirical distribution $\boldsymbol{\nu}$ but also with the length l of the multisample. For instance, if we keep the same $\boldsymbol{\nu}$ but change l to 10, then $T(\boldsymbol{\nu}, 10)$ would consist of all elements of $\mathbb{A}^{10}$ that contain precisely four 1's and six 2's.

The "method of types" consists of addressing the following questions:

- What is the cardinality of $\mathcal{E}(l, n)$? In other words, how many distinct empirical distributions can be generated as $\mathbf{x}_1^l$ varies over $\mathbb{A}^l$? It is not necessary to have a precise count; rather, it suffices to have an upper bound for this cardinality.

- What is the (log) likelihood of each multisample in $T(\boldsymbol{\nu}, l)$, and how is it related to $\boldsymbol{\nu}$?

- For each empirical distribution $\boldsymbol{\nu} \in \mathcal{E}(l, n)$, what is the cardinality of the associated type class $T(\boldsymbol{\nu}, l)$? We require both upper as well as lower bounds on this cardinality.

Let us address each of these questions in order.

Lemma 5.2 *We have that*

$$|\mathcal{E}(l, n)| = \binom{l+n-1}{n-1} = \frac{(l+n-1)!}{l!(n-1)!}. \tag{5.8}$$

Proof. The proof is by induction on n. If $n = 2$, then it is easy to see that $\mathcal{E}(l, 2)$ consists of all distributions of the form $[k/l \;\; (l-k)/l]$ as k varies from 0 to l. Thus $|\mathcal{E}(l, 2)| = l + 1$ for all l, and (5.8) holds for $n = 2$, for all l.

To proceed by induction, suppose (5.8) holds up to $n - 1$, for all l, and suppose $|\mathbb{A}| = n$. Suppose $\boldsymbol{\nu} \in \mathcal{E}(l, n)$. Thus each component of $l\boldsymbol{\nu}$ is an integer, and together they must add up to l. Let k denote the first component of $l\boldsymbol{\nu}$, and note that k can have the values $0, 1, \ldots, l$. If $k = l$ then the next $n - 1$ components of $l\boldsymbol{\nu}$ must all equal zero. Hence there is only one vector $\boldsymbol{\nu} \in \mathcal{E}(l, n)$ with the first component equal to 1. For each $k = 0, \ldots, l - 1$, the next $n - 1$ components of $l\boldsymbol{\nu}$ must all be integers and add up to $l - k$. Thus the next $n - 1$ components of $l\boldsymbol{\nu}$ can have $|\mathcal{E}(l - k, n - 1)|$ possible values. Thus we have established the following recursive relationship:

$$|\mathcal{E}(l, n)| = 1 + \sum_{k=0}^{l-1} |\mathcal{E}(l - k, n - 1)| = 1 + \sum_{k=1}^{l} |\mathcal{E}(k, n - 1)|,$$

after changing the dummy variable of summation. Therefore we need to solve the above recursion with the starting condition $|\mathcal{E}(l, 2)| = l + 1$.

Now let us recall the following property of binomial coefficients:

$$\binom{m}{n} = \binom{m-1}{n} + \binom{m-1}{n-1}.$$

This property can be verified by multiplying both sides by $n!(m-n)!$ and collecting terms. The above equation can be rewritten as

$$\binom{m-1}{n-1} = \binom{m}{n} - \binom{m-1}{n}.$$

Substituting this relationship into the recursive formula for $|\mathcal{E}(l,n)|$, and using the inductive hypothesis gives

$$\begin{aligned}|\mathcal{E}(l,n)| &= 1 + \sum_{k=1}^{l} \binom{k+n-2}{n-2} \\ &= 1 + \sum_{k=1}^{l} \left[\binom{k+n-1}{n-1} - \binom{k+n-2}{n-1} \right].\end{aligned}$$

Note that when $k = 1$, we have that

$$\binom{k+n-2}{n-1} = 1,$$

which cancels the very first term of 1. So this is a telescoping sum, whereby the negative terms and positive terms cancel out, leaving only the very last positive term; that is,

$$|\mathcal{E}(l,n)| = \binom{l+n-1}{n-1}.$$

This completes the proof by induction. □

Lemma 5.3 *Suppose $\boldsymbol{\nu} \in \mathcal{E}(l,n)$, and suppose that the multisample $\mathbf{x}_1^l \in \mathbb{A}^l$ belongs to the corresponding type class $T(\boldsymbol{\nu}, l)$. Then*

$$\log P_\mu^l\{\mathbf{x}\} = -lJ(\boldsymbol{\nu}, \boldsymbol{\mu}), \tag{5.9}$$

where $J(\cdot,\cdot)$ is the loss function defined in (2.30).

Proof. Since $\boldsymbol{\nu} \in \mathcal{E}(l,n)$, every component of $l\boldsymbol{\nu}$ is an integer. Let us define $l_i = l\nu_i$ for each i. Then the multisample $\mathbf{x}_1^l$ contains precisely l_i occurrences of the symbol a_i, for each i. Since the samples are independent of each other, the *order* in which the various symbols occur in $\mathbf{x}_1^l$ does not affect its likelihood. So it follows that

$$P_\mu^l\{\mathbf{x}\} = \prod_{i=1}^{l} \mu_i^{l_i},$$

whence

$$\log P_\mu^l\{\mathbf{x}\} = \sum_{i=1}^{n} l_i \log \mu_i = l \sum_{i=1}^{n} \nu_i \log \mu_i = -lJ(\boldsymbol{\nu}, \boldsymbol{\mu}).$$

□

Now we come to the last of the questions raised above, namely: What is the cardinality of the type class $T(\boldsymbol{\nu}, l)$? Using the same symbols $l_i = l\nu_i$, we see that $|T(\boldsymbol{\nu}, l)|$ is the number of ways of choosing l symbols from $\mathbb{A}$ in such a way that the symbol a_i occurs precisely l_i times. This number equals

$$|T(\boldsymbol{\nu}, l)| = \frac{l!}{\prod_{i=1}^{n} l_1!}.$$

So we derive upper and lower bounds for $|T(\boldsymbol{\nu}, l)|$. As a part of doing this, we also derive another result that may be of independent interest.

Lemma 5.4 *Suppose* $\boldsymbol{\nu}, \boldsymbol{\phi} \in \mathcal{E}(l, n)$. *Then*

$$P_\nu^l(T(\boldsymbol{\phi}, l)) \leq P_\nu^l(T(\boldsymbol{\nu}, l)). \tag{5.10}$$

Remark: The lemma states that for any empirical measure $\boldsymbol{\nu}$, the associated type class has the maximum measure, among all type classes, under the corresponding measure P_ν^l. To illustrate the lemma, suppose $n = 2$ (two possible outputs, call them 0 and 1), and $l = 4$ (four samples). Let $\boldsymbol{\phi} = [2/4 \;\; 2/4]$ and $\boldsymbol{\nu} = [1/4 \;\; 3/4]$. Then the type class $T(\boldsymbol{\phi}, 4)$ consists of all sample paths of length 4 with exactly two occurrences each of 0 and 1, while $T(\boldsymbol{\nu}, 4)$ consists of all sample paths of length 4 that consist of one 0 and three 1's. The cardinalities of the type classes are just the binomial coefficients. Thus $|T(\boldsymbol{\phi}, 4)| = 6$ and $|T(\boldsymbol{\nu}, 4)| = 4$, and $T(\boldsymbol{\phi}, 4)$ is larger than $T(\boldsymbol{\nu}, 4)$. Nevertheless

$$P_\nu^4(T(\boldsymbol{\nu}, 4)) = 4 \times (1/4) \times (3/4)^3 = 27/64,$$

whereas

$$P_\nu^4(T(\boldsymbol{\phi}, 4)) = 6 \times (1/4)^2 \times (3/4)^2 = 54/256 < 27/64.$$

Proof. Since $\boldsymbol{\nu}, \boldsymbol{\phi} \in \mathcal{E}(l, n)$, let us define $l_i = l\nu_i$ as above, and $k_i = l\phi_i$ for all i. Then it follows that

$$P_\nu^l(T(\boldsymbol{\phi}, l)) = |T(\boldsymbol{\phi}, l)| \prod_{i=1}^{n} \nu_i^{k_i} = \frac{l!}{\prod_{i=1}^{n} k_i!} \prod_{i=1}^{n} \nu_i^{k_i} = \frac{l!}{l^l} \prod_{i=1}^{n} \frac{l_i^{k_i}}{k_i!},$$

where we take advantage of the fact that the k_i's add up to l. Now the term $l!/l^l$ is just some integer constant that is independent of both $\boldsymbol{\nu}, \boldsymbol{\phi}$ and can thus be ignored in future calculations. Thus

$$\log P_\nu^l(T(\boldsymbol{\phi}, l)) = \text{const.} + \sum_{i=1}^{n} [k_i \log l_i - \log k_i!].$$

If we note that $\log(k_i!) = \sum_{j=1}^{k_i} \log s$, we can rewrite the above as

$$\log P_\nu^l(T(\boldsymbol{\phi}, l)) = \text{const.} + \sum_{i=1}^{n} \sum_{j=1}^{k_i} [\log l_i - \log j] = \text{const.} + \sum_{i=1}^{n} \sum_{j=1}^{k_i} \log \frac{l_i}{j}.$$

Let us define

$$c(k_1, \ldots, k_n) := \sum_{i=1}^{n} \sum_{j=1}^{k_i} \log \frac{l_i}{j}.$$

The claim is that above quantity is maximized when $k_i = l_i$ for all i.

To prove the claim, let us imagine that there are n bins with capacity $l_1, \ldots, l_n$ respectively. Let us begin with $k_i = 0$ for all i, and then start adding elements one by one to each bin. Then we can see that, if $k_i < l_i$, adding one more element to bin i causes the quantity $c(k_1, \ldots, k_n)$ to increase if $k_i < l_i - 1$, and to stay the same if $k_i = l_i - 1$. This is because $\log(l_i/j) > 0$ whenever $j < l_i$. On the other hand, if $k_i = l_i$, then adding one more element to bin i causes the quantity $c(k_1, \ldots, k_n)$ to decrease, because $\log(l_i/j) < 0$ whenever $j > l_i$. Thus, if $k_i = l_i$, we should not add any more elements to bin i; instead the element should be added to another bin that still has some unused capacity. Continuing in this manner we see that the optimal choice is $k_i = l_i$ for all i. $\square$

Lemma 5.5 *Let $H(\cdot)$ denote the entropy of a distribution, as before. Then*

$$[|\mathcal{E}(l,n)|]^{-1} \exp[lH(\boldsymbol{\nu})] \leq |T(\boldsymbol{\nu}, l)| \leq \exp[lH(\boldsymbol{\nu})], \; \forall \boldsymbol{\nu} \in \mathcal{E}(l,n). \tag{5.11}$$

Proof. Fix $\boldsymbol{\nu} \in \mathcal{E}(l,n)$, and let $\mathbf{x}_1^l \in T(\boldsymbol{\nu}, l)$. Then $\mathbf{x}_1^l$ contains precisely $l_i = l\nu_i$ occurrences of the symbol a_i. Therefore

$$P_\nu^l(\{\mathbf{x}\}) = \prod_{i=1}^n \nu_i^{l_i}, \; \forall \mathbf{x} \in T(\boldsymbol{\nu}, l),$$

$$\log P_\nu^l(\{\mathbf{x}\}) = \sum_{i=1}^n l_i \log \nu_i = l \sum_{i=1}^n \nu_i \log \nu_i = -lH(\boldsymbol{\nu}), \; \forall \mathbf{x} \in T(\boldsymbol{\nu}, l),$$

$$P_\nu^l(\{\mathbf{x}\}) = \exp[-lH(\boldsymbol{\nu})], \; \forall \mathbf{x} \in T(\boldsymbol{\nu}, l),$$

$$P_\nu^l(T(\boldsymbol{\nu}, l)) = |T(\boldsymbol{\nu}, l)| \exp[-lH(\boldsymbol{\nu})]. \tag{5.12}$$

Since P_ν^l is a probability measure on $\mathbb{A}^l$ and $T(\boldsymbol{\nu}, l) \subseteq \mathbb{A}^l$, it follows that

$$1 \geq P_\nu^l(T(\boldsymbol{\nu}, l)) \geq |T(\boldsymbol{\nu}, l)| \exp[-lH(\boldsymbol{\nu})],$$

which is the right inequality in (5.11).

To prove the left inequality in (5.11), observe that the various type classes $T(\boldsymbol{\phi}, l), \boldsymbol{\phi} \in E(l,n)$ partition the sample space $\mathbb{A}^l$. Therefore it follows from Lemma 5.4 that

$$\begin{aligned} 1 = P_\nu^l(\mathbb{A}^l) &= \sum_{\boldsymbol{\phi} \in \mathcal{E}(l,n)} P_\nu^l(T(\boldsymbol{\phi}, l)) \\ &\leq |\mathcal{E}(l,n)| P_\nu^l(T(\boldsymbol{\nu}, l)) \\ &= |\mathcal{E}(l,n)| |T(\boldsymbol{\nu}, l)| \exp[-lH(\boldsymbol{\nu})], \end{aligned}$$

where we make use of (5.10) and (5.12) in the last two steps. This leads to the left inequality in (5.11). $\square$

Theorem 5.6 (Sanov's Theorem for a Finite Alphabet) *The stochastic process $\{\hat{\boldsymbol{\mu}}(\mathbf{x}_1^l)\}$ defined in (5.7) satisfies the large deviation property with the rate function*

$$I(\boldsymbol{\nu}) = D(\boldsymbol{\nu} \| \boldsymbol{\mu}). \tag{5.13}$$

Proof. To show that the function $I(\cdot)$ defined above is indeed the rate function, we apply Definition 5.1. It is clear that the function $I(\cdot)$ is not only lower semi-continuous but is in fact continuous. So it remains only to verify the two inequalities in (5.4).

Suppose $\boldsymbol{\nu} \in \mathcal{E}(l,n)$, and let us ask: What is the probability that the empirical distribution $\hat{\boldsymbol{\mu}}(\mathbf{x}_1^l)$ equals $\boldsymbol{\nu}$? Clearly the answer is

$$\Pr\{\hat{\boldsymbol{\mu}}(\mathbf{x}_1^l) = \boldsymbol{\nu}\} = P_\mu^l(T(\boldsymbol{\nu},l)) = |T(\boldsymbol{\nu},l)| P_\mu^l(\{\mathbf{x}_1^l\}) \text{ where } \mathbf{x}_1^l \in T(\boldsymbol{\nu},l).$$

Now if we use (5.9) and the right inequality in (5.11), we can conclude that

$$\Pr\{\hat{\boldsymbol{\mu}}(\mathbf{x}_1^l) = \boldsymbol{\nu}\} \le \exp[lH(\boldsymbol{\nu}) - lJ(\boldsymbol{\nu},\boldsymbol{\mu})] = \exp[-lD(\boldsymbol{\nu}\|\boldsymbol{\mu})]. \tag{5.14}$$

Similarly, by using (5.9) and the left inequality in (5.11), we can conclude that

$$\Pr\{\hat{\boldsymbol{\mu}}(\mathbf{x}_1^l) = \boldsymbol{\nu}\} \ge [|\mathcal{E}(l,n)|]^{-1} \exp[-lD(\boldsymbol{\nu}\|\boldsymbol{\mu})]. \tag{5.15}$$

Now let $\Gamma \subseteq \mathcal{M}(\mathbb{A})$ be any set of probability distributions on $\mathbb{A}$. Then

$$\begin{aligned}
\Pr\{\hat{\boldsymbol{\mu}}(x_1^l) \in \Gamma\} &= \sum_{\boldsymbol{\nu}\in\mathcal{E}(l,n)\cap\Gamma} \Pr\{\hat{\boldsymbol{\mu}}(x_1^l) = \boldsymbol{\nu}\} \\
&\le |\mathcal{E}(l,n)\cap\Gamma| \sup_{\boldsymbol{\nu}\in\mathcal{E}(l,n)\cap\Gamma} \Pr\{\hat{\boldsymbol{\mu}}(x_1^l) = \boldsymbol{\nu}\} \\
&\le |\mathcal{E}(l,n)| \sup_{\boldsymbol{\nu}\in\mathcal{E}(l,n)\cap\Gamma} \exp[-lD(\boldsymbol{\nu}\|\boldsymbol{\mu})].
\end{aligned}$$

Hence

$$\frac{1}{l}\log \Pr\{\hat{\boldsymbol{\mu}}(x_1^l) \in \Gamma\} \le \frac{1}{l}\log|\mathcal{E}(l,n)| + \sup_{\boldsymbol{\nu}\in\Gamma} -D(\boldsymbol{\nu}\|\boldsymbol{\mu}). \tag{5.16}$$

Now we can make a crude estimate of $|\mathcal{E}(l,n)|$ using (5.8). It is clear that, if $l \ge n$, then $l+n-1 \le 2l$, so that

$$|\mathcal{E}(l,n)| \le \frac{2^{n-1}}{(n-1)!} l^{n-1}, \ \forall l \ge n.$$

This is not a particularly "clever" estimate, but as we shall see below, the constants in front will not matter anyhow when we take the limit as $l \to \infty$. Since $|\mathcal{E}(l,n)|$ is polynomial in l, the first term on the right side of (5.16) approaches zero as $l \to \infty$. Therefore

$$\limsup_{l\to\infty} \frac{1}{l}\log \Pr\{\hat{\boldsymbol{\mu}}(x_1^l) \in \Gamma\} \le \sup_{\boldsymbol{\nu}\in\Gamma} -D(\boldsymbol{\nu}\|\boldsymbol{\mu}) = -\inf_{\boldsymbol{\nu}\in\Gamma} D(\boldsymbol{\nu}\|\boldsymbol{\mu}).$$

However, since Γ is a subset of $\bar{\Gamma}$, it follows that

$$-\inf_{\boldsymbol{\nu}\in\Gamma} D(\boldsymbol{\nu}\|\boldsymbol{\mu}) \le -\inf_{\boldsymbol{\nu}\in\bar{\Gamma}} D(\boldsymbol{\nu}\|\boldsymbol{\mu}).$$

Substituting this into the previous bound shows that

$$\limsup_{l\to\infty} \frac{1}{l}\log \Pr\{\hat{\boldsymbol{\mu}}(x_1^l) \in \Gamma\} \le -\inf_{\boldsymbol{\nu}\in\bar{\Gamma}} D(\boldsymbol{\nu}\|\boldsymbol{\mu}).$$

This establishes the right inequality in (5.4).

To establish the left inequality in (5.4), we first establish the following statement:

$$\sup_{\phi \in \mathcal{M}(\mathbb{A})} \inf_{\nu \in \mathcal{E}(l,n)} \rho(\phi, \nu) \leq \frac{n}{2l} \; \forall l. \tag{5.17}$$

In other words, for every l, every distribution $\phi \in \mathcal{M}(\mathbb{A})$ can be approximated to an accuracy of $n/2l$ or less (in the total variation metric) by some distribution $\nu \in \mathcal{E}(l, n)$. To establish this fact, Let $\phi \in \mathcal{M}(\mathbb{A})$ be arbitrary. For each index $i = 1, \ldots, n$, define $q_i = \lceil l\phi_i \rceil$, $r_i = \lfloor l\phi_i \rfloor$. Then

$$|\phi_i - q_i/l| \leq 1/l, \text{ and } |\phi_i - r_i/l| \leq 1/l, \; \forall i.$$

By choosing s_i equal to either q_i or r_i for each index i, it can be ensured that that $\sum_{i=1}^n s_i = l$. (Clearly the choice is not unique.) Then ν defined by $\nu_i = s_i/l$ belongs to $\mathcal{E}(l, n)$. Moreover, by construction, we have that

$$\rho(\phi, \nu) = \frac{1}{2} \sum_{i=1}^n |\phi_i - s_i/l| \leq \frac{n}{2l}.$$

Now suppose ν is an interior point of Γ. Then there is an open ball $\mathcal{B}(\nu)$ in Γ that contains ν. By (5.17), it follows that there exists a sequence $\{\nu_l\}$ such that $\nu_l \in \mathcal{E}(l, n)$ for all l, and in addition, $\nu_l \to \nu$ as $l \to \infty$. Moreover, it is obvious that $\nu_l \in \mathcal{B}(\nu)$ for sufficiently large l. Hence

$$\Pr\{\hat{\mu}(\mathbf{x}_1^l) \in \Gamma\} \geq \Pr\{\hat{\mu}(\mathbf{x}_1^l) = \nu_l\} \geq [|\mathcal{E}(l,n)|]^{-1} \exp[-lD(\nu_l \| \mu)],$$

where the last step follows from (5.15). Therefore

$$\begin{aligned} \frac{1}{l} \log \Pr\{\hat{\mu}(\mathbf{x}_1^l) \in \Gamma\} &\geq -\frac{1}{l} \log |\mathcal{E}(l,n)| - D(\nu_l \| \mu) \\ &\geq -D(\nu_l \| \mu) + o(1/l) \\ &\to -D(\nu \| \mu) \text{ as } l \to \infty. \end{aligned}$$

Hence it follows that

$$\liminf_{l \to \infty} \frac{1}{l} \log \Pr\{\nu(\mathbf{x}_1^l) \in \Gamma\} \geq -D(\nu \| \mu), \; \forall \nu \in \Gamma^o.$$

Since the above inequality holds for every $\nu \in \Gamma^o$, we can conclude that

$$\liminf_{l \to \infty} \frac{1}{l} \log \Pr\{\nu(\mathbf{x}_1^l) \in \Gamma\} \geq \sup_{\nu \in \Gamma^o} -D(\nu \| \mu) = -\inf_{\nu \in \Gamma^o} D(\nu \| \mu).$$

This is the left inequality in (5.4). Since both of the relationships in (5.4) hold with $I(\nu) = D(\nu \| \mu)$, it follows that $D(\nu \| \mu)$ is the rate function. □

5.3 LARGE DEVIATION PROPERTY FOR MARKOV CHAINS

In this section we extend the results of the previous section from i.i.d. processes to Markov chains. Along the way we introduce an alternate description of Markov chains in terms of "stationary" distributions, and also introduce two very useful notions, namely the entropy rate and the relative entropy rate.

5.3.1 Stationary Distributions

Suppose $\{X_t\}$ is a stationary Markov process assuming values in $\mathbb{A}$. Until now we have been describing such a Markov chain is in terms of two entities: (i) The **stationary distribution** $\boldsymbol{\pi} \in \mathcal{M}(\mathbb{A})$, where

$$\pi_i := \Pr\{X_t = i\}, i \in \mathbb{A}.$$

(ii) The **state transition matrix** $A \in [0,1]^{n \times n}$, where

$$a_{ij} := \Pr\{X_{t+1} = j | X_t = i\}, \ \forall i, j \in \mathbb{A}.$$

Note that each row of A is a probability distribution and belongs to $\mathcal{M}(\mathbb{A})$, since the i-th row of A is the conditional distribution of X_{t+1} given that $X_t = i$. Thus A is a stochastic matrix.

But there is an alternative description of a Markov chain that is more convenient for present purposes, namely the *vector of doublet frequencies.* To motivate this alternate description, we first introduce the notion of a stationary distribution on $\mathbb{A}^k$. Other authors also use the phrase "consistent" distribution.

Suppose $\{X_t\}$ is a stationary stochastic process (not necessarily Markov) assuming values in a finite set $\mathbb{A}$. As before, define

$$X_s^t := X_s X_{s+1} \ldots X_{t-1} X_t.$$

Clearly this notation makes sense only when $s \leq t$. For each integer $k \geq 1$ and each $\mathbf{i} := (i_1, \ldots, i_k) \in \mathbb{A}^k$, let us define the k-tuple frequencies

$$\mu_{\mathbf{i}} := \Pr\{X_{t+1}^{t+k} = i_1 \ldots i_k\}.$$

This probability does not depend on t since the process is stationary. Now note that, for each $(k-1)$-tuple $\mathbf{i} \in \mathbb{A}^{k-1}$, the events

$$\{X_{t+1}^{t+k} = i_1 \ldots i_{k-1} 1\}, \ldots, \{X_{t+1}^{t+k} = i_1 \ldots i_{k-1} n\}$$

are mutually disjoint, and together generate the event

$$\{X_{t+1}^{t+k-1} = \mathbf{i}\}.$$

Thus

$$\mu_{\mathbf{i}} = \sum_{j \in \mathbb{A}} \mu_{\mathbf{i}j}, \ \forall \mathbf{i} \in \mathbb{A}^{k-1}.$$

By entirely analogous reasoning, it also follows that

$$\mu_{\mathbf{i}} = \sum_{j \in \mathbb{A}} \mu_{j\mathbf{i}}, \ \forall \mathbf{i} \in \mathbb{A}^{k-1}.$$

This motivates the next definition.

Definition 5.7 *A distribution* $\boldsymbol{\nu} \in \mathcal{M}(\mathbb{A}^2)$ *is said to be* **stationary** *(or consistent) if*

$$\sum_{j \in \mathbb{A}} \nu_{ij} = \sum_{j \in \mathbb{A}} \nu_{ji}, \ \forall i \in \mathbb{A}. \tag{5.18}$$

For $k \geq 3$, a distribution $\boldsymbol{\nu} \in \mathcal{M}(\mathbb{A}^k)$ is said to be **stationary** *(or consistent) if*

$$\sum_{j \in \mathbb{A}} \nu_{\mathbf{i}j} = \sum_{j \in \mathbb{A}} \nu_{j\mathbf{i}}, \ \forall \mathbf{i} \in \mathbb{A}^{k-1},$$

and in addition, the resulting distribution $\bar{\boldsymbol{\nu}}$ on $\mathbb{A}^{k-1}$ defined by

$$\bar{\boldsymbol{\nu}}_{\mathbf{i}} := \sum_{j \in \mathbb{A}} \nu_{\mathbf{i}j} \ \forall \mathbf{i} \in \mathbb{A}^{k-1} = \sum_{j \in \mathbb{A}} \nu_{j\mathbf{i}} \ \forall \mathbf{i} \in \mathbb{A}^{k-1}$$

is stationary. Equivalently, a distribution $\boldsymbol{\nu} \in \mathcal{M}(\mathbb{A}^k)$ is said to be **stationary** *(or consistent) if*

$$\sum_{\mathbf{j}_1 \in \mathbb{A}^{l_1}} \sum_{\mathbf{j}_2 \in \mathbb{A}^{l_2}} \nu_{\mathbf{j}_1 \mathbf{i} \mathbf{j}_2} = \nu_{\mathbf{i}}, \tag{5.19}$$

for all $\mathbf{i} \in \mathbb{A}^{k-l_1-l_2}$, $l_1 \geq 0, l_2 \geq 0$ such that $l_1 + l_2 \leq k-1$. The set of all stationary distributions on $\mathbb{A}^k$ is denoted by $\mathcal{M}_s(\mathbb{A}^k)$.

Now let us return to Markov chains. Suppose $\{X_t\}$ is a stationary Markov chain assuming values in the finite set $\mathbb{A}$. Define the vector $\boldsymbol{\mu} \in \mathcal{M}(\mathbb{A}^2)$ by

$$\mu_{ij} = \Pr\{X_t X_{t+1} = ij\}, \ \forall i, j \in \mathbb{A}.$$

Then, as per the above discussion, $\boldsymbol{\mu} \in \mathcal{M}_s(\mathbb{A}^2)$. The vector $\boldsymbol{\mu}$ is called the **vector of doublet frequencies**. The claim is that the doublet frequency vector $\boldsymbol{\mu}$ captures all the relevant information about the Markov chain, and does so in a more natural way. The stationary distribution of the Markov chain is given by

$$\bar{\mu}_i := \sum_{j \in \mathbb{A}} \mu_{ij} = \sum_{j \in \mathbb{A}} \mu_{ji},$$

while the state transition matrix A is given by

$$a_{ij} = \frac{\mu_{ij}}{\bar{\mu}_i}.$$

Dividing by $\bar{\mu}_i$ can be justified by observing that if $\bar{\mu}_i = 0$ for some index i, then the corresponding element i can simply be dropped from the set $\mathbb{A}$. With these definitions, it readily follows that $\bar{\boldsymbol{\mu}}$ is a row eigenvector of A, because

$$(\bar{\boldsymbol{\mu}} A)_j = \sum_{i=1}^{n} \bar{\mu}_i a_{ij} = \sum_{i=1}^{n} \mu_{ij} = \bar{\mu}_j.$$

Note that the above reasoning breaks down if $\boldsymbol{\mu} \in \mathbb{S}_{n^2}$ but $\boldsymbol{\mu} \notin \mathcal{M}_s(\mathbb{A}^2)$.

More generally, suppose $\{X_t\}$ is an s-step Markov chain, so that

$$\Pr\{X_t = v | X_0^{t-1} = \mathbf{u}_0^{t-1}\} = \Pr\{X_t = v | X_{t-s}^{t-1} = \mathbf{u}_{t-s}^{t-1}\} \ \forall t.$$

Then the process is completely characterized by its $(s+1)$-tuple frequencies

$$\mu_{\mathbf{i}} := \Pr\{X_t^{t+s} = \mathbf{i}\}, \ \forall \mathbf{i} \in \mathbb{A}^{s+1}.$$

The probability distribution $\boldsymbol{\mu}$ is stationary and thus belongs to $\mathcal{M}_s(\mathbb{A}^{s+1})$. Now an s-step Markov chain assuming values in $\mathbb{A}$ can also be viewed as a conventional (one-step) Markov chain over the state space $\mathbb{A}^s$. Moreover, if the current state is $i\mathbf{j}$ where $i \in \mathbb{A}, \mathbf{j} \in \mathbb{A}^{s-1}$, then a transition is possible only to a state of the form $\mathbf{j}k, k \in \mathbb{A}$. Thus, even though the state transition matrix has dimension $n^s \times n^s$, each row of the transition matrix can have at most n nonzero elements. The entry in row $i\mathbf{j}$ and column $\mathbf{j}k$ equals

$$\Pr\{X_t = k | X_{t-s}^{t-1} = i\mathbf{j}\} = \frac{\mu_{i\mathbf{j}k}}{\bar{\mu}_{i\mathbf{j}}},$$

where as before we define

$$\bar{\mu}_{\mathbf{l}} := \sum_{i \in \mathbb{A}} \mu_{i\mathbf{l}} = \sum_{i \in \mathbb{A}} \mu_{\mathbf{l}i}, \ \forall \mathbf{l} \in \mathbb{A}^s.$$

5.3.2 Entropy and Relative Entropy Rates

In this subsection, we introduce two important notions called the entropy rate of a Markov process, and the relative entropy rate between two Markov processes. The reader is cautioned that the terminology is not completely standard. The definition of the entropy of a Markov process over a finite alphabet is given by Shannon [117]. His definition is extended from Markov processes to arbitrary stationary ergodic processes[4] by McMillan [96] and Breiman [22]. A closed-form formula for the entropy of a Markov process is given in [109]. The Kullback-Leibler divergence rate between hidden Markov models is defined in [72], and a closed-form formula for this rate between Markov processes is also given in [109]. So as to harmonize the various terms in use, in the present book we use the phrases "entropy rate" and "relative entropy rate." In particular, since the Kullback-Leibler divergence is often referred to as relative entropy, it is not too much of a stretch to refer to the Kullback-Leibler divergence rate as the relative entropy rate. However, the phrase "entropy rate" is used only to make it sound similar to "relative entropy rate," and "entropy of a process" is the more common phrase.

Suppose $\{X_t\}_{t \geq 0}$ is a stochastic process assuming values in a finite alphabet $\mathbb{A}$. We will call this process "stationary" if, for every pair of nonnegative integers t, τ, the joint distribution of $\mathcal{X}_0^t$ is the same as that of $X_\tau^{t+\tau}$. For such processes, it is possible to define an entropy, or as we prefer to call it, an entropy rate.

Definition 5.8 *Suppose $\{X_t\}_{t \geq 0}$ is a stationary stochastic process assuming values in a finite alphabet $\mathbb{A}$. Then the* **entropy rate** *of the process is denoted by $H_r(\{X_t\})$, and is defined as*

$$H_r(\{X_t\}) = \lim_{t \to \infty} H(X_t | X_0^{t-1}), \tag{5.20}$$

where the conditional entropy $H(\cdot|\cdot)$ is defined in Definition 2.10.

[4]We will not define these terms, as we will not study this general case in this book.

The next theorem shows that the entropy rate is well-defined and also gives an alternate characterization.

Theorem 5.9 *Suppose* $\{X_t\}_{t \geq 0}$ *is a stationary stochastic process assuming values in a finite alphabet* $\mathbb{A}$. *Then there exists a constant* $c \geq 0$ *such that*

$$H(X_t|X_0^{t-1}) \downarrow c \text{ as } t \to \infty,$$

and this constant is the entropy rate of the process. An alternate expression for c *is*

$$c = \lim_{t \to \infty} \frac{H(X_0^t)}{t}. \tag{5.21}$$

Proof. Recall that conditioning on more random variables cannot increase the conditional entropy. Therefore, for each $t \geq 1$, $\tau \geq 0$, we have

$$H(X_{t+\tau}|X_0^{t+\tau-1}) \leq H(X_{t+\tau}|X_\tau^{t+\tau-1}) = H(X_t|X_0^{t-1}),$$

where the last step follows from the stationarity of the process. This shows that the sequence of real numbers $\{H(X_t|X_0^{t-1})\}$ is nonincreasing. Since the sequence is clearly bounded below by zero, it has a limit, which can be denoted by c. This proves the first assertion.

To prove the second assertion, note from the definition of conditional entropy that

$$H(X_t|X_0^{t-1}) = H(X_0^t) - H(X_0^{t-1}).$$

Therefore

$$\sum_{\tau=1}^{t} H(X_\tau|X_0^{\tau-1}) = \sum_{\tau=1}^{t} [H(X_0^\tau) - H(X_0^{\tau-1})] = H(X_0^t) - H(X_0),$$

because the summation on the right is a "telescoping" sum where the second part of one term cancels the first part of the next term. Therefore

$$\frac{1}{t} \sum_{\tau=1}^{t} H(X_\tau|X_0^{\tau-1}) = \frac{H(X_0^t)}{t} - \frac{H(X_0)}{t}.$$

Now, if the sequence $\{H(X_t|X_0^{t-1})\}$ converges to the constant c, so does the sequence of the averages of the terms. In other words,

$$\begin{aligned} c &= \lim_{t \to \infty} \frac{1}{t} \sum_{\tau=1}^{t} H(X_\tau|X_0^{\tau-1}) \\ &= \lim_{t \to \infty} \frac{H(X_0^t)}{t} - \lim_{t \to \infty} \frac{H(X_0)}{t} \\ &= \lim_{t \to \infty} \frac{H(X_0^t)}{t}, \end{aligned}$$

because $H(X_0)$ is just some constant, as a result of which the ratio $H(X_0)/t$ approaches zero as $t \to \infty$. This establishes (5.21). □

The reader is advised that Theorem 5.9 is a very simple result that is very easy to prove. A far more significant result in this direction is known as the

Shannon-McMillan-Breiman theorem, or as the Asymptotic Equipartition Property (AEP), and states that the log likelihood ratio of *each sample path* converges to the entropy rate c in probability and almost surely. This advanced result is not needed here, and the interested reader is referred to [30].

Next we define the concept of the relative entropy rate.

Definition 5.10 *Suppose $\{X_t\}_{t\geq 0}$ and $\{Y_t\}_{t\geq 0}$ are two stationary stochastic processes assuming values in a common finite alphabet $\mathbb{A}$. Then the* **relative entropy rate**!*For stationary processes between them is denoted by $D_r(\{X_t\}\|\{Y_t\})$ and is defined as the limit*

$$\lim_{t\to\infty} \frac{D(X_0^t\|Y_0^t)}{t}, \tag{5.22}$$

if the limit exists. Otherwise it is left as undefined.

Note that, unlike the entropy rate of a stationary process, which always exists, the relative entropy rate may or may not exist (i.e., be well-defined).

Next we define the entropy rate and relative entropy rate for stationary distributions.

Suppose $\boldsymbol{\nu}, \boldsymbol{\mu} \in \mathcal{M}_s(\mathbb{A}^k)$ are *stationary* distributions on $\mathbb{A}^k$ for some finite set $\mathbb{A}$ and some integer $k \geq 2$. We define $\bar{\boldsymbol{\mu}} \in \mathcal{M}_s(\mathbb{A}^{k-1})$ by

$$\bar{\mu}_{\mathbf{i}} := \sum_{j\in\mathbb{A}} \mu_{\mathbf{i}j} = \sum_{j\in\mathbb{A}} \mu_{j\mathbf{i}}, \ \forall \mathbf{i} \in \mathbb{A}^{k-1}.$$

The overbar serves to remind us that $\bar{\boldsymbol{\mu}}$ is "reduced by one dimension" from $\boldsymbol{\mu}$. Because $\boldsymbol{\mu}$ is a stationary distribution, it does not matter whether the reduction is on the first component or the last. The symbol $\bar{\boldsymbol{\nu}}$ is defined similarly.

Definition 5.11 *Suppose $\boldsymbol{\nu}, \boldsymbol{\mu} \in \mathcal{M}_s(\mathbb{A}^k)$ for some integer $k \geq 2$. Then*

$$H_r(\boldsymbol{\mu}) := H(\boldsymbol{\mu}) - H(\bar{\boldsymbol{\mu}}) \tag{5.23}$$

is called the **entropy rate** *of $\boldsymbol{\mu}$, while*

$$D_r(\boldsymbol{\nu}\|\boldsymbol{\mu}) := D(\boldsymbol{\nu}\|\boldsymbol{\mu}) - D(\bar{\boldsymbol{\nu}}\|\bar{\boldsymbol{\mu}}) \tag{5.24}$$

is called the **relative entropy rate** *between $\boldsymbol{\nu}$ and $\boldsymbol{\mu}$.*

Note that $\bar{\boldsymbol{\mu}}, \bar{\boldsymbol{\nu}}$ are marginal distributions of $\boldsymbol{\mu}, \boldsymbol{\nu}$ respectively on $\mathbb{A}^{k-1}$. Thus it readily follows that both $H_r(\cdot)$ and $D_r(\cdot\|\cdot)$ are nonnegative-valued.

At this point, the symbols H_r and D_r are doing double duty. It is now shown that both usages coincide. In other words, while the entropy rate defined in Definition 5.8 and relative entropy rate defined in Definition 5.10 are applicable to arbitrary stationary stochastic processes, they reduce to the quantities in Definition 5.11 in the case of Markov processes.

Theorem 5.12 *Suppose $\{X_t\}$ is a Markov process over $\mathbb{A}$ with doublet frequency $\boldsymbol{\nu} \in \mathcal{M}_s(\mathbb{A}^2)$, stationary distribution $\bar{\boldsymbol{\nu}} \in \mathbb{S}_n$, and transition matrix $A \in [0,1]^{n \times n}$. Then the entropy rate of the Markov process is given by*

$$H_r(\{X_t\}) = H(X_1|X_0).$$

Moreover

$$H_r(\{X_t\}) = H_r(\bar{\boldsymbol{\nu}}) = \sum_{i=1}^{n} \bar{\nu}_i H(\mathbf{a}_i), \tag{5.25}$$

where $\mathbf{a}_i \in \mathbb{S}_n$ denotes the i-th row of the matrix A.

Proof. First, note that, because the process is Markov, the conditional probability distribution of X_t given X_0^t is the same as that of X_t given only X_{t-1}. Therefore

$$H(X_t|X_0^{t-1}) = H(X_t|X_{t-1}) = H(X_1|X_0),$$

where the last step follows from the stationarity of the process. So the sequence of real numbers $\{H(X_t|X_0^{t-1})\}$ converges in one step and the limit is $H(X_1|X_0)$. This shows that the entropy rate of the Markov process is $H(X_1|X_0)$. Next, from (2.19), it follows that

$$H(X_1|X_0) = H((X_1, X_0)) - H(X_0) = H(\boldsymbol{\nu}) - H(\bar{\boldsymbol{\nu}}) = H_r(\boldsymbol{\nu})$$

because of the definition of H_r. It remains only to show that $H_r(\boldsymbol{\nu})$ is given by the summation in (5.25). We reason as follows:

$$\begin{aligned} H_r(\boldsymbol{\nu}) &= H(\boldsymbol{\nu}) - H(\bar{\boldsymbol{\nu}}) \\ &= -\sum_{i=1}^{n}\sum_{j=1}^{n} \nu_{ij} \log \nu_{ij} + \sum_{i=1}^{n} \bar{\nu}_i \log \bar{\nu}_i \\ &= \sum_{i=1}^{n}\sum_{j=1}^{n} \nu_{ij} \log \frac{\bar{\nu}_i}{\nu_{ij}} \\ &= -\sum_{i=1}^{n}\sum_{j=1}^{n} \bar{\nu}_i a_{ij} \log a_{ij} \\ &= \sum_{i=1}^{n} \bar{\nu}_i H(\mathbf{a}_i), \end{aligned}$$

which is the same as (5.25). □

Theorem 5.13 *Suppose $\{X_t\}$, $\{Y_t\}$ are Markov processes over a finite set $\mathbb{A}$, with doublet frequencies $\boldsymbol{\nu}, \boldsymbol{\mu} \in \mathcal{M}_s(\mathbb{A}^2)$, stationary distributions $\bar{\boldsymbol{\nu}}, \bar{\boldsymbol{\mu}} \in \mathbb{S}_n$ respectively, and transition matrices $A, B \in [0,1]^{n \times n}$ respectively. Then the relative entropy rate between the two processes is given by*

$$D_r(\{X_t\}\|\{Y_t\}) = D_r(\boldsymbol{\nu}\|\boldsymbol{\mu}). \tag{5.26}$$

Moreover

$$D_r(\boldsymbol{\nu}\|\boldsymbol{\mu}) = \sum_{i=1}^{n} \bar{\nu}_i D(\mathbf{a}_i\|\mathbf{b}_i), \tag{5.27}$$

where $\mathbf{a}_i, \mathbf{b}_i \in \mathbb{S}_n$ denote the i-th rows of A and B respectively.

The proof of Theorem 5.13 depends on a lemma that is of interest in its own right, because it is not restricted to Markov processes.

Lemma 5.14 *Given stationary stochastic processes* $\{X_t\}, \{Y_t\}$, *define for all* $t \geq 1$

$$\alpha_t = \frac{D(X_0^t \| Y_0^t)}{t}, \tag{5.28}$$

$$\beta_t = D(X_0^{t+1} \| Y_0^{t+1}) - D(X_0^t \| Y_0^t). \tag{5.29}$$

Then the following statements are equivalent.

- *The relative entropy rate* $D_r(\{X_t\} \| \{Y_t\})$ *is well-defined and equals* d.
- *The sequence* $\{\alpha_t\}$ *converges to* d *as* $t \to \infty$.
- *The sequence* $\{\beta_t\}$ *converges in the Césaro sense to* d.

Proof. (Of Lemma 5.14): Recall that a sequence $\{\beta_t\}$ is said to "converge in the Césaro sense" if the sequence of averages

$$\gamma_t = \frac{1}{t} \sum_{\tau=0}^{t-1} \beta_\tau$$

converges. Now observe that for all $t \geq 1$, we have

$$\begin{aligned} \sum_{\tau=0}^{t-1} \beta_\tau &= \sum_{\tau=0}^{t-1} [D(X_0^{\tau+1} \| Y_0^{\tau+1}) - D(X_0^\tau \| Y_0^\tau)] \\ &= D(X_0^t \| Y_0^t) - D(X_0 \| Y_0), \end{aligned}$$

because the summation is a telescoping sum whereby the second part in one term cancels the first part in the next term. Therefore

$$\alpha_t = \frac{1}{t} D(X_0^t \| Y_0^t) = \gamma_t + \frac{1}{t} D(X_0 \| Y_0).$$

As $t \to \infty$, the second term on the right approaches zero. Therefore $\{\alpha_t\}$ converges if and only if $\{\gamma_t\}$ converges, which is in turn equivalent to the convergence of $\{\beta_t\}$ in the Césaro sense. □

Proof. (Of Theorem 5.13): Observe that, by definition, we have that

$$\bar{\nu}_i = \sum_{i=1}^n \nu_{ij} = \sum_{i=1}^n \nu_{ji}, \bar{\mu}_i = \sum_{i=1}^n \mu_{ij} = \sum_{i=1}^n \mu_{ji},$$

and that

$$a_{ij} = \frac{\nu_{ij}}{\bar{\nu}_i}, b_{ij} = \frac{\mu_{ij}}{\bar{\mu}_i}.$$

From Lemma 5.14, we know that the relative entropy rate between the two Markov processes $\{X_t\}$ and $\{Y_t\}$ is well-defined if and only if the sequence

$\{\beta_t\}$ defined in (5.29) converges in the Césaro sense. Now, from the chain rule for relative entropy, namely Theorem 2.23, it follows that

$$D(X_0^{t+1}\|Y_0^{t+1}) = D(X_0^t\|Y_0^t) + \sum_{\mathbf{i}_0^t \in \mathbb{A}^t} \Pr\{X_0^t = \mathbf{i}_0^t\} \cdot D(X_{t+1}|X_0^t = \mathbf{i}_0^t \| Y_{t+1}|Y_0^t = \mathbf{i}_0^t).$$

Therefore, in general it is true that

$$\beta_t = D(X_0^{t+1}\|Y_0^{t+1}) - D(X_0^t\|Y_0^t) = \sum_{\mathbf{i} \in \mathbb{A}^t} \Pr\{X_0^t = \mathbf{i}_0^t\} \cdot D(X_{t+1}|X_0^t = \mathbf{i}_0^t \| Y_{t+1}|Y_0^t = \mathbf{i}_0^t).$$

However, since both $\{X_t\}$ and $\{Y_t\}$ are Markov processes, it follows that the conditional distribution of $(X_{t+1}|X_0^t = \mathbf{i}_0^t)$ is the same as that of $(X_{t+1}|X_t = i_t)$. A similar statement holds for Y_{t+1}. Therefore

$$\beta_t = \sum_{\mathbf{i}_0^t \in \mathbb{A}^t} \Pr\{X_0^t = \mathbf{i}_0^t\} \cdot D(X_{t+1}|X_t = i_t \| Y_{t+1}|Y_t = i_t) = \sum_{\mathbf{i}_0^{t-1} \in \mathbb{A}^{t-1}} \sum_{i_t \in \mathbb{A}} \Pr\{X_0^t = \mathbf{i}_0^t\} \cdot D(X_{t+1}|X_t = i_t \| Y_{t+1}|Y_t = i_t).$$

This can be rearranged as

$$\beta_t = \sum_{i_t \in \mathbb{A}} \left[\sum_{\mathbf{i}_0^{t-1} \in \mathbb{A}^{t-1}} \Pr\{X_0^t = \mathbf{i}_0^t\} \right] \cdot D(X_{t+1}|X_t = i_t \| Y_{t+1}|Y_t = i_t).$$

Now observe that

$$\sum_{\mathbf{i}_0^{t-1} \in \mathbb{A}^{t-1}} \Pr\{X_0^t = \mathbf{i}_0^t\} = \Pr\{X_t = i_t\}.$$

Therefore

$$\beta_t = \sum_{i_t \in \mathbb{A}} \Pr\{X_t = i_t\} \cdot D(X_{t+1}|X_t = i_t \| Y_{t+1}|Y_t = i_t).$$

By the stationarity of the process, the quantities are being summed are all independent of t, and it follows that

$$\beta_t = \sum_{i \in \mathbb{A}} \bar{\nu}_i D(\mathbf{a}_i \| \mathbf{b}_i), \ \forall t \geq 1.$$

Therefore the sequence $\{\beta_t\}$ converges in just one step, and the relative entropy rate is the right side of the above equation. This is precisely the desired conclusion. □

5.3.3 The Rate Function for Doubleton Frequencies

Suppose $\{X_t\}$ is a Markov process assuming values in a finite set $\mathbb{A}$. Given an observation $\mathbf{x}_1^l = x_1 \ldots x_l$, we can form an empirical distribution $\phi(\mathbf{x}_1^l) \in \mathbb{S}_n$ in analogy with (5.1); that is,

$$\phi_j(\mathbf{x}_1^l) := \frac{1}{l} \sum_{t=1}^{l} I_{\{x_t = j\}}, \ \forall j \in \mathbb{A}. \tag{5.30}$$

Thus $\boldsymbol{\phi}$ is an approximation to the stationary distribution $\boldsymbol{\pi}$ of the Markov chain.

As shown earlier, the Markov process $\{X_t\}$ is completely characterized by its vector of doublet frequencies $\boldsymbol{\mu} \in \mathcal{M}_s(\mathbb{A}^2)$. If $\{X_t\}$ is a Markov chain, so is the stochastic process consisting of doublets $\{(X_t, X_{t+1})\}$. So, given a sample path $\mathbf{x}_1^l$, we can estimate the vector of doublet frequencies using this sample path. Since this point is germane to the subsequent discussion, it is worth describing how precisely the doublet frequency vector is estimated. Given the sample path $\mathbf{x}_1^l = x_1 \ldots x_l$, we could define

$$\theta_{ij}(\mathbf{x}_1^l) := \frac{1}{l-1} \sum_{t=1}^{l-1} I_{\{X_t X_{t+1} = ij\}}. \tag{5.31}$$

This procedure produces a vector $\boldsymbol{\theta} \in \mathbb{S}_{n^2}$ which is a measure on $\mathbb{A}^2$, and can be interpreted as an empirical estimate for $\boldsymbol{\mu}$, the true but unknown vector of doublet frequencies. The difficulty however is that the distribution $\boldsymbol{\theta}$ is *not stationary* in general. If we define $\bar{\boldsymbol{\theta}} \in \mathbb{S}_n$ by

$$\bar{\theta}_i := \frac{1}{l-1} \sum_{t=1}^{l-1} I_{\{X_t = i\}},$$

then it is certainly true that

$$\bar{\theta}_i = \sum_{j=1}^{n} \theta_{ij}.$$

However, in general

$$\sum_{j=1}^{n} \theta_{ji} \neq \bar{\theta}_i.$$

Hence $\boldsymbol{\theta} \in \mathbb{S}_{n^2}$ is *not* a stationary distribution in general. Moreover, there is no simple relationship between $\bar{\boldsymbol{\theta}} \in \mathbb{S}_n$ and $\boldsymbol{\phi} \in \mathbb{S}_n$ defined in (5.30).

On the other hand, if $x_l = x_1$ so that the sample path is a cycle, then $\boldsymbol{\theta} \in \mathcal{M}_s(\mathbb{A}^2)$. This suggests that we should use only cyclic sample paths to construct empirical estimates of doublet frequencies, or to carry the argument a bit farther, that we must *create cycles*, artificially if necessary, in the sample path. Accordingly, given a sample path $\mathbf{x}_1^l = x_1 \ldots x_l$, we construct the empirical estimate $\boldsymbol{\nu} = \boldsymbol{\nu}(\mathbf{x}_1^l)$ as follows:

$$\nu_{ij}(\mathbf{x}_1^l) := \frac{1}{l} \sum_{t=1}^{l} I_{\{x_t x_{t+1} = ij\}}, \ \forall i, j \in \mathbb{A}, \tag{5.32}$$

where x_{l+1} is taken as x_1. If we compare (5.32) with (5.31), we see that we have in effect augmented the original sample path $\mathbf{x}_1^l$ by adding a "ghost" transition from x_l back to x_1 so as to create a cycle, and used this artificial sample path of length $l+1$ to construct the empirical estimate. The advantage of doing so is that the resulting vector $\boldsymbol{\nu}$ is *always stationary*, unlike $\boldsymbol{\theta}$ in (5.31) which may not be stationary in general.

It should be intuitively obvious that $\boldsymbol{\nu}(\mathbf{x}_1^l)$ is stationary, but this is shown formally.

Lemma 5.15 *The measure $\boldsymbol{\nu}(\mathbf{x}_1^l) \in \mathcal{M}(\mathbb{A}^2)$ defined in (5.32) is stationary and thus belongs to $\mathcal{M}_s(\mathbb{A}^2)$. Moreover, its one-dimensional marginals equal $\boldsymbol{\phi}(\mathbf{x}_1^l)$ as defined in (5.30).*

Proof. The claim is that

$$\sum_{j \in \mathbb{A}} \nu_{ji} = \sum_{j \in \mathbb{A}} \nu_{ij} = \phi_i, \ \forall i \in \mathbb{A}, \tag{5.33}$$

thus showing that $\boldsymbol{\nu}$ is stationary, and that both of its one-dimensional marginals are $\boldsymbol{\phi}$ as defined in (5.30). To establish (5.33), observe that $\sum_{j \in \mathbb{A}} \nu_{ij}$ is obtained by counting the number of times that i occurs as the first symbol in $x_1x_2, x_2x_3, \ldots, x_{l-1}x_l, x_lx_1$, and then dividing by l. Similarly the quantity $\sum_{j \in \mathbb{A}} \nu_{ji}$ is obtained by counting the number of times that i occurs as the second symbol in $x_1x_2, x_2x_3, \ldots, x_{l-1}x_l, x_lx_1$, and then dividing by l. Now, for $2 \leq t \leq l-1$, x_t is the first symbol in x_tx_{t+1}, and the second symbol in $x_{t-1}x_t$. Next, x_1 is the first symbol in x_1x_2 and the second symbol in the ghost transition x_lx_1. Similarly x_l is the second symbol in $x_{l-1}x_l$ and the first symbol in the ghost transition x_lx_1. Thus $\boldsymbol{\nu}$ is stationary. It is easy to verify that the one-dimensional marginals of $\boldsymbol{\nu}$ are indeed $\boldsymbol{\phi}$. □

To derive the rate function for this situation, we again use the method of types. For a given integer l, the sample space of all possible sample paths is clearly $\mathbb{A}^l$. With each sample path $\mathbf{x}_1^l \in \mathbb{A}^l$, we associate a corresponding empirical distribution $\hat{\boldsymbol{\mu}}(\mathbf{x}_1^l)$ defined as $\boldsymbol{\nu}(\mathbf{x}_1^l)$ of (5.32). We again define $\mathbf{x}_1^l, \mathbf{y}_1^l \in \mathbb{A}^l$ to be equivalent if they lead to the same empirical distribution, that is, if $\hat{\boldsymbol{\mu}}(\mathbf{x}_1^l) = \hat{\boldsymbol{\mu}}(\mathbf{y}_1^l)$. Let us define $\mathcal{E}(l, n, 2)$ to be the subset of $\mathcal{M}_s(\mathbb{A}^2)$ that can be generated as empirical measures from a sample path of length l over an alphabet of size n. Note that earlier we had introduced the symbol $\mathcal{E}(l, n)$ for the set of all empirical measures in $\mathbb{S}_n$ that can be generated from a sample path of length l over an alphabet of size n. So it is clear that $\mathcal{E}(l, n, 2) \subseteq \mathcal{E}(l, n^2)$. However, not every empirical measure in $\mathcal{E}(l, n^2)$ will be *stationary*. This is why we introduce a new symbol $\mathcal{E}(l, n, 2)$.

As before, for each $\boldsymbol{\nu} \in \mathcal{E}(l, n, 2)$, define

$$T(\boldsymbol{\nu}, l) := \{\mathbf{x}_1^l \in \mathbb{A}^l : \hat{\boldsymbol{\mu}}(\mathbf{x}_1^l) = \boldsymbol{\nu}\}.$$

Then $T(\boldsymbol{\nu}, l) \subseteq \mathbb{A}^l$ is once again called the **type class** of $\boldsymbol{\nu}$. We will once again address the following questions:

- What is the cardinality of $\mathcal{E}(l, n, 2)$? In other words, how many distinct *stationary* empirical measures $\hat{\boldsymbol{\mu}}(\mathbf{x}_1^l)$ can be generated as $\mathbf{x}_1^l$ varies over $\mathbb{A}^l$?
- For a given $\boldsymbol{\nu} \in \mathcal{E}(l, n, 2)$, what is the cardinality of the associated type class $T(\boldsymbol{\nu}, l)$?
- What is the (log) likelihood of each sample path in $T(\boldsymbol{\nu}, l)$, and how is it related to $\boldsymbol{\nu}$?

We have seen in Section 5.2 that, if the process is i.i.d., and we estimate one-dimensional marginal using (5.1), then every sample path in $T(\boldsymbol{\nu}, l)$ has exactly the same (log) likelihood. However, if the process is not i.i.d., then this statement is no longer true: Different sample paths in $T(\boldsymbol{\nu}, l)$ can have different (log) likelihoods. Nevertheless, it is possible to adapt the arguments from Section 5.2 to derive a rate function for the present situation.

We state at once the main result of this subsection. The proof is given in stages.

Theorem 5.16 *Suppose $\{X_t\}$ is a stationary Markov chain assuming values in a finite alphabet $\mathbb{A}$. Let $\boldsymbol{\mu} \in \mathcal{M}_s(\mathbb{A}^2)$ denote the vector of doublet frequencies corresponding to this Markov chain, and let $\boldsymbol{\nu}(\mathbf{x}_1^l) \in \mathcal{M}_s(\mathbb{A}^2)$ denote the empirical distribution constructed as in (5.32). Suppose $\mu_{ij} > 0 \ \forall i, j \in \mathbb{A}$. Then the $\mathcal{M}_s(\mathbb{A}^2)$-valued process $\{\boldsymbol{\nu}(\mathbf{x}_1^l)\}$ satisfies the LDP with the rate function*

$$I(\boldsymbol{\nu}) := D_c(\boldsymbol{\nu} \| \boldsymbol{\mu}) = D(\boldsymbol{\nu} \| \boldsymbol{\mu}) - D(\bar{\boldsymbol{\nu}} \| \bar{\boldsymbol{\mu}}). \tag{5.34}$$

The proof of the theorem is given through a couple of preliminary lemmas. Each $\boldsymbol{\nu} \in \mathcal{E}(l, n, 2)$ is of the form $\nu_{ij} = l_{ij}/l$ for some integer l_{ij}. Moreover, the corresponding reduced distribution $\bar{\boldsymbol{\nu}}$ over $\mathbb{A}$ is given by $\bar{\nu}_i = \bar{l}_i/l$ where

$$\bar{l}_i = \sum_{j=1}^{n} l_{ij} = \sum_{j=1}^{n} l_{ji}, \ \forall i.$$

Throughout the proof, l denotes the length of the sample path and $l_{ij}, \bar{l}_i$ denote these integers.

Lemma 5.17 *With all notation as above, we have*

$$|\mathcal{E}(l, n, 2)| \leq (l+1)^{n^2}. \tag{5.35}$$

Proof. Suppose $\boldsymbol{\nu} \in \mathcal{E}(l, n, 2)$. Then each component ν_{ij} has $l+1$ possible values, namely $0/l, 1/l, \ldots, (l-1)/l, l/l$. Thus the maximum number of possible vectors in $\mathcal{E}(l, n, 2)$ is $(l+1)^{n^2}$. □

Lemma 5.18 *For all $\boldsymbol{\psi} \in \mathcal{M}_s(\mathbb{A}^2)$, we have that*

$$\min_{\boldsymbol{\nu} \in \mathcal{E}(l,n,2)} \|\boldsymbol{\psi} - \boldsymbol{\nu}\|_1 \leq \frac{2n^2}{l} \tag{5.36}$$

for all sufficiently large l.

The proof of this lemma uses some methods that are not discussed in this book, and are not needed elsewhere. Hence the proof is not given. The interested reader is referred to [29, Lemma 5.2]. The bound (5.36) can be compared to (5.17).

The next lemma is much more crucial.

Lemma 5.19 *Suppose $\boldsymbol{\nu} \in \mathcal{E}(l, n, 2)$. Then the cardinality of the type class $T(\boldsymbol{\nu}, l)$ is bounded by*

$$(2l)^{-2n^2} \exp[l H_r(\boldsymbol{\nu})] \leq |T(\boldsymbol{\nu}, l)| \leq l \exp[l H_r(\boldsymbol{\nu})]. \tag{5.37}$$

Proof. Suppose $\boldsymbol{\nu} \in \mathcal{E}(l,n,2)$. As we have done elsewhere, define $\bar{\boldsymbol{\nu}} \in \mathbb{S}_n$ as the reduced version of $\boldsymbol{\nu}$, and $l_{ij} = l\nu_{ij}, \bar{l}_i = l\bar{\nu}_i$ to be associated integers. With the distribution $\boldsymbol{\nu}$, we can associate a directed graph $\mathcal{G}(\boldsymbol{\nu})$ with n nodes and l edges by placing l_{ij} edges from node i to node j. Because $\boldsymbol{\nu}$ is a stationary distribution, it is clear that every node in the graph $\mathcal{G}(\boldsymbol{\nu})$ has equal in-degree and out-degree. Therefore the graph $\mathcal{G}(\boldsymbol{\nu})$ is a union of cycles. In other words, it is possible to start at a node, and trace out a path that ends at the same node. Note that it is permissible for the path to pass through any of the n nodes one or more times. Indeed if $l > n$, this is unavoidable. The cardinality of the associated type class $T(\boldsymbol{\nu},l,2)$ is precisely equal to the number of different ways of tracing out such a path. While every $\boldsymbol{\nu} \in \mathcal{M}_s(\mathbb{A}^2)$ corresponds to a graph where every node has equal in- and out-degrees, not every $\boldsymbol{\nu} \in \mathcal{M}_s(\mathbb{A}^2)$ (or equivalently, not every graph with this property) belongs $\mathcal{E}(l,n,2)$. For example, the distribution

$$\boldsymbol{\nu}_{ij} = \frac{\bar{l}_i}{l}\delta_{ij}, \ \forall i,j \in \mathbb{A},$$

with $\bar{l}_i > 0$ for all i, where δ is the Kronecker delta, belongs to $\mathcal{M}_s(\mathbb{A}^2)$ but not to $\mathcal{E}(l,n,2)$. In order for $\boldsymbol{\nu}$ to belong to $\mathcal{E}(l,n,2)$ and not just $\mathcal{M}_s(\mathbb{A}^2)$, a necessary and sufficient condition is that the graph $\mathcal{G}(\boldsymbol{\nu})$ should consist of one strongly connected component. This explains why the "diagonal" distribution above fails to belong to $\mathcal{E}(l,n,2)$, because in this case there is more than one strongly connected component.

Next we introduce an alternate way of describing a graph $\mathcal{G}(\boldsymbol{\nu})$ associated with a $\boldsymbol{\nu} \in \mathcal{M}_s(\mathbb{A}^2)$. Recall that $\{a_1,\ldots,a_n\}$ denote the elements of the state space $\mathbb{A}$, ordered in some arbitrary fashion. Given a sample path $\mathbf{x}_1^l$ and the associated empirical estimate $\boldsymbol{\nu}(\mathbf{x}_1^l)$ constructed as in (3.1), let us define n sequences $S(1)$ through $S(n)$ as follows: The set $S(i)$ has cardinality $\bar{l}_i$, and consists of all the symbols that follow a_i in the sample path, in that order. To illustrate, suppose $n=3$, $\mathbb{A} = \{a,b,c\}$, $l=10$, and[5]

$$\mathbf{x}_1^l = abaccbacbc.$$

Then, with the ghost transition from c back to a added, the three sets are defined as

$$S(a) = bcc, S(b) = aac, S(c) = cbba.$$

It is easy to see that, given the n sets $S(1)$ through $S(n)$, we can reconstruct the associated cycle; however, we would not know the starting point. In other words, given the sets $S(1)$ through $S(n)$, there are at most n different corresponding sample paths, corresponding to choosing one of the n nodes as the starting point of the cycle. Moreover, if $\boldsymbol{\theta}$ belongs to the same type class as $\boldsymbol{\nu}$, then each $S(i,\boldsymbol{\theta})$ is a permutation of the corresponding set $S(i,\boldsymbol{\nu})$.

This suggests a way of finding an upper bound for $|T(\boldsymbol{\nu},l,2)|$. Given a sample path $\mathbf{x}_1^l$ and the associated distribution $\boldsymbol{\nu}$, in order to enumerate all elements of $T(\boldsymbol{\nu},l,2)$, there are two things we can do. First, we can permute

[5]This is clearer than writing $A = \{1,2,3\}$ or $\mathbb{A} = \{a_1,a_2,a_3\}$.

the elements of $S(i)$ for $i = 1, \ldots, n$, which leads to $\prod_{i=1}^{n} |S(i)|!$ variants of the sample path that generate the same empirical distribution. Second, for each such permutation we can do a cyclic shift of the sample path and thus change the starting point. Note all permutations of each $S(i)$ and/or not all cyclic shifts of the starting point lead to a valid sample path, or to distinct sample paths. But the combination of these two actions provides *an upper bound* on the cardinality of $|T(\boldsymbol{\nu}, l, 2)|$. Specifically, it follows that

$$|T(\boldsymbol{\nu}, l, 2)| \leq n \frac{\prod_{i \in \mathbb{A}} \bar{l}_i!}{\prod_{i \in \mathbb{A}} \prod_{j \in \mathbb{A}} l_{ij}!} \leq l \frac{\prod_{i \in \mathbb{A}} \bar{l}_i!}{\prod_{i \in \mathbb{A}} \prod_{j \in \mathbb{A}} l_{ij}!} \quad \forall l \geq n. \tag{5.38}$$

The denominator term arises because each of the l_{ij} edges are indistinguishable for each i, j, so we need to divide by the number of ways in which these can be permuted.

Finding a lower bound for $|T(\boldsymbol{\nu}, l, 2)|$ is a bit more involved. To obtain a lower bound on the number of paths, we use the following argument from [34], which is explained more clearly in [68, p. 17]. Pick any one cycle that spans each node in the connected component exactly once, and then delete all of these edges from the graph. Delete the corresponding elements from the sets $S(1)$ through $S(n)$. This has the effect of reducing the cardinality of each set $S(i)$ by exactly one. Then all possible permutations of these reduced sets will result in valid paths, because it is not possible to get "stuck" at any node—the edges of the deleted cycle serve as an escape route. Thus there at least $\prod_{i \in \mathbb{A}} (\bar{l}_i - 1)!$ permutations that will result in paths. Therefore

$$\frac{\prod_{i \in \mathbb{A}} (\bar{l}_i - 1)!}{\prod_{i \in \mathbb{A}} \prod_{j \in \mathbb{A}} l_{ij}!} \leq |T(\boldsymbol{\nu}, l, 2)|,$$

or equivalently

$$\frac{1}{\prod_{i \in \mathbb{A}} \bar{l}_i} \frac{\prod_{i \in \mathbb{A}} \bar{l}_i!}{\prod_{i \in \mathbb{A}} \prod_{j \in \mathbb{A}} l_{ij}!} \leq |T(\boldsymbol{\nu}, l, 2)|. \tag{5.39}$$

Again, the denominator term arises because each of the l_{ij} edges are indistinguishable for each i, j, so we need to divide by the number of ways in which these can be permuted. Combining (5.38) and (5.39) leads to

$$\frac{1}{\prod_{i \in \mathbb{A}} \bar{l}_i} \frac{\prod_{i \in \mathbb{A}} \bar{l}_i!}{\prod_{i \in \mathbb{A}} \prod_{j \in \mathbb{A}} l_{ij}!} \leq |T(\boldsymbol{\nu}, l, 2)| \leq l \frac{\prod_{i \in \mathbb{A}} \bar{l}_i!}{\prod_{i \in \mathbb{A}} \prod_{j \in \mathbb{A}} l_{ij}!} \quad \forall l \geq n. \tag{5.40}$$

To complete the proof of the theorem, we express both sides of (5.40) in terms of $H_r(\boldsymbol{\nu})$. For this purpose, we introduce a symbol for the combinatorial index when there are n possible outcomes. Suppose $m_1, \ldots, m_n$ are integers with $m := \sum_{i=1}^{n} m_i$. For notational convenience let $\mathbf{m}$ denote the vector $[m_i]$. Then the integer

$$C_{\mathbf{m}}^{m} := \frac{m!}{\prod_{i=1}^{n} m_i!}$$

represents the number of distinct ways of assigning m labels to the n elements of $\mathbb{A}$ in such a way that the i-th element is assigned precisely m_i labels. If

$n = 2$ so that $m_2 = m - m_1$, then it is customary to write just $C^m_{m_1}$ or $C^m_{m_2}$. Define the associated probability distribution

$$\boldsymbol{\zeta} := [m_i/m, i = 1, \ldots, n] \in \mathbb{S}_n.$$

Then a classic result found for example in [32, Lemmas 1 and 2] states that

$$\left(C^{m+n-1}_{n-1}\right)^{-1} \exp[H(\boldsymbol{\zeta})] \leq C^m_{\mathbf{m}} \leq \exp[H(\boldsymbol{\zeta})]. \tag{5.41}$$

Since we are interested in the case where m approaches infinity, we observe that, for all $m \geq n$, we have

$$C^{m+n-1}_{n-1} = \frac{(m-n-1)!}{(n-1)!m!} = \frac{\prod_{i=1}^{n-1}(m+n-i)}{(n-1)!} \leq (2m)^{n-1}.$$

As a result we can rewrite (5.41) as

$$(2m)^{-(n-1)} \exp[H(\boldsymbol{\zeta})] \leq C^m_{\mathbf{m}} \leq \exp[H(\boldsymbol{\zeta})], \ \forall m \geq n. \tag{5.42}$$

This bound is less sharp than that in (5.41), but is easier to work with. As we shall see, so long as we get the exponential term exactly right, the polynomial terms in front don't really matter.

Since $\sum_{j \in \mathbb{A}} l_{ij} = \bar{l}_i$ for all $i \in \mathbb{A}$, we can see lots of occurrences of the combinatorial parameter in (5.40). By observing that $\bar{l}_i \leq l$ for all i, we can rewrite (5.40) as

$$\frac{1}{l^n} \prod_{i \in \mathbb{A}} C^{\bar{l}_i}_{l_{i1}, \ldots, l_{in}} \leq |T(l, n, s+1)| \leq l \prod_{i \in \mathbb{A}} C^{\bar{l}_i}_{l_{i1}, \ldots, l_{in}}. \tag{5.43}$$

Now we make use of the upper and lower bounds in (5.41). For this purpose, let us define

$$\bar{\boldsymbol{\zeta}}_i := [l_{i1}/\bar{l}_i \ldots l_{in}/\bar{l}_i] \in \mathbb{S}_n \ \forall \in \mathbb{A}.$$

Then it follows from (5.41) that, for all $i \in \mathbb{A}$,

$$\frac{1}{2^{n-1}\bar{l}_i^{n-1}} \exp(\bar{l}_i H(\bar{\boldsymbol{\zeta}}_i)) \leq C^{\bar{l}_i}_{l_{i1}, \ldots, l_{in}} \leq \exp(\bar{l}_i H(\bar{\boldsymbol{\zeta}}_i)). \tag{5.44}$$

When we substitute these bounds in (5.43), we need to compute two quantities, namely

$$l^n \prod_{i \in \mathbb{A}} 2^{n-1} \bar{l}_i^{n-1},$$

and

$$\prod_{i \in \mathbb{A}} \exp(\bar{l}_i H(\bar{\boldsymbol{\zeta}}_i)) = \exp\left(\sum_{i \in \mathbb{A}} \bar{l}_i H(\bar{\boldsymbol{\zeta}}_i)\right).$$

The first term is easy. Since $\bar{l}_i \leq l$ for all i, the first product is bounded by $(2^{n-1}l^n)^n \leq (2l)^{n^2}$. As for the second term, we get

$$\bar{l}_i H(\bar{\boldsymbol{\zeta}}_i) = -\bar{l}_i \sum_{j \in \mathbb{A}} \frac{l_{ij}}{\bar{l}_i} [\log l_{ij} - \log \bar{l}_i] = \bar{l}_i \log \bar{l}_i - \sum_{j \in \mathbb{A}} l_{ij} \log l_{ij},$$

where we use the fact that $\sum_{j \in \mathbb{A}} l_{ij} = \bar{l}_i$. Therefore

$$\sum_{i \in \mathbb{A}} \bar{l}_i H(\bar{\boldsymbol{\zeta}}_i) = - \sum_{i \in \mathbb{A}} \sum_{j \in \mathbb{A}} l_{ij} \log l_{ij} + \sum_{i \in \mathbb{A}} \bar{l}_i \log \bar{l}_i$$
$$= l[H(\boldsymbol{\zeta}) - H(\bar{\boldsymbol{\zeta}})] = l H_c(\boldsymbol{\zeta}). \tag{5.45}$$

Substituting from (5.44) and (5.45) into (5.43) leads to the desired bound, namely

$$(2l)^{-n^2} \exp(l H_r(\boldsymbol{\zeta})) \leq |T(\boldsymbol{\zeta}, l, 2)| \leq l \exp(l H_r(\boldsymbol{\zeta})).$$

□

At last we come to the proof of the main theorem.

Proof. (Of Theorem 5.16): Suppose we have a sample path $\mathbf{x}_1^l$. Let us compute its likelihood in terms of the properties of the corresponding empirical distribution $\boldsymbol{\nu}(\mathbf{x}_1^l)$. We have

$$\Pr\{X_1^l = \mathbf{x}_1^l\} = \Pr\{X_1 = x_1\} \cdot \prod_{t=1}^{l-1} \Pr\{X_{t+1} = x_{t+1} | X_t = x_t\}.$$

Hence[6]

$$\Pr\{X_1^l = \mathbf{x}_1^l\} = \bar{\mu}(x_1) \cdot \prod_{t=1}^{l-1} \frac{\mu(x_t x_{t+1})}{\bar{\mu}(x_t)}$$
$$= \bar{\mu}(x_1) \cdot \prod_{t=1}^{l} \frac{\mu(x_t x_{t+1})}{\bar{\mu}(x_t)} \cdot \frac{\bar{\mu}(x_l)}{\mu(x_l x_1)}$$
$$= \frac{\bar{\mu}(x_1)\bar{\mu}(x_l)}{\mu(x_l x_1)} \cdot \prod_{t=1}^{l} \frac{\mu(x_t x_{t+1})}{\bar{\mu}(x_t)}, \tag{5.46}$$

where as before we take $x_{l+1} = x_1$. Now, since $\mu_{ij} > 0$ for all i, j, there exist constants $\underline{c}$ and $\bar{c}$ such that

$$\underline{c} \leq \frac{\bar{\mu}_i \bar{\mu}_j}{\mu_{ij}} \leq \bar{c}, \ \forall i, j.$$

Of course these constants depend on $\boldsymbol{\mu}$, but the point is that they do not depend on the empirical measure $\boldsymbol{\nu}(\mathbf{x}_1^l)$.

Next we examine the product term in (5.46). We have

$$\log \left[\prod_{t=1}^{l} \frac{\mu(x_t x_{t+1})}{\bar{\mu}(x_t)} \right] = \sum_{t=1}^{l} [\log \mu(x_t x_{t+1}) - \log \bar{\mu}(x_t)].$$

When we do the above summation, we observe that the pair $x_t x_{t+1}$ occurs exactly $l_{ij} = l[\boldsymbol{\nu}(\mathbf{x}_1^l)]_{ij}$ times, while x_t occurs exactly $\bar{l}_i = l[\bar{\boldsymbol{\nu}}(\mathbf{x}_1^l)]_i$ times.

[6]In the interests of clarity, in the proof we write $\mu(x_s x_t)$ instead of $\mu_{x_s x_t}$, and $\bar{\mu}(x_t)$ instead of $\bar{\mu}_{x_t}$. However, we continue to use the subscript notation if the arguments are simple indices such as i and j.

Therefore

$$\log \left[\prod_{t=1}^{l} \frac{\mu(x_t x_{t+1})}{\bar{\mu}(x_t)} \right] = l \sum_{i \in \mathbb{A}} \sum_{j \in \mathbb{A}} \nu_{ij} \log \mu_{ij} - l \sum_{i \in \mathbb{A}} \bar{\nu}_i \log \bar{\mu}_i$$
$$= -l[J(\boldsymbol{\nu}, \boldsymbol{\mu}) - J(\bar{\boldsymbol{\nu}}, \bar{\boldsymbol{\mu}})],$$

where we write $\boldsymbol{\nu}$ and $\bar{\boldsymbol{\nu}}$ for the more precise $\boldsymbol{\nu}(\mathbf{x}_1^l)$ and $\bar{\boldsymbol{\nu}}(\mathbf{x}_1^l)$. Substituting this into (5.46) shows that the likelihood of each sample path can be bounded as follows:

$$\log \Pr\{X_1^l = \mathbf{x}_1^l\} + l[J(\boldsymbol{\nu}, \boldsymbol{\mu}) - J(\bar{\boldsymbol{\nu}}, \bar{\boldsymbol{\mu}})] \in [\log \underline{c}, \log \bar{c}]. \tag{5.47}$$

In large deviation theory, the quantity of interest is the log of the likelihood that a particular empirical estimate will occur, normalized by the length of the observation. Accordingly, let us denote the empirical distribution generated by a sample path as $\hat{\boldsymbol{\mu}}(\mathbf{x}_1^l)$, and define

$$\delta(l, \boldsymbol{\nu}) := \frac{1}{l} \log \Pr\{\hat{\boldsymbol{\mu}}(\mathbf{x}_1^l) = \boldsymbol{\nu}\}.$$

Now we know from (5.47) that the log likelihood of each sample path within the type class $T(\boldsymbol{\nu}, l)$ looks like $l[J(\boldsymbol{\nu}, \boldsymbol{\mu}) - J(\bar{\boldsymbol{\nu}}, \bar{\boldsymbol{\mu}})]$, and we know from (5.37) that $\log |T(\boldsymbol{\nu})|$ looks like $lH_r(\boldsymbol{\nu})$. Combining these two facts leads to

$$\begin{aligned}
\delta(l, \boldsymbol{\nu}) &\leq H_r(\boldsymbol{\nu}) - J(\boldsymbol{\nu}, \boldsymbol{\mu}) + J(\bar{\boldsymbol{\nu}}, \bar{\boldsymbol{\mu}}) + o(1/l) \\
&= H(\boldsymbol{\nu}) - H(\bar{\boldsymbol{\nu}}) - J(\boldsymbol{\nu}, \boldsymbol{\mu}) + J(\bar{\boldsymbol{\nu}}, \bar{\boldsymbol{\mu}}) + o(1/l) \\
&= -D(\boldsymbol{\nu} \| \boldsymbol{\mu}) + D(\bar{\boldsymbol{\nu}} \| \bar{\boldsymbol{\mu}}) + o(1/l) \\
&= -D_r(\boldsymbol{\nu} \| \boldsymbol{\mu}) + o(1/l).
\end{aligned} \tag{5.48}$$

Similarly we get

$$\delta(l, \boldsymbol{\nu}) \geq -D_r(\boldsymbol{\nu} \| \boldsymbol{\mu}) + o(1/l). \tag{5.49}$$

The remainder of the proof is entirely analogous to that of Theorem 5.6. Let $\Gamma \subseteq \mathcal{M}_s(\mathbb{A}^2)$ be any set of stationary distributions on $\mathbb{A}^2$. Then

$$\begin{aligned}
\Pr\{\hat{\boldsymbol{\mu}}(\mathbf{x}_1^l) \in \Gamma\} &= \sum_{\boldsymbol{\nu} \in \mathcal{E}(l,n,2) \cap \Gamma} \Pr\{\hat{\boldsymbol{\mu}}(\mathbf{x}_1^l) = \boldsymbol{\nu}\} \\
&\leq |\mathcal{E}(l, n, 2) \cap \Gamma| \sup_{\boldsymbol{\nu} \in \mathcal{E}(l,n,2) \cap \Gamma} \Pr\{\hat{\boldsymbol{\mu}}(\mathbf{x}_1^l) = \boldsymbol{\nu}\}.
\end{aligned}$$

Hence

$$\frac{1}{l} \log \Pr\{\hat{\boldsymbol{\mu}}(\mathbf{x}_1^l) \in \Gamma\} \leq \frac{1}{l} \log |\mathcal{E}(l, n, 2)| + \sup_{\boldsymbol{\nu} \in \Gamma} \delta(l, \boldsymbol{\nu}).$$

Since $|\mathcal{E}(l, n, 2)|$ is polynomial in l, the first term approaches zero as $l \to \infty$. Next, from (5.49) it follows that $\delta(l, \boldsymbol{\nu})$ approaches $-D_r(\boldsymbol{\nu} \| \boldsymbol{\mu})$ as $l \to \infty$. Combining these two facts shows that

$$\limsup_{l \to \infty} \frac{1}{l} \log \Pr\{\hat{\boldsymbol{\mu}}(\mathbf{x}_1^l) \in \Gamma\} \leq \sup_{\boldsymbol{\nu} \in \Gamma} -D_r(\boldsymbol{\nu} \| \boldsymbol{\mu}) = -\inf_{\boldsymbol{\nu} \in \Gamma} D_r(\boldsymbol{\nu} \| \boldsymbol{\mu}).$$

This establishes the right inequality in (5.4).

To establish the left inequality, suppose $\boldsymbol{\nu}$ is an interior point of Γ. Then there is an open ball $\mathcal{B}(\boldsymbol{\nu})$ in $\mathcal{M}_s(\mathbb{A}^2)$ that contains $\boldsymbol{\nu}$. Now it follows from Lemma 5.18 that there exists a sequence of elements $\{\boldsymbol{\nu}_l\}$ such that $\boldsymbol{\nu}_l \in \mathcal{E}(l,n,2) \cap \Gamma$ for sufficiently large l and such that $\boldsymbol{\nu}_l \to \boldsymbol{\nu}$. Hence

$$\Pr\{\hat{\boldsymbol{\mu}}(x_1^l) \in \Gamma\} \geq \Pr\{\hat{\boldsymbol{\mu}}(x_1^l) = \boldsymbol{\nu}_l\},$$

$$\begin{aligned}\frac{1}{l}\log \Pr\{\hat{\boldsymbol{\mu}}(x_1^l) \in \Gamma\} &\geq \delta(l,\boldsymbol{\nu}_l)\\ &\geq -D_r(\boldsymbol{\nu}_l\|\boldsymbol{\mu}) + o(1/l)\\ &\to -D_r(\boldsymbol{\nu}\|\boldsymbol{\mu}) \text{ as } k \to \infty.\end{aligned}$$

Hence it follows that

$$\liminf_{l\to\infty} \frac{1}{l}\log \Pr\{\boldsymbol{\nu}(\mathbf{x}_1^l) \in \Gamma\} \geq -D_r(\boldsymbol{\nu}\|\boldsymbol{\mu}),\ \forall \boldsymbol{\nu} \in \Gamma^o.$$

Since the above inequality holds for every $\boldsymbol{\nu} \in \Gamma^o$, we can conclude that

$$\liminf_{l\to\infty} \frac{1}{l}\log \Pr\{\boldsymbol{\nu}(\mathbf{x}_1^l) \in \Gamma\} \geq \sup_{\boldsymbol{\nu}\in\Gamma^o} -D_r(\boldsymbol{\nu}\|\boldsymbol{\mu}) = -\inf_{\boldsymbol{\nu}\in\Gamma^o} D_r(\boldsymbol{\nu}\|\boldsymbol{\mu}).$$

This establishes that the relationships in (5.4) hold with $I(\boldsymbol{\nu}) = D_r(\boldsymbol{\nu}\|\boldsymbol{\mu})$. Since $I(\cdot)$ is clearly continuous (and not just lower semi-continuous), this shows that $D_r(\boldsymbol{\nu}\|\boldsymbol{\mu})$ is the rate function. □

It may be remarked that when the samples $\{x_t\}$ come from an i.i.d. process, we have exact formulas for both the size of each type class, and the likelihood of each sample within a type class. In the case where the samples come from a Markov process, we have only bounds. However, the "correction terms" in these bounds approach zero as the number of samples approaches infinity, thus allowing us to deduce the rate function for doubleton frequencies in a straightforward fashion.

Theorem 5.16 deals with conventional one-step Markov chains. However, the methods can be extended readily to multi-step processes. In the interests of brevity, only the final results are stated, and the interested reader is referred to [135] for details.

Suppose $\{X_t\}$ is an s-step Markov process, which is completely characterized by the $(s+1)$-tuple frequency vector

$$\mu_{\mathbf{i}} := \Pr\{X_t^{t+s} = \mathbf{i}\},\ \forall \mathbf{i} \in \mathbb{A}^{s+1}.$$

Note that the frequency vector $\boldsymbol{\mu}$ is stationary and thus belongs to $\mathcal{M}_s(\mathbb{A}^{s+1})$. Since an s-step Markov process over $\mathbb{A}$ can be viewed as a conventional (one-step) Markov process over the state space $\mathbb{A}^s$, we can identify the stationary distribution

$$\bar{\mu}_{\mathbf{i}} := \sum_{j\in\mathbb{A}} \mu_{\mathbf{i}j} = \sum_{j\in\mathbb{A}} \mu_{j\mathbf{i}},\ \forall \mathbf{i} \in \mathbb{A}^s,$$

while the transition probabilities are given by

$$\Pr\{X_t = j | X_{t-s}^{t-1} = \mathbf{i}\} = \frac{\mu_{\mathbf{i}j}}{\mu_{\mathbf{i}}}.$$

Suppose $\mathbf{x}_1^l$ is a sample path of length l of an s-step Markov chain. To construct a *stationary* empirical measure on the basis of this sample path, we define the augmented sample path $\tilde{\mathbf{x}}_1^l := x_1 \ldots x_l x_1 \ldots x_s = \mathbf{x}_1^l \cdot \mathbf{x}_1^s \in \mathbb{A}^{l+s}$. Here the symbol $\cdot$ denotes the concatenation of two strings. The above augmentation is the s-step generalization of adding a single ghost transition from x_l to x_1 in the case of one-step Markov chains. In this case we add s ghost transitions. Then we define

$$\nu_{\mathbf{i}} := \frac{1}{l} \sum_{t=1}^{l} I_{\{x_t^{t+s} = \mathbf{i}\}}, \ \forall \mathbf{i} \in \mathbb{A}^{s+1}. \tag{5.50}$$

Compare (5.50) to (5.32). Then the resulting empirical measure $\boldsymbol{\nu}(\mathbf{x}_1^l)$ belongs to $\mathcal{M}_s(\mathbb{A}^{s+1})$. For this empirical measure we can state the following result.

Theorem 5.20 *Suppose $\{X_t\}$ is a stationary s-step Markov assuming values in the finite set $\mathbb{A}$, with the $(s+1)$-tuple frequency vector $\boldsymbol{\mu} \in \mathcal{M}_s(\mathbb{A}^{s+1})$. Define $\boldsymbol{\nu}(\mathbf{x}_1^l) \in \mathcal{M}_s(\mathbb{A}^{s+1})$ as in (5.50). Then the $\mathcal{M}_s(\mathbb{A}^{s+1})$-valued process $\{\boldsymbol{\nu}(\mathbf{x}_1^l)\}$ satisfies the LDP with the rate function*

$$I(\boldsymbol{\nu}) := D_r(\boldsymbol{\nu} \| \boldsymbol{\mu}) = D(\boldsymbol{\nu} \| \boldsymbol{\mu}) - D(\bar{\boldsymbol{\nu}} \| \bar{\boldsymbol{\mu}}). \tag{5.51}$$

5.3.4 The Rate Function for Singleton Frequencies

In this subsection, we first introduce a very important technique known as the "contraction principle," which permits us to derive rate functions for *functions* of the empirically estimated frequencies. Using the contraction principle, we then derive the rate function for singleton frequencies of a Markov chain. The contraction principle directly leads to a very appealing formula. By applying duality theory, we then derive another formula that is equivalent to this one.

Theorem 5.21 (The Contraction Principle) *Suppose the stochastic process $\{\hat{\boldsymbol{\mu}}(\mathbf{x}_1^l)\}$ assuming values in $\mathbb{S}_m$ satisfies the large deviation property with the rate function $I(\cdot) : \mathbb{S}_m \to \mathbb{R}_+$. Suppose that $f : \mathbb{S}_m \to \mathbb{S}_k$ is continuous. Then the stochastic process $\{f[\hat{\boldsymbol{\mu}}(\mathbf{x}_1^l)]\}$ satisfies the large deviation property with the rate function $J(\cdot) : \mathbb{S}_k \to \mathbb{R}_+$ defined by*

$$J(\boldsymbol{\theta}) := \inf_{\boldsymbol{\nu} \in f^{-1}(\boldsymbol{\theta})} I(\boldsymbol{\nu}), \ \forall \boldsymbol{\theta} \in \mathbb{S}_k. \tag{5.52}$$

Remarks:

1. The rate function for the new stochastic process $\{f[\hat{\boldsymbol{\mu}}(\mathbf{x}_1^l)]\}$ has a very intuitive interpretation. The rate function of $\boldsymbol{\theta} \in \mathbb{S}_k$ is the "slowest" value among the original rate function $I(\boldsymbol{\nu})$ as $\boldsymbol{\nu}$ varies over all preimages of $\boldsymbol{\theta}$, that is, over all $\boldsymbol{\nu} \in \mathbb{S}_m$ that map into $\boldsymbol{\theta}$.

2. In the general case, where the stochastic process $\{X_l\}$ assumes values in $\mathbb{R}^m$, and $f : \mathbb{R}^m \to \mathbb{R}^k$, it is necessary to add the assumption that

the original rate function $I(\cdot)$ is "good," that is, all level sets of $I(\cdot)$ are compact. See for example [41, p. 126, Theorem 4.2.1]. However, in the restricted situation being studied here, where the rate function has a compact set (namely $\mathbb{S}_m$) as its domain, we need not worry about this additional condition.

Proof. The proof consists of two steps. First, it is shown that the function $J(\cdot)$ is lower semi-continuous. Second, instead of establishing the two relationships in (5.4), we instead establish (5.2) and (5.3). As pointed out earlier, for lower semi-continuous functions, (5.4) is equivalent to (5.2) and (5.3).

To show that the function $J(\cdot)$ is lower semi-continuous, we begin by observing that, since the original rate function $I(\cdot)$ is lower semi-continuous and the set $\mathbb{S}_m$ is compact, the infimum in (5.52) is actually a minimum.[7] Thus for each $\boldsymbol{\theta} \in \mathbb{S}_k$, there exists a $\boldsymbol{\nu} \in \mathbb{S}_m$ such that $f(\boldsymbol{\nu}) = \boldsymbol{\theta}$ and $J(\boldsymbol{\theta}) = I(\boldsymbol{\nu})$. Now suppose $\{\boldsymbol{\theta}_i\}$ is a sequence in $\mathbb{S}_k$ that converges to $\boldsymbol{\theta}^* \in \mathbb{S}_k$. The objective is to show that

$$J(\boldsymbol{\theta}^*) \leq \liminf_{i \to \infty} I(\boldsymbol{\theta}_i). \tag{5.53}$$

Towards this end, let us choose, for each index i, a $\boldsymbol{\nu}_i \in \mathbb{S}_m$ such that $f(\boldsymbol{\nu}_i) = \boldsymbol{\theta}_i$ and $I(\boldsymbol{\nu}_i) = J(\boldsymbol{\theta}_i)$. Now, since $\mathbb{S}_m$ is compact[8] it follows that $\{\boldsymbol{\nu}_i\}$ contains a convergent subsequence. Let us renumber this subsequence again as $\{\boldsymbol{\nu}_i\}$, and let $\boldsymbol{\nu}^* \in \mathbb{S}_m$ denote its limit. Now, since f is a continuous map, it follows that

$$f(\boldsymbol{\nu}^*) = \lim_{i \to \infty} f(\boldsymbol{\nu}_i) = \lim_{i \to \infty} \boldsymbol{\theta}_i = \boldsymbol{\theta}^*.$$

Hence $\boldsymbol{\nu}^* \in f^{-1}(\boldsymbol{\theta}^*)$. The definition of $J(\cdot)$ plus the lower semi-continuity of $I(\cdot)$ together imply that

$$J(\boldsymbol{\theta}^*) = \inf_{\boldsymbol{\nu} \in f^{-1}(\boldsymbol{\theta}^*)} I(\boldsymbol{\nu}) \leq I(\boldsymbol{\nu}^*) \leq \liminf_{i \to \infty} I(\boldsymbol{\nu}_i) = \liminf_{i \to \infty} J(\boldsymbol{\theta}_i).$$

Hence (5.53) is established and $J(\cdot)$ is shown to be lower semi-continuous.

Next, suppose $\Omega \subseteq \mathbb{S}_k$ is an open set. Since the map f is continuous, the preimage $f^{-1}(\Omega)$ is also open. Therefore

$$\Pr\{f(\hat{\boldsymbol{\mu}}_l) \in \Omega\} = \Pr\{\hat{\boldsymbol{\mu}}_l \in f^{-1}(\Omega)\}.$$

As a consequence,

$$\begin{aligned}
\liminf_{l \to \infty} \frac{1}{l} \log \Pr\{f(\hat{\boldsymbol{\mu}}_l) \in \Omega\} &= \liminf_{l \to \infty} \frac{1}{l} \log \Pr\{\hat{\boldsymbol{\mu}}_l \in f^{-1}(\Omega)\} \\
&\geq - \inf_{\boldsymbol{\nu} \in f^{-1}(\Omega)} I(\boldsymbol{\nu}) \\
&= - \inf_{\boldsymbol{\theta} \in \Omega} J(\boldsymbol{\theta}).
\end{aligned}$$

[7] In the case of a general stochastic process assuming values in $\mathbb{R}^m$, we invoke the "goodness" of $I(\cdot)$.

[8] In the general case, we invoke the fact that the original rate function $I(\cdot)$ is a "good" rate function.

This establishes that (5.2) holds with I replaced by J and Γ replaced by Ω. The proof of (5.3) is entirely similar and follows upon noting that if $\Omega \subseteq \mathbb{S}_k$ is a closed set, then so is $f^{-1}(\Omega)$. □

To apply the contraction principle to derive the rate function for singleton frequencies, let us define a map $\mathbf{f} : \mathcal{M}_s(\mathbb{A}^2) \to \mathbb{S}_n$ by

$$[\mathbf{f}(\boldsymbol{\nu})]_i := \sum_{j \in \mathbb{A}} \nu_{ij} = \sum_{j \in \mathbb{A}} \nu_{ji}.$$

Thus $\mathbf{f}$ maps a stationary distribution $\boldsymbol{\nu}$ on $\mathbb{A}^2$ onto its one-dimensional marginal $\bar{\boldsymbol{\nu}}$. Moreover, if we construct $\boldsymbol{\nu}(\mathbf{x}_1^l)$ for a sample $\mathbf{x}_1^l$ using the formula (5.32), then the corresponding $\mathbf{f}[\boldsymbol{\nu}(\mathbf{x}_1^l)]$ is the empirical distribution $\boldsymbol{\phi}(\mathbf{x}_1^l)$ of singleton frequencies defined in (5.30). Now, by invoking the contraction principle, we can readily conclude the following:

Theorem 5.22 *The $\mathbb{S}_n$-valued process $\{\boldsymbol{\phi}(\mathbf{x}_1^l)\}$ satisfies the LDP with the rate function*

$$J(\boldsymbol{\phi}) := \inf_{\boldsymbol{\nu} \in \mathcal{M}_s(\mathbb{A}^2)} D_r(\boldsymbol{\nu} \| \boldsymbol{\mu}) \text{ s.t. } \bar{\boldsymbol{\nu}} = \boldsymbol{\phi}. \tag{5.54}$$

Recall that

$$D_r(\boldsymbol{\nu} \| \boldsymbol{\mu}) = D(\boldsymbol{\nu} \| \boldsymbol{\mu}) - D(\bar{\boldsymbol{\nu}} \| \bar{\boldsymbol{\mu}}).$$

Hence we can also write

$$J(\boldsymbol{\phi}) = \left[\inf_{\boldsymbol{\nu} \in \mathcal{M}_s(\mathbb{A}^2)} D(\boldsymbol{\nu} \| \boldsymbol{\mu}) \text{ s.t. } \bar{\boldsymbol{\nu}} = \boldsymbol{\phi} \right] - D(\boldsymbol{\phi} \| \bar{\boldsymbol{\mu}}). \tag{5.55}$$

The problem of minimizing $D(\boldsymbol{\nu} \| \boldsymbol{\mu})$ where $\boldsymbol{\nu}, \boldsymbol{\mu} \in \mathcal{M}_s(\mathbb{A}^2)$ subject to the constraint that $\bar{\boldsymbol{\nu}} = \boldsymbol{\phi}$ is a special case of the following more general problem: Suppose $\mathbb{A}, \mathbb{B}$ are finite sets (not necessarily of the same size), and $\boldsymbol{\mu}$ is a distribution on $\mathbb{A} \times \mathbb{B}$. Suppose $\boldsymbol{\phi}, \boldsymbol{\psi}$ are distributions on $\mathbb{A}, \mathbb{B}$ respectively. Then the problem is:

$$\min_{\boldsymbol{\mu}} D(\boldsymbol{\nu} \| \boldsymbol{\mu}) \text{ s.t. } \boldsymbol{\nu}_{\mathbb{A}} = \boldsymbol{\phi}, \boldsymbol{\nu}_{\mathbb{B}} = \boldsymbol{\psi}. \tag{5.56}$$

In the somewhat uninteresting case where $\boldsymbol{\mu}$ is itself a product measure of the form $\boldsymbol{\mu}_{\mathbb{A}} \times \boldsymbol{\mu}_{\mathbb{B}}$, the solution is easy: $\boldsymbol{\nu} = \boldsymbol{\phi} \times \boldsymbol{\psi}$. But in general no closed-form solution is available.

Theorem 5.22 gives the rate function as the infimum of a convex minimization problem. The reformulation (5.55) makes it obvious that the objective function is convex in $\boldsymbol{\nu}$ since $-D(\boldsymbol{\phi} \| \bar{\boldsymbol{\mu}})$ is just an additive constant. Now by using duality theory, we obtain an alternate formula for the rate function for singleton frequencies.

Theorem 5.23 *Suppose $\boldsymbol{\phi} \in \mathbb{S}_n$ and $\boldsymbol{\mu} \in \mathcal{M}_s(\mathbb{A}^2)$. Then*

$$\left\{ \inf_{\boldsymbol{\nu} \in \mathcal{M}_s(\mathbb{A}^2)} D(\boldsymbol{\nu} \| \boldsymbol{\mu}) \text{ s.t. } \bar{\boldsymbol{\nu}} = \boldsymbol{\phi} \right\} = \left\{ D(\boldsymbol{\phi} \| \bar{\boldsymbol{\mu}}) + \sup_{\mathbf{u} > \mathbf{0}} \sum_{i=1}^{n} \phi_i \log \frac{u_i}{(\mathbf{u}A)_i} \right\}, \tag{5.57}$$

where as before $a_{ij} = \mu_{ij}/\bar{\mu}_i$ *and* $A = [a_{ij}]$ *is the state transition matrix of the Markov chain associated with the doublet frequency vector* $\boldsymbol{\mu}$*. Therefore an alternate formula for the rate function* $J(\cdot)$ *is*

$$J(\boldsymbol{\phi}) = \sup_{\mathbf{u}>\mathbf{0}} \sum_{i=1}^{n} \phi_i \log \frac{u_i}{(\mathbf{u}A)_i}. \tag{5.58}$$

Proof. Since $D(\boldsymbol{\nu}\|\boldsymbol{\mu})$ is a convex function of $\boldsymbol{\nu}$ and the constraint $\bar{\boldsymbol{\nu}} = \boldsymbol{\phi}$ is linear, it follows that the value of this infimum is the same as the supremum of the dual problem; in other words, there is no duality gap.

To formulate the dual problem, we study the more general problem posed in (5.56), and persist with it as long as we can. The problem can be written out as

$$\inf_{\boldsymbol{\nu}\in\mathbb{S}_{nm}} \sum_{i=1}^{n}\sum_{j=1}^{m} \nu_{ij} \log \frac{\nu_{ij}}{\mu_{ij}} \text{ s.t. } \sum_{j=1}^{m} \nu_{ij} = \phi_i \ \forall i, \text{ and } \sum_{i=1}^{n} \nu_{ij} = \psi_j, \ \forall j. \tag{5.59}$$

Right at the very end we will put $n = m$ and $\boldsymbol{\phi} = \boldsymbol{\psi}$, which will incidentally automatically ensure that $\boldsymbol{\nu} \in \mathcal{M}_s(\mathbb{A}^2)$.

The Lagrangian of the above constrained problem is

$$\begin{aligned} L(\boldsymbol{\nu}, \boldsymbol{\alpha}, \boldsymbol{\beta}) = &\sum_{i=1}^{n}\sum_{j=1}^{m} \nu_{ij} \log \frac{\nu_{ij}}{\mu_{ij}} \\ &+ \sum_{i=1}^{n} \left[\phi_i - \sum_{j=1}^{m} \nu_{ij} \right] \alpha_i + \sum_{j=1}^{m} \left[\psi_j - \sum_{i=1}^{n} \nu_{ij} \right] \beta_j, \end{aligned}$$

where $\boldsymbol{\alpha}, \boldsymbol{\beta}$ are the vectors of Lagrange multipliers. Then

$$\frac{\partial L}{\partial \nu_{ij}} = \log \frac{\nu_{ij}}{\mu_{ij}} + 1 - \alpha_i - \beta_j.$$

Thus, at the optimum, we have

$$\log \frac{\nu_{ij}^*}{\mu_{ij}} = \alpha_i + \beta_j - 1,$$

or

$$\nu_{ij}^* = \mu_{ij} \exp(\alpha_i + \beta_j - 1).$$

Thus

$$\begin{aligned} L^*(\boldsymbol{\alpha}, \boldsymbol{\beta}) :=& \inf_{\boldsymbol{\nu}} L(\boldsymbol{\nu}, \boldsymbol{\alpha}, \boldsymbol{\beta}) \\ =& \sum_{i=1}^{n}\sum_{j=1}^{m} \nu_{ij}^*(\alpha_i + \beta_j - 1) + \sum_{i=1}^{n} \left[\phi_i - \sum_{j=1}^{m} \nu_{ij}^* \right] \alpha_i + \sum_{j=1}^{m} \left[\psi_j - \sum_{i=1}^{n} \nu_{ij}^* \right] \beta_j, \\ =& -\sum_{i=1}^{n}\sum_{j=1}^{m} \nu_{ij}^* + \sum_{i=1}^{n} \phi_i \alpha_i + \sum_{j=1}^{m} \psi_j \beta_j \\ =& -\sum_{i=1}^{n}\sum_{j=1}^{m} \mu_{ij} e^{\alpha_i + \beta_j - 1} + \sum_{i=1}^{n} \phi_i \alpha_i + \sum_{j=1}^{m} \psi_j \beta_j. \end{aligned}$$

By duality theory, the infimum in (5.59) is the unconstrained supremum of $L^*(\boldsymbol{\alpha}, \boldsymbol{\beta})$ with respect to $\boldsymbol{\alpha}, \boldsymbol{\beta}$.

Next, let us reparametrize the problem. We have

$$L^*(\boldsymbol{\alpha}, \boldsymbol{\beta}) = -\sum_{i=1}^{n} \sum_{j=1}^{m} \mu_{ij} e^{\alpha_i + \beta_j - 1} + \sum_{i=1}^{n} \phi_i \alpha_i + \sum_{j=1}^{m} \psi_j \beta_j.$$

Now define

$$\exp(\alpha_i) =: v_i, \alpha_i = \log v_i, \exp(\beta_j - 1) =: w_j, \beta_j = \log w_j + 1,$$

and observe that since $\boldsymbol{\alpha}, \boldsymbol{\beta}$ are unconstrained, the corresponding vectors $\mathbf{v}, \mathbf{w}$ are constrained to be strictly positive; that is, $\mathbf{v} > \mathbf{0}, \mathbf{w} > \mathbf{0}$. In a bit of sloppy notation, we continue to refer to the resulting function as $L^*(\mathbf{v}, \mathbf{w})$. Now

$$L^*(\mathbf{v}, \mathbf{w}) = -\sum_{i=1}^{n} \sum_{j=1}^{m} \mu_{ij} v_i w_j + \sum_{i=1}^{n} \phi_i \log v_i + \sum_{j=1}^{m} \psi_j \log w_j + \sum_{=1}^{m} \psi_j.$$

Next, observe that

$$\sup_{\mathbf{v} > \mathbf{0}, \mathbf{w} > \mathbf{0}} L^*(\mathbf{v}, \mathbf{w}) = \sup_{\mathbf{v} > \mathbf{0}} \sup_{\mathbf{w} > \mathbf{0}} L^*(\mathbf{v}, \mathbf{w}).$$

So let us fix $\mathbf{v} > \mathbf{0}$ and define

$$L^{**}(\mathbf{v}) := \sup_{\mathbf{w} > \mathbf{0}} L^*(\mathbf{v}, \mathbf{w}).$$

To compute $L^{**}(\mathbf{v})$, note that

$$\frac{\partial L}{\partial w_j} = -\sum_{i=1}^{n} \mu_{ij} v_i + \frac{\psi_j}{w_j}.$$

Hence at the optimum we have

$$w_j^* = \frac{\psi_j}{\sum_{i=1}^{n} \mu_{ij} v_i}.$$

Thus

$$\begin{aligned} L^{**}(\mathbf{v}) &= L^*(\mathbf{v}, \mathbf{w}^*) \\ &= -\sum_{j=1}^{m} \psi_j + \sum_{i=1}^{n} \phi_i \log v_i + \sum_{j=1}^{m} \psi_j \log \frac{\psi_j}{\sum_{i=1}^{m} v_i \mu_{ij}} + \sum_{j=1}^{m} \psi_j. \end{aligned}$$

Now observe that the first and last terms on the right side cancel out. At last let us use the facts that $n = m$ and $\boldsymbol{\phi} = \boldsymbol{\psi}$. Thus, after making these substitutions, and interchanging the indices i and j in the last summation, we get

$$L^{**}(\mathbf{v}) = \sum_{i=1}^{n} \phi_i \log \frac{\phi_i v_i}{\sum_{j=1}^{m} v_j \mu_{ji}}.$$

Let us now make one last change of variables by defining

$$v_i = u_i / \bar{\mu}_i, u_i = v_i \bar{\mu}_i, \ \forall i.$$

With this change of variables, $\mathbf{u}$ is also constrained to be a strictly positive vector. Also

$$v_j \mu_{ji} = v_j \frac{\mu_{ji}}{\bar{\mu}_j} = u_j a_{ji},$$

where a_{ij} are the elements of the state transition matrix of the Markov chain. Next,

$$\sum_{j=1}^{n} v_j \mu_{ji} = \sum_{j=1}^{n} u_j a_{ji} = (\mathbf{u}A)_i, \ \forall i.$$

And finally, retaining the same symbol L^{**}, we get

$$\begin{aligned} L^{**}(\mathbf{u}) &= \sum_{i=1}^{n} \phi_i \left[\log(\phi_i / \bar{\mu}_i) + \log \frac{u_i}{(\mathbf{u}A)_i} \right] \\ &= D(\boldsymbol{\phi} \| \bar{\boldsymbol{\mu}}) + \sum_{i=1}^{n} \phi_i \log \frac{u_i}{(\mathbf{u}A)_i}. \end{aligned}$$

Therefore the solution to the original minimization problem is

$$D(\boldsymbol{\phi} \| \bar{\boldsymbol{\mu}}) + \inf_{\mathbf{u} > \mathbf{0}} \sum_{i=1}^{n} \phi_i \log \frac{u_i}{(\mathbf{u}A)_i}.$$

This proves (5.57). □

Chapter Six

Hidden Markov Processes: Basic Properties

In this chapter, we study a special type of stochastic process that forms the main focus of this book, called a "hidden" Markov process (HMP). Some authors also use the expression "hidden Markov model (HMM)." In this book we prefer to say "a process $\{Y_t\}$ *is* a hidden Markov process" or "a process $\{Y_t\}$ *has* a hidden Markov model." We use the two expressions interchangeably.

The chapter is organized as follows: In Section 6.1 we present three distinct types of HMMs, and show that they are all equivalent from the standpoint of their expressive power or modeling ability. In Section 6.2 we study various issues related to the computation of likelihoods in a HMM.

In the remainder of the book, the acronyms HMP for hidden Markov process and HMM for hidden Markov model are used freely.

6.1 EQUIVALENCE OF VARIOUS HIDDEN MARKOV MODELS

In this section we formulate three distinct types of hidden Markov models, and then show that they are all equivalent from the standpoint of their expressive power. This discussion becomes relevant because each of these models appears in the literature. Hence it is important to realize that, while these models may *appear* to be different, in fact each of the models can be transformed to any of the other two.

6.1.1 Three Different-Looking Models

Definition 6.1 *Suppose $\{Y_t\}_{t=1}^{\infty}$ is a stationary stochastic process assuming values in a finite set $\mathbb{M} = \{1, \dots, m\}$.*[1] *We say that $\{Y_t\}$ has a* **Type 1 hidden Markov model**, *or a* **HMM of the deterministic function of a Markov chain type**, *if there exists a stationary Markov process $\{X_t\}_{t=0}^{\infty}$ over a finite state space $\mathbb{N} = \{1, \dots, n\}$ and a function $f : \mathbb{N} \to \mathbb{M}$ such that $Y_t = f(X_t)$.*[2]

---

[1] As always, we really mean to say that Y_t assumes values in a finite set $\{y_1, \dots, y_m\}$ consisting of just abstract labels. We write the set as $\{1, \dots, m\}$ in the interests of simplifying the notation. However, these elements should be viewed as just labels and not as integers.

[2] Here again, we really mean that X_t assumes values in a finite set $\{x_1, \dots, x_n\}$.

From a historical perspective HMMs of Type 1 are the earliest to be introduced into the literature; see [20, 60]. Note that we must have $n \geq m$ in order for the above definition to make sense. If $n < m$, then some elements of the set $\mathbb{M}$ cannot be images of any element of $\mathbb{N}$, and can therefore be deleted from the output space. Observe also that the requirement that $\{X_t\}$ must have a *finite* state space is crucial. Carlyle [27] has shown that *every* stochastic process $\{Y_t\}$ over a finite output space can be expressed in the form $\{f(X_t)\}$ where $\{X_t\}$ is a Markov process whose state space is *countably infinite.*

Definition 6.2 *Suppose $\{Y_t\}_{t=1}^{\infty}$ is a stationary stochastic process assuming values in a finite set $\mathbb{M} = \{y_1, \ldots, y_m\}$. We say that $\{Y_t\}$ has a* **hidden Markov model of Type 2**, *or a* **HMM of the random function of a Markov chain type**, *if there exist a finite integer n, a pair of matrices $A \in [0,1]^{n \times n}$, $B \in [0,1]^{n \times m}$, and a probability distribution $\pi \in \mathbb{S}_n$, such that the following properties hold:*

1. *A and B are both stochastic matrices. Thus each row of A and each row of B add up to one, or equivalently*

$$A\mathbf{e}_n = \mathbf{e}_n, B\mathbf{e}_m = \mathbf{e}_n. \tag{6.1}$$

2. *π is a stationary distribution of A; that is, $\pi A = \pi$.*

3. *Suppose $\{X_t\}$ is a homogeneous Markov chain on the state space $\mathbb{N} = \{1, \ldots, n\}$ with the initial distribution π and state transition matrix A. Thus*

$$\Pr\{X_0 = i\} = \pi_i, \text{ and } \Pr\{X_{t+1} = j | X_t = i\} = a_{ij}, \forall i, j, t. \tag{6.2}$$

   Suppose that, at each time instant t, the random variable Z_t is selected according to the rule

$$\Pr\{Z_t = u | X_t = j\} = b_{ju}, \forall j \in \mathbb{N}, u \in \mathbb{M}, t \geq 0. \tag{6.3}$$

   Then $\{Z_t\}$ has the same law as $\{Y_t\}$.

Thus, in a Type 2 HMM, the current output Y_t can be viewed as a "random" function of the current state X_t, according to the rule (6.3).[3] The Type 2 HMM was apparently first introduced into the literature in [12]. Note that, in contrast to a Type 1 HMM, the Type 2 HMM remains meaningful even if $m > n$.

In the engineering world, the expressions HMP and HMM are invariably reserved for a Type 2 HMP or HMM. According to [50], a Type 1 HMM is referred to as a "Markov source" in the world of communication theory.

---

[3]Note that the distinction between Z_t and Y_t is out of respect to the niceties of mathematical expression and is perhaps pedantic. We really cannot say "Y_t *is* generated by the rule (6.3)." Instead we say "a random variable Z_t selected according to the rule is indistinguishable from Y_t." Having said that, we will hereafter ignore the distinction.

Definition 6.3 *Suppose $\{Y_t\}_{t=1}^{\infty}$ is a stationary stochastic process assuming values in a finite set $\mathbb{M} = \{y_1, \ldots, y_m\}$. We say that $\{Y_t\}$ is a* **hidden Markov model of Type 3**, *or a* **HMM of the joint Markov process type** *if there exist a finite set $\mathbb{N} = \{1, \ldots, n\}$ and a stationary stochastic process $\{X_t\}$ assuming values in $\mathbb{N}$ such that the following properties hold:*

1. *The joint process $\{(X_t, Y_t)\}$ is Markov.*

2. *In addition*

$$\Pr\{(X_t, Y_t)|(X_{t-1}, Y_{t-1})\} = \Pr\{(X_t, Y_t)|X_{t-1}\}. \tag{6.4}$$

In other words

$$Pr\{(X_t, Y_t) = (j, u)|(X_{t-1}, Y_{t-1}) = (i, v)\}$$
$$= Pr\{(X_t, Y_t) = (j, u)|X_{t-1} = i\}, \ \forall i, j \in \mathbb{N}, u, v \in \mathbb{M}. \tag{6.5}$$

Note that in a Type 3 HMM, the associated process $\{X_t\}$ is Markov by itself. The distinction between a Type 2 HMM and a Type 3 HMM is brought out clearly by comparing (6.3) and (6.4). In a Type 2 HMM, the current output Y_t is a "random function" of the *current* state X_t, whereas in a Type 3 HMM, the current output Y_t is a "random function" of the *previous* state X_{t-1}. Of course, in a Type 1 HMM, the current output Y_t is a "deterministic function" of the current state X_t, in contrast with a Type 2 HMM where Y_t is a "random function" of the current state X_t.

Whereas Type 1 and Type 2 HMMs are historical and widely used in the literature, the Type 3 HMM is somewhat nonstandard and appears to have been introduced in [6]. But a Type 3 HMM has two significant advantages over the other types of HMMs. First, as shown in the next subsection, a Type 3 HMM in general requires a smaller state space compared to the other two types of HMMs. Second, when we study realization theory in Chapter 7, the proofs become very streamlined if we use a Type 3 HMM.

6.1.2 Equivalence between the Three Models

The objective of this subsection is to show that in fact all three types of HMMs are the same in terms of their expressive power. However, when it comes to the "economy" of the model as measured by the size of the state space of the associated Markov process, the Type 3 HMM is the most economical whereas the Type 1 HMM is the least economical. A Type 2 HMM lies in between.

Theorem 6.4 *The following statements are equivalent:*

(i) The process $\{Y_t\}$ has a Type 1 HMM (that is, a HMM of the "deterministic function of a Markov chain" type).

(ii) The process $\{Y_t\}$ has a Type 2 HMM (that is, a HMM of the "random function of a Markov chain" type).

(iii) The process $\{Y_t\}$ has a Type 3 HMM (that is, a HMM of the "joint Markov process" type).

Proof. **(i) $\Rightarrow$ (ii).** Clearly every deterministic function of a Markov chain is also a "random" function of the same Markov chain, with every element of B equal to zero or one. Precisely, since both $\mathbb{N}$ and $\mathbb{M}$ are finite sets, the function f simply induces a partition of the state space $\mathbb{N}$ into m subsets $\mathbb{N}_1, \ldots, \mathbb{N}_m$, where $\mathbb{N}_u := \{i \in \mathbb{N} : f(i) = u\}$. Thus two states in $\mathbb{N}_u$ are indistinguishable through the measurement process $\{Y_t\}$. Now set $b_{ju} = 1$ if $j \in \mathbb{N}_u$ and zero otherwise.

(ii) $\Rightarrow$ (iii). If $\{Y_t\}$ is modeled as a Type 2 HMM with $\{X_t\}$ as the associated Markov chain, then the joint process $\{(X_t, Y_t)\}$ is Markov. Indeed, if we define $(X_t, Y_t) \in \mathbb{N} \times \mathbb{M}$, then it readily follows from the HMM conditions that

$$\Pr\{(X_{t+1}, Y_{t+1}) = (j, u) | (X_t, Y_t) = (i, v)\} = a_{ij} b_{ju},$$

and is therefore independent of v. Now define

$$M^{(u)} := [a_{ij} b_{ju}] \in [0, 1]^{n \times n}.$$

Then the process $\{(X_t, Y_t)\}$ is Markov, and its state transition matrix is given by

$$\begin{bmatrix} M^{(1)} & M^{(2)} & \ldots & M^{(m)} \\ \vdots & \vdots & \vdots & \vdots \\ M^{(1)} & M^{(2)} & \ldots & M^{(m)} \end{bmatrix}.$$

Finally, note that the probability that $(X_{t+1}, Y_{t+1}) = (j, u)$ depends only on X_t but not on Y_t. Hence the joint process $\{(X_t, Y_t)\}$ satisfies all the conditions required of the Type 3 HMM.

(iii) $\Rightarrow$ (i). Suppose $\{Y_t\}$ has a Type 3 HMM, and suppose X_t is a Markov process such that the joint process $\{(X_t, Y_t)\}$ is also Markov. Then clearly $Y_t = f[(X_t, Y_t)]$ for a suitable function f. Hence this is also a Type 1 HMM. □

Up to now we have considered only the "expressive power" of the various HMM types. However, this is only part of the problem of stochastic modeling. An equally, if not more, important issue is the "economy" of the representation, that is, the size of the state space of the associated Markov chain. To study this issue, let us use the shorthand expression "$\{Y_t\}$ has a Type 1 (or 2 or 3) HMM of size n" if the Markov process $\{X_t\}$ associated with the HMM of the appropriate type evolves over a set $\mathbb{N}$ of cardinality n. Then the next result is almost a direct consequence of the proof of Theorem 6.4, and shows that a Type 3 HMM is the most economical, while a Type 1 HMM is the least economical.

Corollary 6.5 *Suppose $\{Y_t\}_{t=1}^{\infty}$ is a stationary stochastic process assuming values in a finite set $\mathbb{M} = \{y_1, \ldots, y_m\}$.*

1. *Suppose $\{Y_t\}$ has a Type 1 HMM of size n. Then it has a Type 2 HMM of size n.*

2. *Suppose $\{Y_t\}$ has a Type 2 HMM of size n. Then it has a Type 3 HMM of size n.*

The next lemma discusses the extent to which the above statements can be reversed.

Lemma 6.6 *Suppose $\{Y_t\}_{t=1}^{\infty}$ is a stationary stochastic process assuming values in a finite set $\mathbb{M} = \{y_1, \ldots, y_m\}$.*

(i) Suppose a process $\{Y_t\}$ has a HMM of the random function of a Markov chain type, and let $\{X_t\}$ denote the associated Markov chain. Let A and B denote respectively the state transition matrix and output matrix of the HMM. Then Y_t is a deterministic function of X_t if and only if every element of the matrix B is either zero or one.

(ii) Suppose a process $\{Y_t\}$ has a HMM of the joint Markov process type, and let $\{X_t\}$ denote the associated Markov chain. Define

$$m_{ij}^{(u)} := \Pr\{(X_t, Y_t) = (j, u)|X_{t-1} = i\}, \ \forall i, j \in \mathbb{N}, u \in \mathbb{M}, \tag{6.6}$$

$$M^{(u)} := [m_{ij}^{(u)}, i, j \in \mathbb{N}], \ \forall u \in \mathbb{M}, \tag{6.7}$$

and

$$a_{ij} := \sum_{u \in \mathbb{M}} m_{ij}^{(u)}, \ \forall i, j. \tag{6.8}$$

Then Y_t is a random function of X_t (and not just X_{t-1}) if and only if the following consistency conditions hold: If $a_{ij} \neq 0$, then the ratio $m_{ij}^{(u)}/a_{ij}$ is independent of i.

Proof. The first statement is obvious. Let us consider the second statement. Suppose the process $\{Y_t\}$ has a joint Markov process type of HMM, and let $\{(X_t, Y_t)\}$ be the associated Markov process. Define the matrices $M^{(u)}$ as in (6.6). Then we already know that Y_t is a random function of X_{t-1}. The aim is to show that Y_t is a random function of X_t (and not just X_{t-1}) if and only if the stated condition holds.

"Only if": By assumption, $\{X_t\}$ is a Markov process. Moreover, we have that

$$\Pr\{X_t = j|X_{t-1} = i\} = \sum_{u \in \mathbb{M}} \Pr\{(X_t, Y_t) = (j, u)|X_{t-1} = i\} = \sum_{u \in \mathbb{M}} m_{ij}^{(u)}.$$

Therefore $A = [a_{ij}]$ is the state transition matrix of the Markov process $\{X_t\}$. Now suppose that Y_t is a random function of X_t, and not just X_{t-1}, and define

$$b_{ju} := \Pr\{Y_t = u|X_t = j\}, \ \forall u \in \mathbb{M}, j \in \mathbb{N}.$$

Then we must have $m_{ij}^{(u)} = a_{ij} b_{ju}$ for all i, j, u. If $a_{ij} = 0$ for some i, j, then perforce $m_{ij}^{(u)} = 0 \ \forall u \in \mathcal{M}$. Suppose $a_{ij} \neq 0$. Then it is clear that

$$b_{ju} = \frac{m_{ij}^{(u)}}{a_{ij}} \ \forall i$$

and is therefore independent of i.

"If": This consists of simply reversing the arguments. Suppose the ratio is indeed independent of i, and define b_{ju} as above. Then clearly $m_{ij}^{(u)} = a_{ij} b_{ju}$ and as a result Y_t is a random function of X_t. □

As a simple example, suppose $n = m = 2$,

$$M^{(1)} = \begin{bmatrix} 0.5 & 0.2 \\ 0.1 & 0.4 \end{bmatrix}, M^{(2)} = \begin{bmatrix} 0.2 & 0.1 \\ 0.1 & 0.4 \end{bmatrix}, A = \begin{bmatrix} 0.7 & 0.3 \\ 0.2 & 0.8 \end{bmatrix}.$$

Then

$$\frac{m_{11}^{(1)}}{a_{11}} = 5/7, \ \frac{m_{21}^{(1)}}{a_{21}} = 1/2 \neq 5/7.$$

Hence, while Y_t is a random function of X_{t-1}, it is *not* a random function of X_t.

6.2 COMPUTATION OF LIKELIHOODS

To place the contents of this section in perspective, let us observe that in the HMM world (and it does not matter what the type of the HMM is), we can identify three entities, namely: the model, the output sequence, and the state sequence. Given any two of these entities, we can seek to determine the third. So we can ask three distinct questions.

1. Given a HMM and a state sequence, what is the likelihood of observing a particular output sequence?

2. Given a HMM and an observed output sequence, what is the most likely state sequence?

3. Given observed output and state sequences, what is the most likely HMM?

All three questions are addressed in this section. In Section 6.2.1 we answer the first question. The second question is answered in Section 6.2.2. The third question is addressed through "realization theory" and turns out to be by far the most difficult and deep question. It is addressed in two stages. First, a standard method known as the Baum-Welch method is given in Section 6.2.3. In the Baum-Welch method, it is assumed that the *size n of the state space of $\{X_t\}$ is known beforehand*, and the emphasis is only on choosing the best possible *parameter set* for the HMM. Clearly this is not always a very realistic problem formulation. Ideally, the size n of the state space of $\{X_t\}$ should be determined by the data that we wish to model, and not be fixed beforehand for our convenience. This is known as the "complete realization problem" and is studied in Chapter 7.

6.2.1 Computation of Likelihoods of Output Sequences

As is common in computer science, let us use the symbol $\mathbb{M}^*$ to denote the set of all *finite* strings over the set $\mathbb{M}$. Thus $\mathbb{M}^*$ consists of all strings of the form $\mathbf{u} = u_1 \ldots u_l$, where each $u_i \in \mathbb{M}$ and l is the length of the string. We let $|\mathbf{u}|$ denote the length of the string. Now it is well known that the set $\mathbb{M}^*$ is countably infinite.[4] The objective of this section is to derive a simple formula for the likelihood of observing a string $\mathbf{u} \in \mathbb{M}^*$. In the literature, this simple formula is referred to as the "forward-backward recursion." Throughout the discussion below, we use the notation introduced earlier, namely, if $\{Y_t\}$ is a stochastic process, then Y_k^l denotes $(Y_k, Y_{k+1}, \ldots, Y_{l-1}, Y_l)$. Naturally, this notation makes sense only if $k \leq l$. If $k = l$, then Y_k^k is just Y_k.

Suppose that a process $\{Y_t\}$ has a Type 3 HMM. Thus there is an associated Markov process $\{X_t\}$ such that the joint process $\{(X_t, Y_t)\}$ is Markov, and in addition (6.4) holds. As in (6.6), let us define the $n \times n$ matrices $M^{(u)} \in [0,1]^{n \times n}$ by

$$m_{ij}^{(u)} := \Pr\{(X_t, Y_t) = (j, u) | X_{t-1} = i\}, \ \forall i, j \in \mathbb{N}, u \in \mathbb{M}.$$

and the stochastic matrix $A \in [0,1]^{n \times n}$ by

$$a_{ij} := \sum_{u \in \mathbb{M}} m_{ij}^{(u)}, \ \forall i, j.$$

Then A is the state transition matrix of the Markov process $\{X_t\}$. If the HMM is of Type 2 rather than Type 3, then as shown in the proof of Theorem 6.4, we have that $m_{ij}^{(u)} = a_{ij} b_{ju}$ for all i, j, u . To complete the specification of the Markov process $\{X_t\}$, we need to specify the stationary distribution π, since in general the matrix A could have more than one stationary distribution. Suppose π is also specified. Then the dynamics of the two processes $\{X_t\}$ and $\{Y_t\}$ are completely specified.

Now suppose $\mathbf{u} \in \mathbb{M}^*$ and that $|\mathbf{u}| = l$. We wish to compute the likelihood

$$\mathbf{f_u} := \Pr\{Y_t^{t+l-1} = \mathbf{u}\}. \tag{6.9}$$

In other words, we wish to compute the probability of observing the sequence of outputs $\mathbf{u} = u_1 \ldots u_l$ in exactly that order. Note that, since the process $\{Y_t\}$ is stationary, the number $f_{\mathbf{u}}$ as defined in (6.9) is independent of t, the time at which the observations start. So we might as well take $t = 1$. We refer to the quantity $f_{\mathbf{u}}$ as the **frequency** of the string $\mathbf{u}$.

To compute the frequency $f_{\mathbf{u}}$ corresponding to a particular $\mathbf{u} \in \mathbb{M}^*$, let us observe that if $\mathbf{i} := (i_0, i_1, \ldots, i_l) \in \mathbb{N}^{l+1}$ is a particular sequence of states,

[4] If we permit strings of *infinite* length, then the resulting set is uncountably infinite.

then

$$\Pr\{(X_0^l, Y_1^l) = (\mathbf{i}, \mathbf{u})\} = \Pr\{X_0 = i_0\} \cdot \prod_{t=1}^{l} \Pr\{(X_t, Y_t) = (i_t, u_t) | X_{t-1} = i_{t-1}\}$$
$$= \pi_{i_0} \prod_{t=1}^{l} m_{i_{t-1} i_t}^{(u_t)}.$$

Now to compute $f_{\mathbf{u}}$, we can observe that

$$f_{\mathbf{u}} = \Pr\{Y_1^l = \mathbf{u}\} = \sum_{\mathbf{i} \in \mathbb{N}^{l+1}} \Pr\{(X_0^l, Y_1^l) = (\mathbf{i}, \mathbf{u})\}.$$

So in principle we could compute the above probability for each sequence $\mathbf{i} \in \mathbb{N}^{l+1}$ and then add them up. This gives the expression

$$f_{\mathbf{u}} = \sum_{(i_0, \ldots, i_l) \in \mathbb{N}^{l+1}} \pi_{i_0} \prod_{t=1}^{l} m_{i_{t-1} i_t}^{(u_t)}. \tag{6.10}$$

If the above equation is interpreted literally and we sum over all possible state sequences, this computation requires $O(n^{l+1})$ computations. Clearly this is a very silly way to compute $f_{\mathbf{u}}$.

Instead let us note that (6.10) can be interpreted as a matrix product. In fact it equals

$$f_{\mathbf{u}} = \pi M^{(u_1)} \cdots M^{(u_l)} \mathbf{e}_n. \tag{6.11}$$

The equivalence of the two formulas (6.10) and (6.11) can be seen easily by expanding the right side of (6.10) as

$$f_{\mathbf{u}} = \sum_{i_0} \sum_{i_1} \cdots \sum_{i_l} \pi_{i_0} m_{i_0 i_1}^{(u_1)} \cdots m_{i_{l-1} i_l}^{(u_l)}.$$

Now if we simply start from the leftmost term in (6.11) and multiply π by $M^{(u_1)}$, and then multiply the resulting row vector by $M^{(u_2)}$, and so on, and finally multiply the resulting row vector by $\mathbf{e}_n$, then the complexity is $O(ln^2)$, since multiplying an $1 \times n$ row vector by an $n \times n$ matrix requires $O(n^2)$ operations, and we need to do this $l + 1$ times.[5]

The very useful formula (6.11) can be given a very nice interpretation, which is sometimes referred to in the HMM literature as the "forward-backward recursion." Observe that, given a sequence $\mathbf{u} \in \mathbb{M}^*$ of length l, the formula (6.11) for $f_{\mathbf{u}}$ can be written as

$$f_{\mathbf{u}} = \alpha(\mathbf{u}, k) \beta(\mathbf{u}, k),$$

where

$$\alpha(\mathbf{u}, k) = \pi M^{(u_1)} \cdots M^{(u_k)} = \pi \prod_{t=1}^{k} M^{(u_t)}, \tag{6.12}$$

[5]Since $l + 1$ and l are of the same order of magnitude, we can write $O(ln^2)$ instead of $O((l + 1)n^2)$.

$$\beta(\mathbf{u}, k) = M^{(u_{k+1})} \cdots M^{(u_l)} \mathbf{e}_n = \prod_{t=k+1}^{l} M^{(u_t)} \mathbf{e}_n. \tag{6.13}$$

These formulas are valid for *every* k between 1 and l, provided we take the empty product as the identity matrix. These formulas can now be given a very simple interpretation in terms of conditional probabilities, which is the basis of the "forward-backward recursion." Observe that $\alpha(\mathbf{u}, k)$ is a $1 \times n$ row vector, whereas $\beta(\mathbf{u}, k)$ is an $n \times 1$ column vector.

Lemma 6.7 *With $\alpha(\mathbf{u}, k)$ and $\beta(\mathbf{u}, k)$ defined as in (6.12) and (6.13) respectively, we have*

$$\alpha_i(\mathbf{u}, k) = \Pr\{X_k = i \wedge (Y_1, \ldots, Y_k) = (u_1, \ldots u_k)\}. \tag{6.14}$$

$$\beta_i(\mathbf{u}, k) = \Pr\{X_k = i \wedge (Y_{k+1}, \ldots, Y_l) = (u_{k+1}, \ldots, u_l)\}. \tag{6.15}$$

The proof is obvious from the formulas (6.12) and (6.13) respectively, and is left as an exercise to the reader.

6.2.2 The Viterbi Algorithm

In this subsection, we study the following question: Suppose we are given a HMM, and an observation $\mathbf{u} \in \mathcal{M}^l$. Thus we know the parameters of an HMM, and are given that $Y_1^l = \mathbf{u}$. The question is: What is the "most likely" state sequence X_0^l, where "most likely" is interpreted in the sense of the maximum *a posteriori* estimate as defined in Section 1.2.4. Thus the problem is: Find

$$\operatorname*{argmax}_{\mathbf{i} \in \mathbb{N}^{l+1}} \Pr\{X_0^l = \mathbf{i} | Y_1^l = \mathbf{u}\}.$$

The first, but very crucial, step is to observe that

$$\Pr\{X_0^l = \mathbf{i} | Y_1^l = \mathbf{u}\} = \frac{\Pr\{X_0^l = \mathbf{i} \wedge Y_1^l = \mathbf{u}\}}{\Pr\{Y_1^l = \mathbf{u}\}}.$$

Now the variable of optimization here is $\mathbf{i} \in \mathbb{N}^{l+1}$, which *does not appear* in the denominator. So we can treat the denominator as a constant and simply maximize the numerator with respect to $\mathbf{i}$. In other words, the problem is to find

$$\operatorname*{argmax}_{\mathbf{i} \in \mathbb{N}^{l+1}} \Pr\{X_0^l = \mathbf{i} \wedge Y_1^l = \mathbf{u}\}.$$

An explicit expression for the right side can be deduced from (6.10), namely

$$\Pr\{X_0^l = \mathbf{i} \wedge Y_1^l = \mathbf{u}\} = \pi_{i_0} \prod_{t=1}^{l} m_{i_{t-1} i_t}^{(u_t)}.$$

Maximizing the right side with respect to $i_0, \ldots, i_l$ is a very difficult problem if we try solve it directly. The well-known **Viterbi algorithm** is a systematic approach that allows us to break down the single (and seemingly intractable) optimization problem into a *sequence* of optimization problems, each of which is tractable.

Towards this end, let us fix a time $t \leq l$, a state $j \in \mathbb{N}$, and define

$$\gamma(t,j;\mathbf{i}) := \Pr\{X_0^{t-1} = \mathbf{i} \wedge X_t = j \wedge Y_1^t = \mathbf{u}_1^t\}$$
$$= \pi_{i_0} \cdot \left[\prod_{s=1}^{t-1} m_{i_{s-1}i_s}^{(u_s)}\right] \cdot m_{i_{t-1}j}^{(u_t)}, \tag{6.16}$$

$$\gamma^*(t,j) := \max_{\mathbf{i} \in \mathbb{N}^t} \gamma(t,j;\mathbf{i}), \tag{6.17}$$

$$I^*(t,j) := \{\mathbf{i} \in \mathbb{N}^t : \gamma(t,j;\mathbf{i}) = \gamma^*(t,j)\}. \tag{6.18}$$

Thus $I^*(t,j)$ consists of the most likely state sequences $\mathbf{i} \in \mathbb{N}^t$ that lead to state j at time t and match the observation history up to time t. The key result is stated next. It is an instance of the "principle of optimality." This principle states that in a sequential optimization problem, any subsequence of an optimal solution is also optimal (under suitable conditions of course).

Theorem 6.8 *Fix t and j, and define $I^*(t,j)$ as in 6.18. Suppose $t \geq 2$ and that $i_0 \dots i_{t-1} \in I^*(t,j)$. Then*

$$i_0 \dots i_{t-2} \in I^*(t-1, i_{t-1}). \tag{6.19}$$

More generally, for any $s \leq t-1$, we have

$$i_0 \dots i_{s-1} \in I^*(s, i_s). \tag{6.20}$$

Proof. Suppose by way of contradiction that

$$i_0 \dots i_{t-2} \notin I^*(t-1, i_{t-1}).$$

This implies that there exists another sequence $j_0 \dots j_{t-2} \in \mathbb{N}^{t-1}$ such that

$$\gamma(t-1, i_{t-1}; i_0 \dots i_{t-2}) < \gamma(t-1, i_{t-1}; j_0 \dots j_{t-2}).$$

Expanding this inequality leads to

$$\pi_{i_0} \left[\prod_{s=1}^{t-2} m_{i_{s-1}i_s}^{(u_s)}\right] \cdot m_{i_{t-2}i_{t-1}}^{(u_{t-1})} < \pi_{j_0} \left[\prod_{s=1}^{t-2} m_{j_{s-1}j_s}^{(u_s)}\right] \cdot m_{j_{t-2}i_{t-1}}^{(u_{t-1})}.$$

Multiplying both sides by $m_{i_{t-1}j}^{(u_t)}$ shows that

$$\gamma(t,j;\mathbf{i}) < \gamma(t,j;j_0 \dots j_{t-2} i_{t-1}),$$

which contradicts the assumption that $i_0 \dots i_{t-2} i_{t-1} \in I^*(t,j)$. Hence (6.19) is true. The proof of (6.20) is entirely similar. □

Theorem 6.9 *The function $\gamma^*(\cdot,\cdot)$ satisfies the recursive relationship*

$$\gamma^*(t,j) = \max_{i \in \mathbb{N}} \left[\gamma^*(t-1,i) \cdot m_{ij}^{(u_t)}\right], t \leq l. \tag{6.21}$$

Proof. Fix $t \leq l$ and $j \in \mathbb{N}$. Suppose $i_0 \ldots i_{t-1} \in I^*(t,j)$ is an optimal state sequence. Then it follows from Theorem 6.8 that

$$\begin{aligned}\gamma^*(t,j) &= \gamma(t,j,i_0 \ldots i_{t-1}) \\ &= \gamma(t,i_{t-1};i_0 \ldots i_{t-2}) \cdot m^{(u_t)}_{i_{t-1}j} \text{ from (6.16)} \\ &= \gamma^*(t-1,i_{t-1}) \cdot m^{(u_t)}_{i_{t-1}j} \end{aligned} \tag{6.22}$$

At this point the only variable of optimization left is i_{t-1}, which we can simply relabel as i. Choosing $i = i_{t-1}$ so as to maximize the right side of (6.22) leads to the recursive relationship (6.21). □

Now we have everything in place to state the Viterbi algorithm.

Step 1. (Initialization) For each $i_1 \in \mathbb{N}$, choose $i_0 \in \mathbb{N}$ so as to maximize the product $\pi_{i_0} m^{(u_1)}_{i_0 i_1}$. In case there is more than one optimal i_0 corresponding to a given i_1, choose any one of the optimal i_0. Thus leads to n optimal trajectories $i_0 i_1$, and n corresponding optimal values $\gamma(i,i_1) = \pi_{i_0} m^{(u_1)}_{i_0 i_1}$, one for each $i_1 \in \mathbb{N}$.

Step 2. (Recursion) Suppose $2 \leq t \leq l$. At time t, for each $i_{t-1} \in \mathbb{N}$ we have an optimal value $\gamma^*(t-1,i_{t-1})$ and an associated optimal trajectory $i_0 \ldots i_{t-2}$. Now, for each $i_t \in \mathbb{N}$, choose $i_{t-1} \in \mathbb{N}$ as the value of i that achieves the maximum in (6.21), with $j = i_t$. If there is more than one value of i_{t-1} that achieves the maximum, choose any one value. Once i_{t-1} is determined, concatenate i_{t-1} to the optimal trajectory up to i_{t-1}, and call it the optimal trajectory up to i_t. The optimal value $\gamma^*(t,i_t)$ is the maximum value in (6.21).

Step 3. (Completion) Let $t = l$. At this stage we have n optimal values $\gamma^*(l,1), \ldots, \gamma^*(l,n)$, and n associated optimal trajectories. Now choose as i_l the value of j that maximizes $\gamma^*(l,j)$ and choose the associated optimal trajectory as the most likely state sequence.

Now let us analyze the computational complexity of the algorithm. At each time t and for each state $j \in \mathbb{N}$, we need to compute the product $\gamma(t-1,i) m^{(u_t)}_{ij}$ for each $i \in \mathbb{N}$, and then find the largest value of the product. This has complexity $O(n)$ for each $j \in \mathbb{N}$, or $O(n^2)$ in all. Since we need to do this l times in all, the overall complexity is $O(ln^2)$, which is the same as the complexity of computing the frequency $f_{\mathbf{u}}$ for a given $\mathbf{u} \in \mathcal{M}^l$. Note that a frontal attack on the problem by enumerating all possible state trajectories would have complexity $O(n^l)$, which would be unacceptable.

6.2.3 The Baum-Welch Algorithm

The objective of this subsection is to present a method universally referred to as the "Baum-Welch algorithm." This algorithm pertains to hidden Markov models of Type 2 as defined in Definition 6.2, wherein the cardinality of the state space of the underlying Markov process is known, but the state transition and emission matrices A and B as defined in (6.2) and (6.3) are not known. It is natural that, in observing the output sequence of a Type

2 HMM, we know the cardinality of the output space, or equivalently, the integer m representing the number of columns of the matrix B. However, it may or may not be reasonable to assume that we also know the cardinality of the state space of the underlying Markov process, that is the integer n. The Baum-Welch algorithm addresses the case where n is known. The so-called "complete realization problem" for hidden Markov models, discussed in Chapter 7, addresses the problem of *determining* the integer n from the statistics of the process $\{Y_t\}$.

The problem addressed by the Baum-Welch algorithm is this: Given an observation sequence $Y_1^l = \mathbf{y}_1^l$, identify the *maximum likelihood estimate* of the matrices A and B. The algorithm consists of a simple recursive technique for constructing a sequence of estimates, call it $\{(A^{(k)}, B^{(k)})\}$, such that the likelihood of the data given the model increases with each iteration. Therefore such a sequence is guaranteed to find at least a local maximum. The proof of the above "hill-climbing" property is given in two papers [12, 13] whose contents are beyond the scope of this book. Thus only the conclusions of these papers are summarized here. The reader will note that Welch is not a coauthor in either of the papers and wonder why his name is associated with the algorithm. Welch explains in [137] that he too had independently come up with the same updating rule as in these publications, and had observed the fact that the likelihood increases with each iteration. However, he could not give a mathematical proof.

The algorithm is as follows: It is assumed that an initial estimate $(A^{(0)}, B^{(0)})$ for the unknown Type 2 HMM is available, where both matrices are strictly positive. Set the counter k to zero.

1. Given the observation sequence $Y_1^l = \mathbf{y}_1^l$, use the Viterbi algorithm to compute the most likely state sequence $\mathbf{x}_1^l$, under the model $(A^{(k)}, B^{(k)})$.

2. Using the state sequence $\mathbf{x}_1^l$, use the method described in Section 4.1.2 for determining the most likely state transition matrix $A^{(k+1)}$.

3. Using the state sequence $\mathbf{x}_1^l$ and the observation sequence $\mathbf{y}_1^l$, determine the most likely emission matrix $B^{(k+1)}$ taking into account the fact that the output sequence Y_1^l is conditionally independent given the state sequence X_1^l. Specifically, for each index $i \in \mathbb{N}$ and each index $u \in \mathbb{M}$, define

$$b_{iu}^{(k+1)} = \frac{c_{iu}}{d_i},$$

where

$$d_i = \sum_{t=1}^{l} I_{\{x_t = i\}}, c_{iu} = \sum_{t=1}^{l} I_{\{x_t = i \wedge y_t = u\}}$$

denote the number of times that state i occurs in the sequence $\mathbf{x}_1^l$ found from the Viterbi algorithm, and the number of times that the pair (i, u) occurs in the sequence $(\mathbf{x}_1^l, \mathbf{y}_1^l)$.

4. Update the counter k and repeat.

As stated above, it can be shown that the likelihood of the observation sequence $\mathbf{y}_1^l$ given the model $(A^{(k)}, B^{(k)})$ is an increasing function of k. See [12, 13] for details.

Since the Baum-Welch algorithm is basically a hill-climbing technique in the parameter space of the A, B matrices, it is guaranteed to converge at least to a local maximizer of the likelihood function. However, as with any hill-climbing algorithm, it is also prone to get stuck in local maxima, especially if the data is being generated by a "true but unknown" HMM that is quite far from the initial guess $(A^{(0)}, B^{(0)})$. On the other hand, if the starting point of the iterations is sufficiently close to the true but unknown HMM, then the algorithm can be expected to converge to the true HMM. Note that the Baum-Welch algorithm is also referred to as the "expectation maximization" algorithm.

Chapter Seven

Hidden Markov Processes: The Complete Realization Problem

In this chapter we continue our study of hidden Markov processes begun in Chapter 6. The focus of study here is the so-called complete realization problem, which was broached in Section 6.2. In that section, we discussed the Baum-Welch algorithm, which attempts to construct a HMM given an output observation sequence of finite length, once the cardinality of the underlying state space is specified. The problem discussed in this chapter is far deeper, and can be stated as follows: Suppose $\mathbb{M} = \{1, \ldots, m\}$ is a finite set[1] and that $\{Y_t\}_{t \geq 0}$ is a stationary stochastic process assuming values in $\mathbb{M}$. We wish to derive necessary and/or sufficient conditions for $\{Y_t\}$ to be a hidden Markov process, where the cardinality of the state space for the underlying Markov process is not specified beforehand; it is required only to be finite. In Chapter 6, three different types of HMMs were introduced. As shown in Theorem 6.4, if the process $\{Y_t\}$ has any one kind of a HMM realization if and only if it has all three. Thus we are free to use whichever HMM we wish. It turns out that the Type 3, or joint Markov model, is the easiest to analyze, and accordingly, that is the one used in this chapter.

It turns out that it is quite easy to prove a universal *necessary* condition for the given process to have a HMM. But this condition is *not sufficient in general.* Unfortunately the demonstration of this fact is rather long. One can in principle derive a "necessary and sufficient condition," but as pointed out by Anderson [6], the "necessary and sufficient condition" is virtually a restatement of the problem to be solved and does not shed any insight into the solution. However, if one adds the requirement that the process $\{Y_t\}$ is also "mixing," then it is possible to present conditions that are "almost necessary and sufficient" for the process to have a HMM realization. That is the main conclusion of this chapter. However, a great many technical details need to be worked out to reach the conclusion. The contents of this chapter are largely based on [133].

[1] As always, we really mean that $\mathbb{M} = \{s_1, \ldots, s_m\}$, a collection of abstract symbols, and we write $\mathbb{M} = \{1, \ldots, m\}$ only to simplify notation.

7.1 FINITE HANKEL RANK: A UNIVERSAL NECESSARY CONDITION

In this subsection we introduce a very useful matrix which we refer to as a "Hankel" matrix, because it has some superficial similarity to a Hankel matrix. Given the set $\mathbb{M}$ in which the stochastic process $\{Y_t\}$ assumes its values, let us define some lexical ordering of the elements in $\mathbb{M}$. The specific order itself does not matter, and the reader can verify that all of the discussion in the present chapter is insensitive to the specific lexical ordering used. For each integer l, the set $\mathbb{M}^l$ has cardinality m^l and consists of l-tuples. These can be arranged either in first-lexical order (flo) or last-lexical order (llo). First-lexical order refers to indexing the first element, then the second, and so on, while last-lexical order refers to indexing the last element, then the next to last, and so on. For example, suppose $m = 2$ and that $\mathbb{M} = \{1, 2\}$ in the natural order. Then

$$\mathbb{M}^3 \text{ in llo } = \{111, 112, 121, 122, 211, 212, 221, 222\},$$

$$\mathbb{M}^3 \text{ in flo } = \{111, 211, 121, 221, 112, 212, 122, 222\}.$$

Given any finite string $\mathbf{u} \in \mathbb{M}^l$, its frequency $f_{\mathbf{u}}$ is defined by

$$f_{\mathbf{u}} := \Pr\{Y_{t+1}^{t+l} = \mathbf{u}\}.$$

Since the process $\{Y_t\}$ is stationary, the above probability is independent of t. Moreover, as seen earlier, the frequency vector is *stationary*; that is

$$f_{\mathbf{u}} = \sum_{v \in \mathbb{M}} f_{\mathbf{u}v} = \sum_{w \in \mathbb{M}} f_{w\mathbf{u}}, \ \forall \mathbf{u} \in \mathbb{M}^*. \tag{7.1}$$

More generally,

$$f_{\mathbf{u}} = \sum_{\mathbf{v} \in \mathbb{M}^r} f_{\mathbf{uv}} = \sum_{\mathbf{w} \in \mathbb{M}^s} f_{\mathbf{wu}}, \ \forall \mathbf{u} \in \mathbb{M}^*. \tag{7.2}$$

Given integers $k, l \geq 1$, the matrix $F_{k,l}$ is defined as

$$F_{k,l} = [f_{\mathbf{uv}}, \mathbf{u} \in \mathbb{M}^k \text{ in flo}, \mathbf{v} \in \mathbb{M}^l \text{ in llo}] \in [0, 1]^{m^k \times m^l}.$$

Thus the rows of $F_{k,l}$ are indexed by an element of $\mathbb{M}^k$ in flo, while the columns are indexed by an element of $\mathbb{M}^l$ in llo. For example, suppose $m = 2$, and $\mathbb{M} = \{1, 2\}$. Then

$$F_{1,2} = \begin{bmatrix} f_{111} & f_{112} & f_{121} & f_{122} \\ f_{211} & f_{212} & f_{221} & f_{222} \end{bmatrix},$$

whereas

$$F_{2,1} = \begin{bmatrix} f_{111} & f_{112} \\ f_{211} & f_{212} \\ f_{121} & f_{122} \\ f_{221} & f_{222} \end{bmatrix}.$$

In general, for a given integer s, the matrices $F_{0,s}, F_{1,s-1}, \ldots, F_{s-1,1}, F_{s,0}$ all contain frequencies of the m^s s-tuples in the set $\mathbb{M}^s$. However, the dimensions of the matrices are different, and the elements are arranged in a different order. Note that by convention $F_{0,0}$ is taken as the 1×1 matrix 1 (which can be thought of as the frequency of occurrence of the empty string).

Given integers $k, l \geq 1$, we define the matrix $H_{k,l}$ as

$$H_{k,l} := \begin{bmatrix} F_{0,0} & F_{0,1} & \ldots & F_{0,l} \\ F_{1,0} & F_{1,1} & \ldots & F_{1,l} \\ \vdots & \vdots & \vdots & \vdots \\ F_{k,0} & F_{k,1} & \ldots & F_{k,l} \end{bmatrix}.$$

Note that $H_{k,l}$ has $1 + m + \ldots + m^k$ rows, and $1 + m + \ldots + m^l$ columns. In general, $H_{k,l}$ is not a "true" Hankel matrix, since it is not constant along backward diagonals. It is not even "block Hankel." However, it resembles a Hankel matrix in the sense that the matrix in the (i, j)-th block consists of frequencies of strings of length $i + j$. Finally, we define H (without any subscripts) to be the infinite matrix of the above form, that is,

$$H := \begin{bmatrix} F_{0,0} & F_{0,1} & \ldots & F_{0,l} & \ldots \\ F_{1,0} & F_{1,1} & \ldots & F_{1,l} & \ldots \\ \vdots & \vdots & \vdots & \vdots & \vdots \\ F_{k,0} & F_{k,1} & \ldots & F_{k,l} & \ldots \\ \vdots & \vdots & \vdots & \vdots & \vdots \end{bmatrix}.$$

Through a mild abuse of language we refer to H as the Hankel matrix associated with the process $\{Y_t\}$.

In this section, it is shown that a process $\{Y_t\}$ has a HMM *only if* the matrix H has finite rank. Taking some liberties with the English language, we refer to this as the "finite Hankel rank condition." Theorem 7.1 below shows that the finite Hankel rank condition is a *universal necessary condition* for a given process to have a HMM realization. However, as shown in the next subsection, the finiteness of the rank of H is only necessary, but not sufficient in general.

Theorem 7.1 *Suppose $\{Y_t\}$ has a Type 3 HMM with the associated $\{X_t\}$ process having n states. Then Rank$(H) \leq n$.*

Proof. The definition of a Type 3 HMM implies that the process $\{X_t\}$ is Markov over a set $\mathbb{N}$ of cardinality n. Define the matrices $M^{(1)}, \ldots, M^{(m)}, A$ as in (6.7) and (6.8) respectively, and let $\boldsymbol{\pi}$ denote the stationary distribution associated with this Markov process. Now suppose $\mathbf{u} \in \mathbb{M}^l$, specifically that $\mathbf{u} = u_1 \ldots u_l$. Then from the sum of products formula (6.11) derived earlier, we have

$$f_{\mathbf{u}} = \sum_{i=1}^{n} \sum_{j_1=1}^{n} \cdots \sum_{j_l=1}^{n} \pi_i m_{ij_1}^{(u_1)} \cdots m_{j_{l-1}j_l}^{(u_l)} = \boldsymbol{\pi} M^{(u_1)} \cdots M^{(u_l)} \mathbf{e}_n. \qquad (7.3)$$

Note that

$$\sum_{l \in \mathbb{M}} M^{(l)} = A, \ \boldsymbol{\pi} \left[\sum_{l \in \mathbb{M}} M^{(l)} \right] = \boldsymbol{\pi}, \text{ and } \left[\sum_{l \in \mathbb{M}} M^{(l)} \right] \mathbf{e}_n = \mathbf{e}_n. \tag{7.4}$$

Thus the sum of the matrices $M^{(u)}$ is the state transition matrix of the Markov chain, and $\boldsymbol{\pi}$ and $\mathbf{e}_n$ are respectively a row eigenvector and a column eigenvector of A corresponding to the eigenvalue 1.

Now let us return to the matrix H. Using (7.3), we see at once that H can be factored as a product KL, where

$$K = \begin{bmatrix} \pi \\ \pi M^{(1)} \\ \vdots \\ \pi M^{(m)} \\ \pi M^{(1)} M^{(1)} \\ \vdots \\ \pi M^{(m)} M^{(m)} \\ \vdots \end{bmatrix},$$

$$L = [\mathbf{e}_n \mid M^{(1)}\mathbf{e}_n | \ldots | M^{(m)}\mathbf{e}_n \mid M^{(1)}M^{(1)}\mathbf{e}_n | \ \ldots \mid M^{(m)}M^{(m)}\mathbf{e}_n | \ldots].$$

In other words, the rows of K consist of $\pi M^{(u_1)} \cdots M^{(u_l)}$ as $\mathbf{u} \in \mathbb{M}^l$ is in flo and l increases, whereas the columns of L consist of $M^{(u_1)} \cdots M^{(u_l)}\mathbf{e}_n$ as $\mathbf{u} \in \mathbb{M}^l$ is in llo and l increases. Now note that the first factor has n columns whereas the second factor has n rows. Hence $\text{Rank}(H) \leq n$. □

It has been shown by Sontag [120] that the problem of deciding whether or not a given "Hankel" matrix has finite rank is undecidable.

7.2 NONSUFFICIENCY OF THE FINITE HANKEL RANK CONDITION

Let us refer to the process $\{Y_t\}$ as "having finite Hankel rank" if $\text{Rank}(H)$ is finite. Thus Theorem 7.1 shows that $\text{Rank}(H)$ being finite is a *necessary* condition for the given process to have a HMM. However, the converse is *not true in general*—it is possible for a process to have finite Hankel rank and yet not have a realization as a HMM. The original example in this direction was given by Fox and Rubin [56]. However, their proof contains an error, in the opinion of this author. In a subsequent paper, Dharmadhikari and Nadkarni [47] quietly and without comment simplified the example of Fox and Rubin and also gave a correct proof (without explicitly pointing out that the Fox-Rubin proof is erroneous). In this section, we present the example of [47] and slightly simplify their proof. It is worth noting that the example crucially depends on rotating a vector by an angle α that is not commensurate with π, that is, α/π is not a rational number. A similar approach is used by Benvenuti and Farina [15, Example 4] to construct a nonnegative impulse response with finite Hankel rank which does not have a finite rank *nonnegative* realization.

Let us begin by choosing numbers $\lambda \in (0, 0.5], \alpha \in (0, 2\pi)$ such that α and π are noncommensurate. In particular, this rules out the possibility that $\alpha = \pi$. Now define

$$h_l := \lambda^l \sin^2(l\alpha/2), \ \forall l \geq 1.$$

Note that we can also write

$$h_l = \lambda^l \frac{(e^{\mathbf{i}l\alpha/2} - e^{-\mathbf{i}l\alpha/2})^2}{4},$$

where (just in this equation) $\mathbf{i}$ denotes $\sqrt{-1}$. Simplifying the expression for h_l shows that

$$h_l = \frac{\lambda^l}{4}(\zeta^l + \zeta^{-l} - 2), \tag{7.5}$$

where $\zeta := e^{\mathbf{i}\alpha}$. Because h_l decays at a geometric rate with respect to l, the following properties are self-evident.

1. $h_i > 0 \ \forall i$. Note that $l\alpha$ can never equal a multiple of π because α and π are noncommensurate.

2. We have that
$$\sum_{i=1}^{\infty} h_i =: \delta < 1. \tag{7.6}$$

3. We have that
$$\sum_{i=1}^{\infty} i h_i < \infty.$$

4. The infinite Hankel matrix
$$\bar{H} := \begin{bmatrix} h_1 & h_2 & h_3 & \cdots \\ h_2 & h_3 & h_4 & \cdots \\ h_3 & h_4 & h_5 & \cdots \\ \vdots & \vdots & \vdots & \ddots \end{bmatrix}$$
has finite rank of 3.

This last property follows from standard linear system theory. Given a sequence $\{h_i\}_{i \geq 1}$, let us define its z-transform $\tilde{h}(\cdot)$ by[2]

$$\tilde{h}(z) := \sum_{i=1}^{\infty} h_i z^{i-1}.$$

[2]Normally in z-transformation theory, the sequence $\{h_i\}$ is indexed starting from $i = 0$, whereas here we have chosen to begin with $i = 1$. This causes the somewhat unconventional-looking definition.

Thanks to an old theorem of Kronecker [86], it is known that the Hankel matrix $\bar{H}$ has finite rank if and only if $\tilde{h}$ is a *rational* function of z, in which case the rank of the Hankel matrix is the same as the degree of the rational function $\tilde{h}(z)$. Now it is a ready consequence of (7.5) that

$$\tilde{h}(z) = \frac{1}{4}\left[\frac{\lambda\zeta}{1-\lambda\zeta z} + \frac{\lambda\zeta^{-1}}{1-\lambda\zeta^{-1}z} - 2\frac{\lambda}{1-\lambda z}\right].$$

Hence the infinite matrix $\bar{H}$ has rank 3.

The counterexample is constructed by defining a Markov process $\{X_t\}$ with a countable state space and another process $\{Y_t\}$ with just two output values such that Y_t is a function of X_t. The process $\{Y_t\}$ satisfies the finite Hankel rank condition; in fact $\text{Rank}(H) \leq 5$. And yet no Markov process with a finite state space can be found such that Y_t is a function of that Markov process. Since we already know from Section 6.4 that the existence of all the three kinds of HMMs is equivalent, this is enough to show that the process $\{Y_t\}$ does not have a joint Markov process type of HMM.

The process $\{X_t\}$ is Markovian with a countable state space $\{0, 1, 2, \ldots\}$. The transition probabilities of the Markov chain are defined as follows:

$$\Pr\{X_{t+1} = 0 | X_t = 0\} = 1 - \delta = 1 - \sum_{i=1}^{\infty} h_i,$$

$$\Pr\{X_{t+1} = i | X_t = 0\} = h_i \text{ for } i = 1, 2, \ldots,$$

$$\Pr\{X_{t+1} = i | X_t = i+1\} = 1 \text{ for } i = 1, 2, \ldots,$$

and all other probabilities are zero. Thus the dynamics of the Markov chain are as follows: If the chain starts in the initial state 0, then it makes a transition to state i with probability h_i, or remains in 0 with the probability $1 - \sum_i h_i = 1 - \delta$. Once the chain moves to the state i, it then successively goes through the states $i-1, i-2, \ldots, 1, 0$. Then the process begins again. Thus the dynamics of the Markov chain consist of a series of cycles beginning and ending at state 0, but where the lengths of the cycles are random, depending on the transition out of the state 0.

Clearly $\{X_t\}$ is a Markov process. Now we define $\{Y_t\}$ to be a function of this Markov process. Let $Y_t = a$ if $X_t = 0$, and let $Y_t = b$ otherwise, i.e., if $X_t = i$ for some $i \geq 1$. Thus the output process $\{Y_t\}$ assumes just two values a and b. Note that in the interests of clarity we have chosen to denote the two output states as a and b instead of 1 and 2. For this process $\{Y_t\}$ we shall show that, first,

$$\text{Rank}(H) \leq \text{Rank}(\bar{H}) + 2 = 5,$$

where H is the Hankel matrix associated with the process $\{Y_t\}$, and second, that there is no Markov process $\{Z_t\}$ with a finite state space such that Y_t is a (deterministic) function of Z_t.

The stationary distribution of the Markov chain is as follows:

$$\pi_0 = g := \left[1 + \sum_{i=1}^{\infty} i h_i\right]^{-1},$$

$$\pi_i = g \sum_{j=i}^{\infty} h_j, i \geq 1.$$

To verify this, note the structure of the state transition matrix A of the Markov chain: State 0 can be reached only from states 0 and 1. Thus column 0 of A has $1-\delta$ in row 0, 1 in row 1, and zeros in all other rows. For $i \geq 1$, state i can be reached only from states 0 and $i+1$. Hence column i has h_i in row 0, 1 in row $i+1$, and zeros elsewhere. As a result

$$(\pi A)_0 = g \left(1 - \delta + \sum_{j=1}^{\infty} h_j \right) = g(1 - \delta + \delta) = g = \pi_0,$$

while for $i \geq 1$,

$$(\pi A)_i = h_i \pi_0 + \pi_{i+1} = g \left[h_i + \sum_{j=i+1}^{\infty} h_j \right] = g \sum_{j=i}^{\infty} h_j = \pi_i.$$

To verify that this is indeed a probability vector, note that

$$\sum_{i=0}^{\infty} \pi_i = g \left[1 + \sum_{i=1}^{\infty} \sum_{j=i}^{\infty} h_j \right] = g \left[1 + \sum_{j=1}^{\infty} \sum_{i=1}^{j} h_j \right] = g \left[1 + \sum_{j=1}^{\infty} j h_j \right] = 1,$$

in view of the definition of g.

Next, let us compute the frequencies of various output strings. Note that if $Y_t = a$, then certainly $X_t = 0$. Hence, if $Y_t = a$, then the conditional probability of Y_{t+1} does not depend on the values of $Y_i, i < t$. Therefore, for arbitrary strings $\mathbf{u}, \mathbf{v} \in \{a, b\}^*$, we have

$$f_{\mathbf{u}a\mathbf{v}} = f_{\mathbf{u}a} \cdot f_{\mathbf{v}|\mathbf{u}a} = f_{\mathbf{u}a} \cdot f_{\mathbf{v}|a}.$$

Hence the infinite matrix $H^{(a)}$ defined by

$$H^{(a)} := [f_{\mathbf{u}a\mathbf{v}}, \mathbf{u}, \mathbf{v} \in \{a, b\}^*]$$

has rank one. In such a case, it is customary to refer to a as a "Markovian state"; see [60].

Next, let us compute the frequencies of strings of the form $ab^l a, ab^l, b^l a$, and b^l. A string of the form $ab^l a$ can occur only of $X_t = 0, X_{t+1} = l, \ldots, X_{t+l} = 1, X_{t+l+1} = 0$. All transitions except the first one have probability one, while the first transition has probability h_l. Finally, the probability that $X_t = 0$ is π_0. Hence

$$f_{ab^l a} = \pi_0 h_l, \ \forall l.$$

Next, note that

$$f_{ab^l} = f_{ab^{l+1}} + f_{ab^l a}.$$

Hence, if we define

$$\pi_0 \gamma_l := f_{ab^l},$$

then γ_l satisfies the recursion

$$\pi_0 \gamma_l = \pi_0 \gamma_{l+1} + \pi_0 h_l.$$

To start the recursion, note that

$$\pi_0 \gamma_1 = f_{ab} = f_a - f_{aa} = \pi_0 - \pi_0(1-\delta) = \pi_0 \delta = \pi_0 \sum_{i=1}^{\infty} h_i.$$

Therefore

$$\pi_0 \gamma_l = \pi_0 \sum_{i=l}^{\infty} h_i, \text{ or } \gamma_l = \sum_{i=l}^{\infty} h_i.$$

Now we compute the frequencies f_{b^l} for all l. Note that

$$f_{b^l} = f_{b^{l+1}} + f_{ab^l} = f_{b^{l+1}} + \pi_0 \gamma_l.$$

Hence if we define $\pi_0 \eta_l := f_{b^l}$, then η_l satisfies the recursion

$$\eta_l = \eta_{l+1} + \gamma_l.$$

To start the recursion, note that

$$f_b = 1 - f_a = 1 - \pi_0.$$

Now observe that

$$\pi_0 = \left[1 + \sum_{i=1}^{\infty} i h_i \right]^{-1},$$

and as a result

$$1 - \pi_0 = \pi_0 \sum_{i=1}^{\infty} i h_i = \pi_0 \sum_{i=1}^{\infty} \sum_{j=1}^{i} h_i = \pi_0 \sum_{j=1}^{\infty} \sum_{i=j}^{\infty} h_i = \pi_0 \sum_{j=1}^{\infty} \gamma_j.$$

Hence

$$f_{b^l} = \pi_0 \eta_l, \text{ where } \eta_l = \sum_{i=l}^{\infty} \gamma_i.$$

Finally, to compute $f_{b^l a}$, note that

$$f_{b^l a} + f_{b^{l+1}} = f_{b^l}.$$

Hence

$$f_{b^l a} = f_{b^l} - f_{b^{l+1}} = \pi_0(\eta_l - \eta_{l+1}) = \pi_0 \gamma_l.$$

Now let us look at the Hankel matrix H corresponding to the process $\{Y_t\}$. We can think of H as the interleaving of two infinite matrices $H^{(a)}$ and $H^{(b)}$, where

$$H^{(a)} = [f_{\mathbf{u}a\mathbf{v}}, \mathbf{u}, \mathbf{v} \in \{a, b\}^*],$$

$$H^{(b)} = [f_{\mathbf{u}b\mathbf{v}}, \mathbf{u}, \mathbf{v} \in \{a, b\}^*].$$

We have already seen that $H^{(a)}$ has rank one, since a is a Markovian state. Hence it follows that

$$\text{Rank}(H) \leq \text{Rank}(H^{(a)}) + \text{Rank}(H^{(b)}) = \text{Rank}(H^{(b)}) + 1.$$

To bound $\text{Rank}(H^{(b)})$, fix integers l, n, and define

$$H_{l,n}^{(b)} := [f_{\mathbf{u}b\mathbf{v}}, \mathbf{u} \in \{a,b\}^l, \mathbf{v} \in \{a,b\}^n].$$

Note that $H_{l,n}^{(b)} \in [0,1]^{2^l \times 2^n}$. It is now shown that

$$\text{Rank}(H_{l,n}^{(b)}) \leq \text{Rank}(\bar{H}) + 1 = 4. \tag{7.7}$$

Since the right side is independent of l, n, it follows that

$$\text{Rank}(H^{(b)}) \leq 4,$$

whence

$$\text{Rank}(H) \leq 5.$$

To prove (7.7), suppose $\mathbf{u} \in \{a,b\}^{l-1}$ is arbitrary. Then

$$f_{\mathbf{u}ab\mathbf{v}} = f_{\mathbf{u}a} \cdot f_{b\mathbf{v}|\mathbf{u}a} = f_{\mathbf{u}a} \cdot f_{b\mathbf{v}|a},$$

because a is a Markovian state. Hence each of the 2^{l-1} rows $[f_{\mathbf{u}ab\mathbf{v}}, \mathbf{u} \in \{a,b\}^{l-1}]$ is a multiple of the row $[f_{b\mathbf{v}|a}]$, or equivalently, of the row $[f_{a^l b\mathbf{v}}]$. Hence $\text{Rank}(H^{(b)})$ is unaffected if we keep only this one row and jettison the remaining $2^{l-1} - 1$ rows. Similarly, as $\mathbf{u}$ varies over $\{a,b\}^{l-2}$, each of the rows $[f_{\mathbf{u}abb\mathbf{v}}]$ is proportional to $[f_{a^{l-2}abb\mathbf{v}}] = [f_{a^{l-1}b^2\mathbf{v}}]$. So we can again retain just the row $[f_{a^{l-1}b^2\mathbf{v}}]$ and discard the rest. Repeating this argument l times shows that $H^{(b)}$ has the same rank as the $(l+1) \times 2^n$ matrix

$$\begin{bmatrix} f_{a^l b\mathbf{v}} \\ f_{a^{l-1}b^2\mathbf{v}} \\ \vdots \\ f_{ab^l\mathbf{v}} \\ f_{b^{l+1}\mathbf{v}} \end{bmatrix}, \mathbf{v} \in \{a,b\}^n.$$

A similar exercise can now be repeated with $\mathbf{v}$. If $\mathbf{v}$ has the form $\mathbf{v} = a\mathbf{w}$, then

$$f_{a^i b^{l+1-i} a\mathbf{w}} = f_{a^i b^{l+1-i} a} \cdot f_{\mathbf{w}|a}.$$

So all 2^{n-1} columns $[f_{a^i b^{l+1-i} a\mathbf{w}}, \mathbf{w}\{a,b\}^{n-1}]$ are proportional to the single column $[f_{a^i b^{l+1-i} a^n}]$. So we can keep just this one column and throw away the rest. Repeating this argument shows that $H^{(b)}$ has the same rank as the $(l+1) \times (n+1)$ matrix

$$\begin{array}{c} \\ a^l \\ a^{l-1}b \\ \vdots \\ ab^{l-1} \\ b^l \end{array} \begin{array}{c} \begin{array}{ccccc} ba^n & b^2a^{n-1} & \cdots & b^n a & b^{n+1} \end{array} \\ \begin{bmatrix} f_{a^l ba^n} & f_{a^l b^2 a^{n-1}} & \cdots & f_{a^l b^n a} & f_{a^l b^{n+1}} \\ f_{a^{l-1}b^2 a^n} & f_{a^{l-1}b^3 a^{n-1}} & \cdots & f_{a^{l-1}b^{n+1}a} & f_{a^{l-1}b^{n+2}} \\ \vdots & \vdots & \vdots & \vdots & \vdots \\ f_{ab^l a^n} & f_{ab^{l+1}a^{n-1}} & \cdots & f_{ab^{l+n-1}a} & f_{ab^{l+n}} \\ f_{b^{l+1}a^n} & f_{b^{l+2}a^{n-1}} & \cdots & f_{b^{l+n}a} & f_{b^{l+n+1}} \end{bmatrix} \end{array}.$$

The structure of this matrix becomes clear if we note that

$$\begin{aligned} f_{a^i b^j a^t} &= f_{a^i} \cdot f_{b^j a^t | a} \\ &= f_{a^i} \cdot f_{b^j a | a} \cdot f_{a^{t-1} | a} \\ &= \pi_0 (1-\delta)^{i-1} \cdot h_j \cdot (1-\delta)^{t-1}. \end{aligned} \tag{7.8}$$

The strings in the last row and column either do not begin with a, or end with a, or both. So let us divide the first row by $\pi_0(1-\delta)^{l-1}$, the second row by $\pi_0(1-\delta)^{l-2}$, etc., the l-th row by π_0, and do nothing to the last row. Similarly, let us divide the first column by $(1-\delta)^{n-1}$, the second column by $(1-\delta)^{n-2}$, etc., the n-th column by $(1-\delta)^0 = 1$, and leave the last column as is. The resulting matrix has the same rank as $H^{(b)}_{l,n}$, and the matrix is

$$\begin{bmatrix} h_1 & h_2 & \ldots & h_n & \times \\ h_2 & h_3 & \ldots & h_{n+1} & \times \\ \vdots & \vdots & \vdots & \vdots & \vdots \\ h_l & h_{l+1} & \ldots & h_{l+n} & \times \\ \times & \times & \ldots & \times & \times \end{bmatrix},$$

where $\times$ denotes a number whose value does not matter. Now the upper left $l \times n$ submatrix is a submatrix of $\bar{H}$; as a result its rank is bounded by 3. This proves (7.7).[3]

To carry on our analysis of this example, we make use of z-transforms. This is not done in [47], but it simplifies the arguments to follow. As shown earlier, the z-transform of the sequence $\{h_i\}$ is given by

$$\tilde{h}(z) = \frac{1}{4}\left[\frac{\lambda\zeta}{1-\lambda\zeta z} + \frac{\lambda\zeta^{-1}}{1-\lambda\zeta^{-1}z} - 2\frac{\lambda}{1-\lambda z}\right] = \frac{\psi_h(z)}{\phi(z)},$$

where

$$\phi(z) := (1-\lambda\zeta)(1-\lambda\zeta^{-1})(1-\lambda), \tag{7.9}$$

and $\psi_h(z)$ is some polynomial of degree no larger than two; its exact form does not matter. Next, recall that

$$\gamma_i = \sum_{j=i}^{\infty} h_j.$$

Now it is an easy exercise to show that

$$\tilde{\gamma}(z) = \frac{\delta - \tilde{h}(z)}{1-z},$$

where, as defined earlier, $\delta = \sum_{i=1}^{\infty} h_i$. Even though we are dividing by $1-z$ in the above expression, in reality $\tilde{\gamma}$ does not have a pole at $z=1$, because $\tilde{h}(1) = \delta$. Hence we can write

$$\tilde{\gamma}(z) = \frac{\psi_\gamma(z)}{\phi(z)},$$

[3] Through better bookkeeping, Dharmadhikari and Nadkarni [47] show that the rank is bounded by 3, not 4. This slight improvement is not worthwhile since all that matters is that the rank is finite.

where again ψ_γ is some polynomial of degree no larger than two, and $\phi(z)$ is defined in (7.9). By entirely similar reasoning, it follows from the expression

$$\eta_i = \sum_{j=i}^{\infty} \gamma_j$$

that

$$\tilde{\eta}(z) = \frac{s - \tilde{\gamma}(z)}{1 - z},$$

where

$$s := \sum_{i=1}^{\infty} \gamma_i = \sum_{i=1}^{\infty} \sum_{j=i}^{\infty} h_j = \sum_{j=1}^{\infty} \sum_{i=1}^{j} h_j = \sum_{j=1}^{\infty} j h_j.$$

Here again, $\tilde{\eta}(\cdot)$ does not have a pole at $z = 1$, and in fact

$$\tilde{\gamma}(z) = \frac{\psi_\eta(z)}{\phi(z)},$$

where ψ_η is also a polynomial of degree no larger than two. The point of all these calculations is to show that each of the quantities γ_l, η_l has the form

$$\gamma_l = c_{0,\gamma} \lambda^l + c_{1,\gamma} \lambda^l \zeta^l + c_{2,\gamma} \lambda^l \zeta^{-l}, \tag{7.10}$$

$$\eta_l = c_{0,\eta} \lambda^l + c_{1,\eta} \lambda^l \zeta^l + c_{2,\eta} \lambda^l \zeta^{-l}, \tag{7.11}$$

for appropriate constants. Note that, even though ζ is a complex number, the constants occur in conjugate pairs so that γ_l, η_l are always real. And as we have already seen from (7.5), we have

$$h_l = -\frac{1}{2}\lambda^l + \frac{1}{4}\lambda^l \zeta^l + \frac{1}{4}\lambda^l \zeta^{-l}.$$

Now the expression (7.11) leads at once to two very important observations.

Observation 1: Fix some positive number ρ, and compute the weighted average

$$\frac{1}{T} \sum_{l=1}^{T} \rho^{-l} \eta_l =: \theta(\rho, T).$$

Then it follows that

1. If $\rho < \lambda$, then $\theta(\rho, T) \to \infty$ as $T \to \infty$.

2. If $\rho > \lambda$, then $\theta(\rho, T) \to 0$ as $T \to \infty$.

3. If $\rho = \lambda$, then $\theta(\rho, T) \to c_{0,\eta}$ as $T \to \infty$, where $c_{0,\eta}$ is the constant in (7.11).

If $\rho \neq \lambda$, then the behavior of $\theta(\rho, T)$ is determined by that of $(\lambda/\rho)^l$. If $\rho = \lambda$, then the averages of the oscillatory terms $(\lambda\zeta/\rho)^l$ and $(\lambda/\rho\zeta)^l$ will both approach zero, and only the first term in (7.11) contributes to a nonzero average.

Observation 2: Let T be any fixed integer, and consider the moving average

$$\frac{1}{T}\sum_{j=l+1}^{l+T} \lambda^{-j}\eta_j =: \theta_l^T.$$

This quantity does not have a limit as $l \to \infty$ if α is not commensurate with π. To see this, take the z-transform of $\{\theta_l^T\}$. This leads to

$$\tilde{\theta}^T(z) = \frac{\beta^T(z)}{\phi(z)},$$

where $\beta^T(z)$ is some high degree polynomial. After dividing through by $\phi(z)$, we get

$$\tilde{\theta}^T(z) = \beta_q^T(z) + \frac{\beta_r^T(z)}{\phi(z)},$$

where β_q^T is the quotient and β_r^T is the remainder (and thus has degree no more than two). By taking the inverse z-transform, we see that the sequence $\{\theta_l^T\}$ is the sum of two parts: The first part is a sequence having finite support (which we can think of as the "transient"), and the second is a sequence of the form

$$c_{0,\theta}\lambda^l + c_{1,\theta}\lambda^l\zeta^l + c_{2,\theta}\lambda^l\zeta^{-l}.$$

From this expression it is clear that if α is noncommensurate with π, then θ_l^T does not have a limit as $l \to \infty$.

These two observations are the key to the concluding part of this very long line of reasoning. Suppose by way of contradiction that the output process $\{Y_t\}$ can be expressed as a function of a Markov process $\{Z_t\}$ with a finite state space. Let $\mathbb{N} = \{1, \ldots, n\}$ denote the state space, and let π, A denote the stationary distribution and state transition matrix of the Markov chain $\{Z_t\}$. Earlier we had used these symbols for the Markov chain $\{X_t\}$, but no confusion should result from this recycling of notation. From Item 2 of Theorem 4.7, it follows that by a symmetric permutation of rows and columns (which corresponds to permuting the labels of the states), A can be arranged in the form

$$A = \begin{bmatrix} P & \mathbf{0} \\ R & Q \end{bmatrix},$$

where the rows of P correspond to the recurring states and those of R to transient states. Similarly, it follows from Item 7 of Theorem 4.7 that the components of π corresponding to transient states are all zero. Hence the corresponding states can be dropped from the set $\mathbb{N}$ without affecting anything. So let us assume that all states are recurrent.

Next, we can partition the state space $\mathbb{N}$ into those states that map into a, and those states that map into b. With the obvious notation, we can partition π as $[\pi_a \ \ \pi_b]$ and the state transition matrix as

$$A = \begin{bmatrix} A_{aa} & A_{ab} \\ A_{ba} & A_{bb} \end{bmatrix}.$$

Moreover, from Theorem 3.9, it follows that we can arrange A_{bb} in the form

$$A_{bb} = \begin{bmatrix} A_{11} & 0 & \dots & 0 \\ A_{21} & A_{22} & \dots & 0 \\ \vdots & \vdots & \vdots & \vdots \\ A_{s1} & A_{s1} & \dots & A_{ss} \end{bmatrix},$$

where s is the number of communicating classes within those states that map into the output b, and each of the diagonal matrices A_{ii} is irreducible. Of course, the fact that each of the diagonal blocks is irreducible does still not suffice to determine π uniquely, but as before we can assume that no component of π is zero, because if some component of π is zero, then we can simply drop that component from the state space.

Now it is claimed that $\rho(A_{bb}) = \lambda$, where $\rho(\cdot)$ denotes the spectral radius. To show this, recall that if B is an irreducible matrix with spectral radius $\rho(B)$, and $\boldsymbol{\theta}, \boldsymbol{\phi}$ are respectively the (unique strictly positive) row eigenvector and column eigenvector corresponding to the eigenvalue $\rho(B)$, then the "ergodic average"

$$\frac{1}{T} \sum_{l=1}^{T} [\rho(B)]^{-l} B^l$$

converges to the rank one matrix $\phi\theta$ as $T \to \infty$. Now from the triangular structure of A_{bb}, it is easy to see that $\rho(A_{bb})$ is the maximum among the numbers $\rho(A_{ii}), i = 1, \dots, s$. If we let θ_i, ϕ_i denote the unique row and column eigenvectors of A_{ii} corresponding to $\rho(A_{ii})$, it is obvious that

$$\frac{1}{T} \sum_{l=1}^{T} [\rho(A_{bb}]^{-l} A_{bb}^l \to \text{ Block Diag } \{\phi_i \theta_i I_{\{\rho(A_{ii}) = \rho(A_{bb})\}}\}. \tag{7.12}$$

In other words, if $\rho(A_{ii}) = \rho(A_{bb})$, then the corresponding term $\phi_i \theta_i$ is present in the block diagonal matrix; if $\rho(A_{ii}) < \rho(A_{bb})$, then the corresponding entry in the block diagonal matrix is the zero matrix. Let D denote the block diagonal in (7.12), and note that at least one of the $\rho(A_{ii})$ equals $\rho(A_{bb})$. Hence at least one of the products $\phi_i \theta_i$ is present in the block diagonal matrix D.

From the manner in which the HMM has been set up, it follows that

$$\eta_l = f_{b^l} = \pi_b A_{bb}^l \mathbf{e}.$$

In other words, the only way in which we can observe a sequence of l symbols b in succession is for all states to belong to the subset of $\mathbb{N}$ that map into the output b. Next, let us examine the behavior of the quantity

$$\frac{1}{T} \sum_{l=1}^{T} \rho^{-l} \eta_l = \frac{1}{T} \sum_{l=1}^{T} \rho^{-l} \pi_b A_{bb}^l \mathbf{e},$$

where $\rho = \rho(A_{bb})$. Now appealing to (7.12) shows that the above quantity has a definite limit as $T \to \infty$. Moreover, since π_b and $\mathbf{e}$ are strictly positive, and the block diagonal matrix D has at least one positive block $\phi_i \theta_i$, it follows that

$$\lim_{T \to \infty} \frac{1}{T} \sum_{l=1}^{T} \rho^{-l} \eta_l = \pi D \mathbf{e} \in (0, \infty).$$

By Observation 1, this implies that $\rho(A_{bb}) = \lambda$.

Finally (and at long last), let us examine those blocks A_{ii} which have the property that $\rho(A_{ii}) = \rho(A_{bb}) = \rho$. Since each of these is an irreducible matrix, it follows from Theorem 3.12 that each such matrix has a unique "period" n_i, which is an integer. Moreover, it follows from Theorem 3.24 that A_{ii} has eigenvalues at $\rho \exp(\mathbf{i} 2\pi j / n_i)$, $j = 1, \ldots, n_i - 1$, and all other eigenvalues of A_{ii} have magnitude strictly less than ρ. This statement applies *only* to those indices i such that $\rho(A_{ii}) = \rho(A_{bb}) = \rho$. Now let N denote the least common multiple of all these integers n_i. Then it is clear that the matrix A_{bb} has a whole lot of eigenvalues of the form $\rho \exp(\mathbf{i} 2\pi j / N)$ for some (though not necessarily all) values of j ranging from 0 to $N - 1$; all other eigenvalues of A have magnitude strictly less than ρ. As a result, the quantity

$$\frac{1}{N} \sum_{l=t+1}^{t+N} A_{bb}^{l}$$

has a definite limit at $t \to \infty$. In turn this implies that the quantity

$$\frac{1}{N} \sum_{l=t+1}^{t+N} \pi_b A_{bb}^{l} \mathbf{e} = \frac{1}{N} \sum_{l=t+1}^{t+N} \eta_l$$

has a definite limit at $t \to \infty$. However, this contradicts Observation 2, since α is noncommensurate with π. This contradiction shows that the stochastic process $\{Y_t\}$ cannot be realized as a function of a finite state Markov chain.

7.3 AN ABSTRACT NECESSARY AND SUFFICIENT CONDITION

In this section we reproduce an abstract necessary and sufficient condition for a given probability law to have an HMM realization, as first presented by Heller [64], with a significantly simplified proof due to Picci [106].

Recall that $\mathbb{M}^*$, the set of all finite strings over $\mathbb{M} = \{1, \ldots, m\}$, is a countable set. We let $\mu(\mathbb{M}^*)$ denote the set of all maps $p : \mathbb{M}^* \to [0, 1]$ satisfying the following two conditions:

$$\sum_{u \in \mathbb{M}} p_u = 1, \tag{7.13}$$

$$\sum_{v \in \mathbb{M}} p_{\mathbf{u}v} = p_{\mathbf{u}}, \ \forall \mathbf{u} \in \mathbb{M}^*. \tag{7.14}$$

Note that by repeated application of (7.14), we can show that

$$\sum_{\mathbf{v} \in \mathbb{M}^l} p_{\mathbf{uv}} = p_{\mathbf{u}}, \ \forall \mathbf{u} \in \mathbb{M}^*. \tag{7.15}$$

By taking $\mathbf{u}$ to be the empty string, so that $p_{\mathbf{u}} = 1$, we get from the above that

$$\sum_{\mathbf{v} \in \mathbb{M}^l} p_{\mathbf{v}} = 1, \ \forall l. \tag{7.16}$$

We can think of $\mu(\mathbb{M}^*)$ as the set of all frequency assignments to strings in $\mathbb{M}^*$ that are **right-stationary** by virtue of satisfying (7.14).

Definition 7.2 *Given $p \in \mu(\mathbb{M}^*)$, the tuple $\{\pi, M^{(1)}, \ldots, M^{(m)}\}$ is called a* **HMM realization** *of p if*

$$\boldsymbol{\pi} \in \mathbb{S}_n, \tag{7.17}$$

$$M^{(u)} \in [0,1]^{n \times n} \ \forall u \in \mathbb{M}, \tag{7.18}$$

$$\left[\sum_{u \in \mathbb{M}} M^{(u)} \right] \mathbf{e}_n = \mathbf{e}_n, \tag{7.19}$$

and finally

$$p_{\mathbf{u}} = \boldsymbol{\pi} M^{(u_1)} \ldots M^{(u_l)} \mathbf{e}_n \ \forall \mathbf{u} \in \mathbb{M}^l. \tag{7.20}$$

Given a frequency distribution $p \in \mu(\mathbb{M}^*)$, for each $u \in \mathbb{M}$ we define the conditional distribution

$$p(\cdot|u) := \mathbf{v} \in \mathbb{M}^* \mapsto \frac{p_{u\mathbf{v}}}{p_u}. \tag{7.21}$$

If by chance $p_u = 0$, we define $p(\cdot|u)$ to equal p. Note that $p(\cdot|u) \in \mu(\mathbb{M}^*)$; that is, $p(\cdot|u)$ is also a frequency assignment map. By applying (7.21) repeatedly, for each $\mathbf{u} \in \mathbb{M}^*$ we can define the conditional distribution

$$p(\cdot|\mathbf{u}) := \mathbf{v} \in \mathbb{M}^* \mapsto \frac{p_{\mathbf{uv}}}{p_{\mathbf{u}}}. \tag{7.22}$$

Again, for each $\mathbf{u} \in \mathbb{M}^*$, the conditional distribution $p(\cdot|\mathbf{u})$ is also a frequency assignment. Clearly conditioning can be applied recursively and the results are consistent. Thus

$$p((\cdot|\mathbf{u})|\mathbf{v}) = p(\cdot|\mathbf{uv}), \ \forall \mathbf{u}, \mathbf{v} \in \mathbb{M}^*. \tag{7.23}$$

It is easy to verify that if p satisfies the right-consistency condition (7.14), then so do all the conditional distributions $p(\cdot|\mathbf{u})$ for all $\mathbf{u} \in \mathbb{M}^*$. Thus, if $p \in \mu(\mathbb{M}^*)$, then $p(\cdot|\mathbf{u}) \in \mu(\mathbb{M}^*)$ for all $\mathbf{u} \in \mathbb{M}^*$.

A set $\mathcal{C} \subseteq \mu(\mathbb{M}^*)$ is said to be **polyhedral** if there exist an integer n and distributions $q^{(1)}, \ldots, q^{(n)} \in \mu(\mathbb{M}^*)$ such that $\mathcal{C}$ is the convex hull of these $q^{(i)}$, that is, every $q \in \mathcal{C}$ is a convex combination of these $q^{(i)}$. A set $\mathcal{C} \subseteq \mu(\mathbb{M}^*)$ is said to be **stable** if

$$q \in \mathcal{C} \Rightarrow q(\cdot|\mathbf{u}) \in \mathcal{C} \; \forall \mathbf{u} \in \mathbb{M}^*. \tag{7.24}$$

In view of (7.23), (7.24) can be replaced by weaker-looking condition

$$q \in \mathcal{C} \Rightarrow q(\cdot|u) \in \mathcal{C} \; \forall u \in \mathbb{M}. \tag{7.25}$$

Now we are ready to state the main result of this section, first proved in [64]. However, the proof below follows [106] with some slight changes in notation.

Theorem 7.3 *A frequency distribution $p \in \mu(\mathbb{M}^*)$ has a HMM realization if and only if there exists a stable polyhedral set $\mathcal{C} \subseteq \mu(\mathbb{M}^*)$ containing p.*

Proof. "If" Suppose $q^{(1)}, \ldots, q^{(n)} \in \mu(\mathbb{M}^*)$ are the generators of the polyhedral set $\mathcal{C}$. Thus every $q \in \mathcal{C}$ is of the form

$$q = \sum_{i=1}^{n} a_i q^{(i)}, a_i \geq 0, \sum_{i=1}^{n} a_i = 1.$$

In general neither the integer n nor the individual distributions $q^{(i)}$ are unique, but this does not matter. Now, since $\mathcal{C}$ is stable, $q(\cdot|u) \in \mathcal{C}$ for all $a \in \mathcal{C}, u \in \mathbb{M}$. In particular, for each i, u, there exist constants $\alpha_{ij}^{(u)}$ such that

$$q^{(i)}(\cdot|u) = \sum_{j=1}^{n} \alpha_{ij}^{(u)} q^{(j)}(\cdot), \alpha_{ij}^{(u)} \geq 0, \sum_{j=1}^{n} \alpha_{ij}^{(u)} = 1.$$

Thus from (7.21) it follows that

$$q_{u\mathbf{v}}^{(i)} = \sum_{j=1}^{n} q_u^{(i)} \alpha_{ij}^{(u)} q_{\mathbf{v}}^{(j)} = \sum_{j=1}^{n} m_{ij}^{(u)} q_{\mathbf{v}}^{(j)}, \tag{7.26}$$

where

$$m_{ij}^{(u)} := q_u^{(i)} \alpha_{ij}^{(u)}, \; \forall i, j, u. \tag{7.27}$$

We can express (7.26) more compactly by using matrix notation. For $\mathbf{u} \in \mathbb{M}^*$, define

$$\mathbf{q_u} := [q_{\mathbf{u}}^{(1)} \ldots q_{\mathbf{u}}^{(n)}]^t \in [0,1]^{n \times 1}.$$

Then (7.26) states that

$$\mathbf{q}_{u\mathbf{v}} = M^{(u)} \mathbf{q_v} \; \forall u \in M, \mathbf{v} \in \mathbb{M}^*,$$

where $M^{(u)} = [m_{ij}^{(u)}] \in [0,1]^{n \times n}$. Moreover, it follows from (7.23) that

$$\mathbf{q_{uv}} = M^{(u_1)} \ldots M^{(u_l)} \mathbf{q_v} \; \forall \mathbf{u}\mathbb{M}^l, \mathbf{v} \in \mathbb{M}^*.$$

If we define

$$M^{(\mathbf{u})} := M^{(u_1)} \ldots M^{(u_l)} \; \forall \mathbf{u} \in \mathbb{M}^l,$$

then the above equation can be written compactly as

$$\mathbf{q}_{\mathbf{uv}} = M^{(\mathbf{u})} M^{(\mathbf{v})} \mathbf{q}_{\mathbf{v}} \ \forall \mathbf{u}, \mathbf{v} \in \mathbb{M}^*. \tag{7.28}$$

By assumption, $p \in \mathcal{C}$. Hence there exist numbers $\pi_1, \ldots, \pi_n$, not necessarily unique, such that

$$p(\cdot) = \sum_{i=1}^{n} \pi_i q^{(i)}(\cdot), \pi_i \geq 0 \ \forall i, \sum_{i=1}^{n} \pi_i = 1. \tag{7.29}$$

We can express (7.29) as

$$p(\cdot) = \boldsymbol{\pi} \mathbf{q}(\cdot).$$

Hence, for all $\mathbf{u}, \mathbf{v} \in \mathbb{M}^*$, it follows from (7.28) that

$$p_{\mathbf{uv}} = \boldsymbol{\pi} \mathbf{q}_{\mathbf{uv}} = \boldsymbol{\pi} M^{(\mathbf{u})} \mathbf{q}_{\mathbf{v}}, \ \forall \mathbf{u}, \mathbf{v} \in \mathbb{M}^*. \tag{7.30}$$

In particular, if we let $\mathbf{v}$ equal the empty string, then $\mathbf{q}_{\mathbf{v}} = \mathbf{e}_n$, and $p_{\mathbf{uv}} = p_{\mathbf{u}}$. Thus (7.30) becomes

$$p_{\mathbf{u}} = \boldsymbol{\pi} M^{(\mathbf{u})} \mathbf{e}_n,$$

which is the same as (7.20).

Next, we verify (7.19) by writing it out in component form. We have

$$\begin{aligned}
\sum_{j=1}^{n} \sum_{u \in \mathbb{M}} m_{ij}^{(u)} &= \sum_{u \in \mathbb{M}} \sum_{j=1}^{n} m_{ij}^{(u)} \\
&= \sum_{u \in \mathbb{M}} q_u^{(i)} \left[\sum_{j=1}^{n} \alpha_{ij}^{(u)} \right] \\
&= \sum_{u \in \mathbb{M}} q_u^{(i)} \text{ because } \sum_{j=1}^{n} \alpha_{ij}^{(u)} = 1 \\
&= 1, \ \forall i \text{ because } q^{(i)} \in \mu(\mathbb{M}^*) \text{ and (7.13).}
\end{aligned}$$

Before leaving the "If" part of the proof, we observe that if the probability distribution $p \in \mu(\mathbb{M}^*)$ is also *left-stationary* by satisfying

$$\sum_{u \in \mathbb{M}} p_{u\mathbf{v}} = p_{\mathbf{v}} \ \forall u \in \mathbb{M}, \mathbf{v} \in \mathbb{M}^*,$$

then *it is possible* to choose the vector $\boldsymbol{\pi}$ such that

$$\boldsymbol{\pi} \left[\sum_{u \in \mathbb{M}} M^{(u)} \right] = \boldsymbol{\pi}. \tag{7.31}$$

To see this, we substitute into (7.20) which has already been established. This gives

$$\boldsymbol{\pi} M^{(\mathbf{v})} \mathbf{e}_n = p_{\mathbf{v}} = \sum_{u \in \mathbb{M}} p_{u\mathbf{v}} = \boldsymbol{\pi} \left[\sum_{u \in \mathbb{M}} M^{(u)} \right] M^{(\mathbf{v})} \mathbf{e}_n, \ \forall \mathbf{v} \in \mathbb{M}^*.$$

Now it is not possible to "cancel" $M^{(\mathbf{v})}\mathbf{e}_n$ from both sides of the above equation. However, it is always possible to choose the coefficient vector $\boldsymbol{\pi}$ so as to satisfy (7.31).

"Only if" Suppose p has a HMM realization $\{\boldsymbol{\pi}, M^{(1)}, \ldots, M^{(m)}\}$. Let n denote the dimension of the matrices $M(u)$ and the vector $\boldsymbol{\pi}$. Define the distributions $q^{(1)}, \ldots, q^{(n)}$ by

$$\mathbf{q}_{\mathbf{u}} = [q_{\mathbf{u}}^{(1)} \ldots q_{\mathbf{u}}^{(n)}]^t := M^{(\mathbf{u})}\mathbf{e}_n, \ \forall \mathbf{u} \in \mathbb{M}^*. \tag{7.32}$$

Thus $q_{\mathbf{u}}^{(i)}$ is the i-th component of the column vector $M^{(\mathbf{u})}\mathbf{e}_n$. First it is shown that each $q^{(i)}$ is indeed a frequency distribution. From (7.32), it follows that

$$\sum_{u \in \mathbb{M}} \mathbf{q}_u = \left[\sum_{u \in \mathbb{M}} M^{(u)}\right] \mathbf{e}_n = \mathbf{e}_n,$$

where we make use of (7.19). Thus each $q^{(i)}$ satisfies (7.13) (with p replaced by $q^{(i)}$). Next, to show that each $q^{(i)}$ is right-stationary, observe that for each $\mathbf{u} \in \mathbb{M}^*, v \in \mathbb{M}$ we have

$$\sum_{v \in \mathbb{M}} \mathbf{q}_{\mathbf{u}v} = M^{(\mathbf{u})} \left[\sum_{v \in \mathbb{M}} M^{(v)}\right] \mathbf{e}_n = M^{(\mathbf{u})}\mathbf{e}_n = \mathbf{q}_{\mathbf{u}}.$$

Thus each $q^{(i)}$ is right-stationary. Finally, to show that the polyhedral set consisting of all convex combinations of $q^{(1)}, \ldots, q^{(n)}$ is stable, observe that

$$q^{(i)}(\mathbf{v}|u) = \frac{q^{(i)}(u\mathbf{v})}{q^{(i)}(u)}.$$

Substituting from (7.31) gives

$$\begin{aligned} q^{(i)}(\mathbf{v}|u) &= \frac{1}{q^{(i)}(u)} M^{(u)} M^{(\mathbf{v})}\mathbf{e}_n \\ &= \mathbf{a}_u^{(i)} M^{(\mathbf{v})}\mathbf{e}_n, \\ &= \mathbf{a}_u^{(i)}\mathbf{q}_{\mathbf{v}}, \end{aligned} \tag{7.33}$$

where

$$\mathbf{a}_u^{(i)} := \left[\frac{m_{ij}^{(u)}}{q^{(i)}(u)}, j = 1, \ldots, n\right] \in [0,1]^{1 \times n}.$$

Thus each conditional distribution $q^{(i)}(\cdot|u)$ is a linear combination of $q^{(1)}(\cdot)$ through $q^{(n)}(\cdot)$. It remains only to show that $q^{(i)}(\cdot|u)$ is a *convex* combination, that is, that each $\mathbf{a}_u^{(i)} \in \mathbb{R}_+^n$ and that $\mathbf{a}_u^{(i)}\mathbf{e}_n = 1$. The first is obvious from the definition of the vector $\mathbf{a}^{(i)}$. To establish the second, substitute $\mathbf{v}$ equal to the empty string in (7.33). Then $q^{(i)}(\mathbf{v}|u) = 1$ for all i, u, and $\mathbf{q}_{\mathbf{v}} = \mathbf{e}_n$. Substituting these into (7.32) shows that

$$1 = \mathbf{a}_u^{(i)}\mathbf{e}_n,$$

as desired. Thus the polyhedral set $\mathcal{C}$ consisting of all convex combinations of the $q^{(i)}$ is stable. Finally, it is obvious from (7.20) that p is a convex combination of the $q^{(i)}$ and thus belongs to $\mathcal{C}$. □

7.4 EXISTENCE OF REGULAR QUASI-REALIZATIONS

In this section, we study processes whose Hankel rank is finite, and show that it is *always* possible to construct a "quasi-realization" of such a process. Moreover, any two *regular* quasi-realizations of a finite Hankel rank process are related through a similarity transformation.

Definition 7.4 *Suppose a process* $\{Y_t\}$ *has finite Hankel rank* r. *Suppose* $n \geq r$, $\mathbf{x}$ *is a row vector in* $\mathbb{R}^n$, $\mathbf{y}$ *is a column vector in* $\mathbb{R}^n$, *and* $C^{(u)} \in \mathbb{R}^{n \times n}\ \forall u \in \mathbb{M}$. *Then we say that* $\{n, \mathbf{x}, \mathbf{y}, C^{(u)}, u \in \mathbb{M}\}$ *is a* **quasi-realization** *of the process if three conditions hold. First,*

$$f_{\mathbf{u}} = \mathbf{x} C^{(u_1)} \ldots C^{(u_l)} \mathbf{y}\ \forall \mathbf{u} \in \mathbb{M}^*, \tag{7.34}$$

where $l = |\mathbf{u}|$. *Second,*

$$\mathbf{x} \left[\sum_{u \in \mathbb{M}} C^{(u)} \right] = \mathbf{x}. \tag{7.35}$$

Third,

$$\left[\sum_{u \in \mathbb{M}} C^{(u)} \right] \mathbf{y} = \mathbf{y}. \tag{7.36}$$

We say that $\{n, \mathbf{x}, \mathbf{y}, C^{(u)}, \mathbf{u} \in \mathbb{M}\}$ *is a* **regular quasi-realization** *of the process if* $n = r$, *the rank of the Hankel matrix.*

The formula (7.34) is completely analogous to (7.3). Similarly, (7.35) and (7.36) are analogous to (7.4). The only difference is that the various quantities are not required to be nonnegative. This is why we speak of a "quasi-realization" instead of a true realization. With this notion, it is possible to prove the following powerful statements:

1. Suppose the process $\{Y_t\}$ has finite Hankel rank, say r. Then the process always has a regular quasi-realization.

2. Suppose the process $\{Y_t\}$ has finite Hankel rank r. Suppose further that $\{\boldsymbol{\theta}_1, \boldsymbol{\phi}_1, D_1^{(u)}, u \in \mathbb{M}\}$ and $\{\boldsymbol{\theta}_2, \boldsymbol{\phi}_2, D_2^{(u)}, u \in \mathbb{M}\}$ are two regular quasi-realizations of this process. Then there exists a nonsingular matrix T such that

$$\boldsymbol{\theta}_2 = \boldsymbol{\theta}_1 T^{-1}, D_2^{(u)} = T D_1^{(u)} T^{-1}\ \forall u \in \mathbb{M}, \boldsymbol{\phi}_2 = T \boldsymbol{\phi}_1.$$

These two statements are formally stated and proven as Theorem 7.8 and Theorem 7.9 respectively.

The results of this section are not altogether surprising. Given that the infinite matrix H has finite rank, it is clear that there *must exist* recursive relationships between its various elements. Earlier work, most notably [43, 28], contains some such recursive relationships. However, the present formulas are the cleanest, and also the closest to the conventional formula

(7.3). Note that Theorem 7.8 is more or less contained in the work of Erickson [51]. In [70], the authors generalize the work of Erickson by studying the relationship between two quasi-realizations, *without* assuming that the underlying state spaces have the same dimension. In this case, in place of the similarity transformation above, they obtain "intertwining" conditions of the form $D_2^{(u)} T = T D_1^{(u)}$, where the matrix T may now be rectangular. In the interests of simplicity, in the present book we do not study this more general case. Moreover, the above formulas are the basis for the construction of a "true" (as opposed to quasi) HMM realization in subsequent sections.

Some notation is introduced to facilitate the subsequent proofs. Suppose k, l are integers, and $I \subseteq \mathbb{M}^k, J \subseteq \mathbb{M}^l$; thus every element of I is a string of length k, while every element of J is a string of length l. Specifically, suppose $I = \{\mathbf{i}_1, \ldots, \mathbf{i}_{|I|}\}$, and $J = \{\mathbf{j}_1, \ldots, \mathbf{j}_{|J|}\}$. Then we define

$$F_{I,J} := \begin{bmatrix} f_{\mathbf{i}_1\mathbf{j}_1} & f_{\mathbf{i}_1\mathbf{j}_2} & \cdots & f_{\mathbf{i}_1\mathbf{j}_{|J|}} \\ f_{\mathbf{i}_2\mathbf{j}_1} & f_{\mathbf{i}_2\mathbf{j}_2} & \cdots & f_{\mathbf{i}_2\mathbf{j}_{|J|}} \\ \vdots & \vdots & \vdots & \vdots \\ f_{\mathbf{i}_{|I|}\mathbf{j}_1} & f_{\mathbf{i}_{|I|}\mathbf{j}_2} & \cdots & f_{\mathbf{i}_{|I|}\mathbf{j}_{|J|}} \end{bmatrix}. \tag{7.37}$$

Thus $F_{I,J}$ is a submatrix of $F_{k,l}$ and has dimension $|I| \times |J|$. This notation is easily reconciled with the earlier notation. Suppose k, l are integers. Then we can think of $F_{k,l}$ as shorthand for $F_{\mathbb{M}^k,\mathbb{M}^l}$. In the same spirit, if I is a subset of $\mathbb{M}^k$ and l is an integer, we use the "mixed" notation $F_{I,l}$ to denote $F_{I,\mathbb{M}^l}$. This notation can be extended in an obvious way to the case where either k or l equals zero. If $l = 0$, we have that $\mathbb{M}^0 := \{\emptyset\}$. In this case

$$F_{I,0} := [f_{\mathbf{i}} : \mathbf{i} \in I] \in \mathbb{R}^{|I| \times 1}.$$

Similarly if $J \subseteq \mathbb{M}^l$ for some integer l, then

$$F_{0,J} := [f_{\mathbf{j}} : \mathbf{j} \in J] \in \mathbb{R}^{1 \times |J|}.$$

Finally, given any string $\mathbf{u} \in \mathbb{M}^*$, we define

$$F_{k,l}^{(\mathbf{u})} := [f_{\mathbf{iuj}}, \mathbf{i} \in \mathbb{M}^k \text{ in flo}, \mathbf{j} \in \mathbb{M}^l \text{ in llo}], \tag{7.38}$$

$$F_{I,J}^{(\mathbf{u})} := [f_{\mathbf{iuj}}, \mathbf{i} \in I, \mathbf{j} \in J]. \tag{7.39}$$

Lemma 7.5 *Suppose H has finite rank. Then there exists a smallest integer k such that*

$$Rank(F_{k,k}) = Rank(H).$$

Moreover, for this k, we have

$$Rank(F_{k,k}) = Rank(H_{k+l,k+s}), \ \forall l, s \geq 0. \tag{7.40}$$

Proof. We begin by observing that, for every pair of integers k, l, we have

$$\text{Rank}(H_{k,l}) = \text{Rank}(F_{k,l}). \tag{7.41}$$

To see this, observe that the row indexed by $\mathbf{u} \in \mathbb{M}^{k-1}$ in $F_{k-1,s}$ is the sum of the rows indexed by $v\mathbf{u}$ in $F_{k,s}$, for each s. This follows from (7.1).

Similarly each row in $F_{k-2,s}$ is the sum of m rows in $F_{k-1,s}$ and thus of m^2 rows of $F_{k,s}$, and so on. Thus it follows that every row of $F_{t,s}$ for $t < k$ is a sum of m^{k-t} rows of $F_{k,s}$. Therefore

$$\text{Rank}(H_{k,l}) = \text{Rank}([F_{k,0} \;\; F_{k,1} \;\; \ldots F_{k,l}]).$$

Now repeat the same argument for the columns of this matrix. Every column of $F_{k,t}$ is the sum of m^{k-t} columns of $F_{k,l}$. This leads to the desired conclusion (7.41).

To complete the proof, observe that, since $H_{l,l}$ is a submatrix of $H_{l+1,l+1}$, we have that

$$\text{Rank}(H_{1,1}) \leq \text{Rank}(H_{2,2}) \leq \ldots \leq \text{Rank}(H).$$

Now at each step, there are only two possibilities: Either $\text{Rank}(H_{l,l}) < \text{Rank}(H_{l+1,l+1})$, or else $\text{Rank}(H_{l,l}) = \text{Rank}(H_{l+1,l+1})$. Since $\text{Rank}(H)$ is finite, the first possibility can only occur finitely many times. Hence there exists a smallest integer k such that

$$\text{Rank}(H_{k,k}) = \text{Rank}(H).$$

We have already shown that $\text{Rank}(H_{k,k}) = \text{Rank}(F_{k,k})$. Finally, since $H_{k+l,k+s}$ is a submatrix of H and contains $H_{k.k}$ as a submatrix, the desired conclusion (7.40) follows. □

Note: Hereafter, the symbol k is used *exclusively* for this integer and nothing else. Similarly, hereafter the symbol r is used exclusively for the (finite) rank of the Hankel matrix H and nothing else.

Now consider the matrix $F_{k,k}$, which is chosen so as to have rank r. Thus there exist sets $I, J \subseteq \mathbb{M}^k$, such that $|I| = |J| = r$ and $F_{I,J}$ has rank r. (Recall the definition of the matrix $F_{I,J}$ from (7.37).) In other words, the index sets I, J are chosen such that $F_{I,J}$ is any full rank nonsingular submatrix of $F_{k,k}$. Of course the choice of I and J may not be unique. However, once I, J are chosen, there exist *unique* matrices $U \in \mathbb{R}^{m^k \times r}, V \in \mathbb{R}^{r \times m^k}$ such that $F_{k,k} = UF_{I,J}V$. Hereafter, the symbols U, V are used only for these matrices and nothing else.

The next lemma shows that, once the index sets I, J are chosen (thus fixing the matrices U and V), the relationship $F_{k,k} = UF_{I,J}V$ can be extended to strings of *arbitrary* lengths.

Lemma 7.6 *With the various symbols defined as above, we have*

$$F_{k,k}^{(\mathbf{u})} = UF_{I,J}^{(\mathbf{u})}V, \; \forall \mathbf{u} \in \mathbb{M}^*. \tag{7.42}$$

This result can be compared to [6, Lemma 1].

Proof. For notational convenience only, let us suppose I, J consist of the first r elements of $\mathbb{M}^k$. The more general case can be handled through more messy notation. The matrix U can be partitioned as follows:

$$U = \begin{bmatrix} I_r \\ \bar{U} \end{bmatrix}.$$

This is because $F_{I,k}$ is a submatrix of $F_{k,k}$. (In general we would have to permute the indices so as to bring the elements of I to the first r positions.) Now, by the rank condition and the assumption that $F_{k,k} = UF_{I,J}V = UF_{I,k}$, it follows that

$$\begin{bmatrix} I_r & \mathbf{0} \\ -\bar{U} & I_{m^k-r} \end{bmatrix} H_{k,.} = \begin{bmatrix} F_{I,k} & F_{I,.} \\ \mathbf{0} & F_{\mathbb{M}^k \setminus I,.} - \bar{U}F_{I,.} \end{bmatrix},$$

where

$$F_{I,.} = [F_{I,k+1} \;\; F_{I,k+2} \ldots], \text{ and } F_{\mathbb{M}^k \setminus I,.} = [F_{\mathbb{M}^k \setminus I,k+1} \;\; F_{\mathbb{M}^k \setminus I,k+2} \ldots].$$

This expression allows us to conclude that

$$F_{\mathbb{M}^k \setminus I,.} = \bar{U}F_{I,.}. \tag{7.43}$$

Otherwise the $(2,2)$-block of the above matrix would contain some nonzero element, which would in turn imply that $\text{Rank}(H_{k,.}) > r$, a contradiction. Now the above relationship implies that

$$F_{k,k}^{(\mathbf{u})} = UF_{I,k}^{(\mathbf{u})}, \; \forall \mathbf{u} \in \mathbb{M}^*.$$

Next, as with U, partition V as $V = [I_r \;\; \bar{V}]$. (In general, we would have to permute the columns to bring the elements of J to the first positions.) Suppose $N > k$ is some integer. Observe that $F_{N,k}$ is just $[F_{k,k}^{(\mathbf{u})}, \mathbf{u} \in \mathbb{M}^{N-k} \text{ in flo}]$. Hence

$$\begin{bmatrix} I_r & \mathbf{0} \\ -\bar{U} & I_{m^k-r} \end{bmatrix} H_{.,k} = \begin{bmatrix} F_{I,k}^{(\mathbf{u})}, \mathbf{u} \in \mathbb{M}^* \text{ in flo} \\ \mathbf{0} \end{bmatrix}$$

$$= \left[\begin{array}{c} F_{I,k} \\ \mathbf{0} \\ \hline F_{I,k}^{(\mathbf{u})}, \mathbf{u} \in \mathbb{M}^* \setminus \emptyset \text{ in flo} \\ \mathbf{0} \end{array}\right].$$

Now post-multiply this matrix as shown below:

$$\begin{bmatrix} I_r & \mathbf{0} \\ -\bar{U} & I_{m^k-r} \end{bmatrix} H_{.,k} \begin{bmatrix} I_r & -\bar{V} \\ \mathbf{0} & I_{m^k-r} \end{bmatrix}$$

$$= \begin{bmatrix} F_{I,J} & \mathbf{0} \\ \mathbf{0} & \mathbf{0} \\ F_{I,J}^{(\mathbf{u})} & F_{I,M^k \setminus J}^{(\mathbf{u})} - F_{I,J}^{(\mathbf{u})}\bar{V}, \mathbf{u} \in \mathbb{M}^* \text{in flo} \\ \mathbf{0} & \mathbf{0} \end{bmatrix}.$$

So if $F_{I,M^k \setminus J}^{(\mathbf{u})} - F_{I,J}^{(\mathbf{u})}\bar{V} \neq 0$ for some $\mathbf{u} \in \mathbb{M}^*$, then $\text{Rank}(H_{.,k})$ would exceed $\text{Rank}(F_{I,J})$, which is a contradiction. Thus it follows that

$$F_{I,M^k \setminus J}^{(\mathbf{u})} = F_{I,J}^{(\mathbf{u})}\bar{V}, \; \forall \mathbf{u} \in \mathbb{M}^*. \tag{7.44}$$

The two relationships (7.43) and (7.44) can together be compactly expressed as (7.42), which is the desired conclusion. □

Lemma 7.7 *Choose unique matrices* $\bar{D}^{(u)}, u \in \mathbb{M}$, *such that*

$$F_{I,J}^{(u)} = F_{I,J}\bar{D}^{(u)}, \ \forall u \in \mathbb{M}. \tag{7.45}$$

Then for all $\mathbf{u} \in \mathbb{M}^*$, *we have*

$$F_{I,J}^{(\mathbf{u})} = F_{I,J}\bar{D}^{(u_1)} \ldots \bar{D}^{(u_l)}, \ \textit{where } l = |\mathbf{u}|. \tag{7.46}$$

Choose unique matrices $D^{(u)}, u \in \mathbb{M}$, *such that*

$$F_{I,J}^{(u)} = D^{(u)}F_{I,J}, \ \forall u \in \mathbb{M}. \tag{7.47}$$

Then for all $\mathbf{u} \in \mathbb{M}^*$, *we have*

$$F_{I,J}^{(\mathbf{u})} = D^{(u_1)} \ldots D^{(u_l)}F_{I,J}, \ \textit{where } l = |\mathbf{u}|. \tag{7.48}$$

This result can be compared to [6, Theorem 1].

Proof. We prove only (7.46), since the proof of (7.48) is entirely similar. By the manner in which the index sets I, J are chosen, we have

$$\text{Rank}[F_{I,J} \ \ F_{I,J}^{(u)}] = \text{Rank}[F_{I,J}], \ \forall u \in \mathbb{M}.$$

Hence there exist unique matrices $\bar{D}^{(u)}, u \in \mathbb{M}$ such that (7.45) holds. Now suppose $\mathbf{v}$ is any *nonempty* string in $\mathbb{M}^*$. Then, since $F_{I,J}$ is a maximal rank submatrix of H, it follows that

$$\text{Rank}\begin{bmatrix} F_{I,J} & F_{I,J}^{(u)} \\ F_{I,J}^{(\mathbf{v})} & F_{I,J}^{(\mathbf{v}u)} \end{bmatrix} = \text{Rank}\begin{bmatrix} F_{I,J} \\ F_{I,J}^{(\mathbf{v})} \end{bmatrix}, \ \forall u \in \mathbb{M}.$$

Now post-multiply the matrix on the left side as shown below:

$$\begin{bmatrix} F_{I,J} & F_{I,J}^{(u)} \\ F_{I,J}^{(\mathbf{v})} & F_{I,J}^{(\mathbf{v}u)} \end{bmatrix}\begin{bmatrix} I & -\bar{D}^{(u)} \\ \mathbf{0} & I \end{bmatrix} = \begin{bmatrix} F_{I,J} & \mathbf{0} \\ F_{I,J}^{(\mathbf{v})} & F_{I,J}^{(\mathbf{v}u)} - F_{I,J}^{(\mathbf{v})}\bar{D}^{(u)} \end{bmatrix}.$$

This shows that

$$F_{I,J}^{(\mathbf{v}u)} = F_{I,J}^{(\mathbf{v})}\bar{D}^{(u)}, \ \forall \mathbf{v} \in \mathbb{M}^*, \ \forall u \in \mathbb{M}. \tag{7.49}$$

Otherwise, the (2,2)-block of the matrix on the right side would contain a nonzero element and would therefore have rank larger than that of $F_{I,J}$, which would be a contradiction. Note that if $\mathbf{v}$ is the empty string in (7.49), then we are back to the definition of the matrix $\bar{D}^{(u)}$. Now suppose $\mathbf{u} \in \mathbb{M}^*$ has length l and apply (7.49) recursively. This leads to the desired formula (7.46). □

Suppose $\mathbf{u} \in \mathbb{M}^*$ has length l. Then it is natural to define

$$\bar{D}^{(\mathbf{u})} := \bar{D}^{(u_1)} \ldots \bar{D}^{(u_l)}, \ D^{(\mathbf{u})} := D^{(u_1)} \ldots D^{(u_l)}.$$

With this notation let us observe that the matrices $D^{(\mathbf{u})}$ and $\bar{D}^{(\mathbf{u})}$ "intertwine" with the matrix $F_{I,J}$. That is,

$$F_{I,J}\bar{D}^{(\mathbf{u})} = D^{(\mathbf{u})}F_{I,J}, \ \text{and} \ F_{I,J}^{-1}D^{(\mathbf{u})} = \bar{D}^{(\mathbf{u})}F_{I,J}^{-1}. \tag{7.50}$$

This follows readily from the original relationship

$$F_{I,J}\bar{D}^{(u)} = D^{(u)}F_{I,J} (= F_{I,J}^{(u)}) \ \forall u \in \mathbb{M}$$

applied recursively.

Finally we come to the main theorem about quasi-realizations. We begin by formalizing the notion.

Note that a regular quasi-realization in some sense completes the analogy with the formulas (7.3) and (7.4).

Theorem 7.8 *Suppose the process $\{Y_t\}$ has finite Hankel rank, say r. Then the process always has a regular quasi-realization. In particular, choose the integer k as in Lemma 7.5, and choose index sets $I, J \subseteq \mathbb{M}^k$ such that $|I| = |J| = r$ and $F_{I,J}$ has rank r. Define the matrices $U, V, D^{(u)}, \bar{D}^{(u)}$ as before. The following two choices are regular quasi-realizations. First, let*

$$\mathbf{x} = \boldsymbol{\theta} := F_{0,J}F_{I,J}^{-1}, \ \mathbf{y} = \boldsymbol{\phi} := F_{I,0}, \ C^{(u)} = D^{(u)} \ \forall u \in \mathbb{M}. \tag{7.51}$$

Second, let

$$\mathbf{x} = \bar{\boldsymbol{\theta}} := F_{0,J}, \ \mathbf{y} = \bar{\boldsymbol{\phi}} := F_{I,J}^{-1}F_{I,0}, \ C^{(u)} = \bar{D}^{(u)} \ \forall u \in \mathbb{M}. \tag{7.52}$$

This result can be compared to [6, Theorems 1 and 2].

Proof. With all the spade work done already, the proof is very simple. For any string $\mathbf{u} \in \mathbb{M}^*$, it follows from (7.47) that

$$F_{I,J}^{(\mathbf{u})} = D^{(u_1)} \ldots D^{(u_l)}F_{I,J}, \text{ where } l = |\mathbf{u}|.$$

Next, we have from (7.42) that

$$F_{k,k}^{(\mathbf{u})} = UF_{I,J}^{(\mathbf{u})}V, \ \forall \mathbf{u} \in \mathbb{M}^*.$$

Now observe that, by definition, we have

$$f_{\mathbf{u}} = \sum_{\mathbf{i} \in \mathbb{M}^k} \sum_{\mathbf{j} \in \mathbb{M}^k} f_{\mathbf{iuj}} = \mathbf{e}_{m^k}^t F_{k,k}^{(\mathbf{u})} \mathbf{e}_{m^k} = \mathbf{e}_{m^k}^t UD^{(u_1)} \ldots D^{(u_l)}F_{I,J}V\mathbf{e}_{m^k},$$

where $\mathbf{e}_{m^k}$ is the column vector with m^k ones. Hence (7.34) is satisfied with the choice

$$n = r, \boldsymbol{\theta} := \mathbf{e}_{m^k}^t U, \boldsymbol{\phi} := F_{I,J}V\mathbf{e}_{m^k}, C^{(u)} = D^{(u)} \ \forall u \in \mathbb{M},$$

and the matrices $D^{(u)}$ as defined in (7.47). Since $D^{(\mathbf{u})}F_{I,J} = F_{I,J}\bar{D}^{(\mathbf{u})}$, we can also write

$$f_{\mathbf{u}} = \mathbf{e}_{m^k}^t UF_{I,J}\bar{D}^{(u_1)} \ldots \bar{D}^{(u_l)}V\mathbf{e}_{m^k}.$$

Hence (7.34) is also satisfied with the choice

$$n = r, \bar{\boldsymbol{\theta}} := \mathbf{e}_{m^k}^t UF_{I,J}, \bar{\boldsymbol{\phi}} := V\mathbf{e}_{m^k}, C^{(u)} = \bar{D}^{(u)} \ \forall u \in \mathbb{M},$$

and the matrices $\bar{D}^{(u)}$ as defined in (7.46).

Next, we show that the vectors $\boldsymbol{\theta}, \boldsymbol{\phi}, \bar{\boldsymbol{\theta}}, \bar{\boldsymbol{\phi}}$ can also be written as in (7.51) and (7.52). For this purpose, we proceed as follows:

$$\boldsymbol{\theta} = \mathbf{e}_{m^k}^t U = \mathbf{e}_{m^k}^t UF_{I,J}F_{I,J}^{-1} = \mathbf{e}_{m^k}^t F_{k,J}F_{I,J}^{-1} = F_{0,J}F_{I,J}^{-1}.$$

Therefore

$$\bar{\boldsymbol{\theta}} = \boldsymbol{\theta} F_{I,J} = F_{0,J}.$$

Similarly

$$\boldsymbol{\phi} = F_{I,J} V \mathbf{e}_{m^k} = F_{I,k} \mathbf{e}_{m^k} = F_{I,0},$$

and

$$\bar{\boldsymbol{\phi}} = F_{I,J}^{-1} F_{I,0}.$$

It remains only to prove the eigenvector properties. For this purpose, note that, for each $u \in \mathbb{M}$, we have

$$F_{0,J} \bar{D}^{(u)} = \mathbf{e}_{m^k}^t U F_{I,J} \bar{D}^{(u)} = \mathbf{e}_{m^k}^t U F_{I,J}^{(u)} = F_{0,J}^{(u)}.$$

Now

$$\boldsymbol{\theta} D^{(u)} = F_{0,J} F_{I,J}^{-1} D^{(u)} = F_{0,J} \bar{D}^{(u)} F_{I,J}^{-1} = F_{0,J}^{(u)} F_{I,J}^{-1}.$$

Hence

$$\boldsymbol{\theta} \left[\sum_{u \in \mathbb{M}} D^{(u)} \right] = \sum_{u \in \mathbb{M}} \boldsymbol{\theta} D^{(u)} = \sum_{u \in \mathbb{M}} F_{0,J}^{(u)} F_{I,J}^{-1} = F_{0,J} F_{I,J}^{-1} = \boldsymbol{\theta},$$

since

$$\sum_{u \in \mathbb{M}} F_{0,J}^{(u)} = F_{0,J}.$$

As for $\boldsymbol{\phi}$, we have

$$D^{(u)} \boldsymbol{\phi} = D^{(u)} F_{I,J} V \mathbf{e}_{m^k} = F_{I,J}^{(u)} V \mathbf{e}_{m^k} = F_{I,k}^{(u)} \mathbf{e}_{m^k} = F_{I,0}^{(u)}.$$

Hence

$$\left[\sum_{u \in \mathbb{M}} D^{(u)} \right] \boldsymbol{\phi} = \sum_{u \in \mathbb{M}} D^{(u)} \boldsymbol{\phi} = \sum_{u \in \mathbb{M}} F_{I,0}^{(u)} = F_{I,0} = \boldsymbol{\phi}.$$

This shows that $\{r, \boldsymbol{\theta}, \boldsymbol{\phi}, D^{(u)}\}$ is a quasi-realization. The proof in the case of the barred quantities is entirely similar. We have

$$\bar{\boldsymbol{\theta}} \bar{D}^{(u)} = F_{0,J} \bar{D}^{(u)} = F_{0,J}^{(u)},$$

so

$$\bar{\boldsymbol{\theta}} \left[\sum_{u \in \mathbb{M}} \bar{D}^{(u)} \right] = \sum_{u \in \mathbb{M}} F_{0,J}^{(u)} = F_{0,J} = \bar{\boldsymbol{\theta}}.$$

It can be shown similarly that

$$\left[\sum_{u \in \mathbb{M}} \bar{D}^{(u)} \right] \bar{\boldsymbol{\phi}} = \bar{\boldsymbol{\phi}}.$$

This concludes the proof. □

Next, it is shown that any two "regular" quasi-realizations of the process are related through a similarity transformation.

Theorem 7.9 *Suppose a process* $\{Y_t\}$ *has finite Hankel rank* r*, and suppose* $\{\boldsymbol{\theta}_1, \boldsymbol{\phi}_1, D_1^{(u)}, u \in \mathbb{M}\}$ *and* $\{\boldsymbol{\theta}_2, \boldsymbol{\phi}_2, D_2^{(u)}, u \in \mathbb{M}\}$ *are two regular quasi-realizations of this process. Then there exists a nonsingular matrix* T *such that*

$$\boldsymbol{\theta}_2 = \boldsymbol{\theta}_1 T^{-1}, D_2^{(u)} = T D_1^{(u)} T^{-1} \ \forall u \in \mathbb{M}, \boldsymbol{\phi}_2 = T\boldsymbol{\phi}_1.$$

Proof. Suppose the process has finite Hankel rank, and let r denote the rank of H. Choose the integer k as before, namely, the smallest integer k such that $\text{Rank}(F_{k,k}) = \text{Rank}(H)$. Choose subsets $I, J \subseteq \mathbb{M}^k$ such that $|I| = |J| = r$ and $\text{Rank}(F_{I,J}) = r$. Up to this point, all entities depend only on the process and its Hankel matrix (which depends on the law of the process), and not on the specific quasi-realization. Moreover, the fact that I, J are not unique is not important.

Now look at the matrix $F_{I,J}$, and express it in terms of the two quasi-realizations. By definition,

$$F_{I,J} = \begin{bmatrix} f_{\mathbf{i}_1\mathbf{j}_1} & \cdots & f_{\mathbf{i}_1\mathbf{j}_r} \\ \vdots & \vdots & \vdots \\ f_{\mathbf{i}_r\mathbf{j}_1} & \cdots & f_{\mathbf{i}_r\mathbf{j}_r} \end{bmatrix}.$$

Now, since we are given two quasi-realizations, the relationship (7.34) holds for each quasi-realization. Hence

$$F_{I,J} = \begin{bmatrix} \boldsymbol{\theta}_s D_s^{(\mathbf{i}_1)} \\ \vdots \\ \boldsymbol{\theta}_s D_s^{(\mathbf{i}_r)} \end{bmatrix} [D_s^{(\mathbf{j}_1)}\boldsymbol{\phi}_s \ldots D_s^{(\mathbf{j}_r)}\boldsymbol{\phi}_s], \text{ for } s = 1, 2.$$

Define

$$P_s := \begin{bmatrix} \boldsymbol{\theta}_s D_s^{(\mathbf{i}_1)} \\ \vdots \\ \boldsymbol{\theta}_s D_s^{(\mathbf{i}_r)} \end{bmatrix}, \ Q_s := [D_s^{(\mathbf{j}_1)}\boldsymbol{\phi}_s \ldots D_s^{(\mathbf{j}_r)}\boldsymbol{\phi}_s], \text{ for } s = 1, 2.$$

Then $F_{I,J} = P_1 Q_1 = P_2 Q_2$. Since $F_{I,J}$ is nonsingular, so are P_1, Q_1, P_2, Q_2. Moreover,

$$P_2^{-1} P_1 = Q_2 Q_1^{-1} =: T, \text{ say.}$$

Next, fix $u \in \mathbb{M}$ and consider the $r \times r$ matrix $F_{I,J}^{(u)}$. We have from (7.34) that

$$F_{I,J}^{(u)} = P_1 D_1^{(u)} Q_1 = P_2 D_2^{(u)} Q_2.$$

Hence

$$D_2^{(u)} = P_2^{-1} P_1 D_1^{(u)} Q_1 Q_2^{-1} = T D_1^{(u)} T^{-1}, \ \forall u \in \mathbb{M}.$$

Finally, we can factor the entire matrix H as

$$H = [\boldsymbol{\theta}_s D_s^{(\mathbf{u})}, \mathbf{u} \in \mathbb{M}^* \text{ in flo}][D_s^{(\mathbf{v})}\boldsymbol{\phi}_s, \mathbf{v} \in \mathbb{M}^* \text{ in llo}], \ s = 1, 2,$$

where

$$D^{(\mathbf{u})} := D^{(u_1)} \ldots D^{(u_l)}, \; l = |\mathbf{u}|,$$

and $D^{(\mathbf{v})}$ is defined similarly. Note that the first matrix in the factorization of H has r columns and infinitely many rows, while the second matrix has r rows and infinitely many columns. Thus there exists a nonsingular matrix, say S, such that

$$[\boldsymbol{\theta}_2 D_2^{(\mathbf{u})}, \mathbf{u} \in \mathbb{M}^* \text{ in flo}] = [\boldsymbol{\theta}_1 D_1^{(\mathbf{u})}, \mathbf{u} \in \mathbb{M}^* \text{ in flo}] S^{-1},$$

and

$$[D_2^{(\mathbf{v})} \boldsymbol{\phi}_2, \mathbf{v} \in \mathbb{M}^* \text{ in llo}] = S[D_1^{(\mathbf{v})} \boldsymbol{\phi}_1, \mathbf{v} \in \mathbb{M}^* \text{ in llo}].$$

Choosing $\mathbf{u} = \mathbf{i}_1, \ldots, \mathbf{i}_r$ and $\mathbf{v} = \mathbf{j}_1, \ldots, \mathbf{j}_r$ shows that in fact $S = T$. Finally, choosing $\mathbf{u} = \mathbf{v} = \emptyset$ shows that

$$\boldsymbol{\theta}_2 = \boldsymbol{\theta}_1 T^{-1}, \; \boldsymbol{\phi}_2 = T \boldsymbol{\phi}_1.$$

This concludes the proof. □

We conclude this section with an example from [43] of a regular quasi-realization that does not correspond to a regular realization.

Let $n = 4$, and define the 4×4 "state transition matrix"

$$A = \begin{bmatrix} \lambda_1 & 0 & 0 & 1-\lambda_1 \\ 0 & -\lambda_2 & 0 & 1+\lambda_2 \\ 0 & 0 & -\lambda_3 & 1+\lambda_3 \\ 1-\lambda_1 & c(1+\lambda_2) & -c(1+\lambda_3) & \lambda_1 + c(\lambda_3 - \lambda_2) \end{bmatrix},$$

as well as the "output matrix"

$$B = \begin{bmatrix} 1 & 0 \\ 1 & 0 \\ 1 & 0 \\ 0 & 1 \end{bmatrix}.$$

It is easy to see that $A\mathbf{e}_4 = \mathbf{e}_4$, that is, the matrix A is "stochastic." Similarly $B\mathbf{e}_2 = \mathbf{e}_2$ and so B is stochastic (without quotes). Let $\mathbf{b}_i$ denote the i-th column of B, and let $\text{Diag}(\mathbf{b}_i)$ denote the diagonal 4×4 matrix with the elements of $\mathbf{b}_i$ on the diagonal. Let us define

$$C^{(1)} = A\text{Diag}(\mathbf{b}_1) = \begin{bmatrix} \lambda_1 & 0 & 0 & 0 \\ 0 & -\lambda_2 & 0 & 0 \\ 0 & 0 & -\lambda_3 & 0 \\ 1-\lambda_1 & c(1+\lambda_2) & -c(1+\lambda_3) & 0 \end{bmatrix},$$

$$C^{(2)} = A\text{Diag}(\mathbf{b}_2) = \begin{bmatrix} 0 & 0 & 0 & 1-\lambda_1 \\ 0 & 0 & 0 & 1+\lambda_2 \\ 0 & 0 & 0 & 1+\lambda_3 \\ 0 & 0 & 0 & \lambda_1 + c(\lambda_3 - \lambda_2) \end{bmatrix}.$$

Then $C^{(1)} + C^{(2)} = A$. Note that

$$\mathbf{x} = [0.5 \;\; 0.5c \;\; -0.5c \;\; 0.5]$$

is a "stationary distribution" of A; that is, $\mathbf{x}A = \mathbf{x}$. With these preliminaries, we can define the "quasi-frequencies"

$$f_{\mathbf{u}} = \mathbf{x}C^{(u_1)} \ldots C^{(u_l)}\mathbf{e}_4,$$

where $\mathbf{u} = u_1 \ldots u_l$. Because $\mathbf{x}$ and $\mathbf{e}_4$ are respectively row and column eigenvectors of A corresponding to the eigenvalue one, these quasi-frequencies satisfy the consistency conditions (7.1) and (7.2). Thus, in order to qualify as a quasi-realization, the only thing missing is the property that $f_{\mathbf{u}} \geq 0$ for all strings $\mathbf{u}$.

This nonnegativity property is established in [43] using a Markov chain analogy, and is not reproduced here. All the frequencies will all be nonnegative provided the following inequalities are satisfied:

$$0 < \lambda_i < 1, i = 1, 2, 3; \lambda_1 > \lambda_i, i = 2, 3; 0 < c < 1,$$

$$\lambda_1 + c(\lambda_3 - \lambda_2) > 0; (1 - \lambda_1)^k > c(1 + \lambda_i)^k, i = 2, 3, k = 1, 2.$$

One possible choice (given in [43]) is

$$\lambda_1 = 0.5, \lambda_2 = 0.4, \lambda_3 = 0.3, c = 0.06.$$

Thus the above is a quasi-realization.

To test whether this quasi-realization can be made into a realization (with nonnegative elements), we can make use of Theorem 7.9. *All possible* quasi-realizations of this process can be obtained by performing a similarity transformation on the above quasi-realization. Thus there exists a regular realization (not quasi-realization) of this process if and only if there exists a nonsingular matrix T such that $\mathbf{x}T^{-1}, TC^{(i)}T^{-1}, T\mathbf{e}_4$ are all nonnegative. This can in turn be written as the feasibility of the following set of inequalities:

$$\pi T = \mathbf{x}; TC^{(i)} = M^{(i)}T, i = 1, 2; T\mathbf{e}_4 = \mathbf{e}_4; M^{(i)} \geq \mathbf{0}, i = 1, 2; \pi \geq \mathbf{0}.$$

It can be readily verified that the above set of inequalities is *not* feasible, so that there is no regular realization for this process, only regular quasi-realizations.

As pointed out above, it is possible to check whether a given regular quasi-realization of a stationary process can be converted into a regular realization. There is a related problem that one can examine, namely: Suppose one is given a triplet $\{\mathbf{x}, C^{(u)}, u \in \mathbb{M}, \mathbf{y}\}$ with compatible dimensions. The problem is to determine whether

$$f_{\mathbf{u}} := \mathbf{x}C^{(\mathbf{u})}\mathbf{y} = \mathbf{x}C^{(u_1)} \cdots C^{(u_l)}\mathbf{y} \geq 0 \; \forall \mathbf{u} \in \mathbb{M}^l, \; \forall l.$$

This problem can be viewed as one of deciding whether a given rational power series always has nonnegative coefficients. This problem is known to be undecidable; see [111, Theorem 3.13]. Even if $m = 2$, the above problem is undecidable if $n \geq 50$, where n is the size of the vector $\mathbf{x}$. The arguments of [21] can be adapted to prove this claim. Most likely the problem remains undecidable even if we add the additional requirements that

$$\mathbf{x}\left[\sum_{u \in \mathbb{M}} C^{(u)}\right] = \mathbf{x},$$

$$\left[\sum_{u \in \mathbb{M}} C^{(u)} \right] \mathbf{y} = \mathbf{y},$$

because the above two conditions play no role in determining the nonnegativity or otherwise of the "quasi-frequencies" $f_{\mathbf{u}}$, but serve only to ensure that these quasi-frequencies are stationary.

7.5 SPECTRAL PROPERTIES OF ALPHA-MIXING PROCESSES

In this section, we add the assumption that the finite Hankel rank process under study is also α-mixing, and show that the regular quasi-realizations have an additional property, namely: the matrix that plays the role of the state transition matrix in the HMM has a spectral radius of one, this eigenvalue is simple, and all other eigenvalues have magnitude strictly less than one. This property is referred to as the "quasi-strong Perron property." As a corollary, it follows that if an α-mixing process has a regular realization (and not just a quasi-realization), then the underlying Markov chain is irreducible and aperiodic.

We begin by defining the notion of α-mixing. Suppose the process $\{Y_t\}$ is defined on the probability space (S, Ω), where Ω is a σ-algebra on the set S. For each pair of indices s, t with $s < t$, define Σ_s^t to be the σ-algebra generated by the random variables $Y_s, \ldots, Y_t$, and note that Σ_s^t is a subalgebra of Ω. Then the **α-mixing coefficient** $\alpha(l)$ of the process $\{Y_t\}$ is defined as

$$\alpha(l) := \sup_{A \in \Sigma_0^t, B \in \Sigma_{t+l}^{\infty}} |P(A \cap B) - P(A)P(B)|.$$

The process $\{Y_t\}$ is said to be **α-mixing** if $\alpha(l) \to 0$ as $l \to \infty$. Note that in the definition above, A is an event that depends strictly on the "past" random variables before time t, whereas B is an event that depends strictly on the "future" random variables after time $t + l$. If the future were to be completely independent of the past, we would have $P(A \cap B) = P(A)P(B)$. Thus the α-mixing coefficient measures the extent to which the future is independent of the past.

Remark: As will be evident from the proofs below, actually all that is required is that

$$\sum_{\mathbf{w} \in \mathbb{M}^l} f_{\mathbf{uwv}} \to f_{\mathbf{u}} f_{\mathbf{v}} \text{ as } l \to \infty, \ \forall \mathbf{u}, \mathbf{v} \in \mathbb{M}^k, \tag{7.53}$$

where k is the *fixed* integer arising from the finite Hankel rank condition. It is easy to see that if the process $\{Y_t\}$ is α-mixing, then (7.53) follows.

Since the process assumes values in a finite alphabet, (7.53) is equivalent to the condition

$$\max_{A \in \Sigma_1^k, B \in \Sigma_{l+k+1}^{2k}} |P(A \cap B) - P(A)P(B)| \to 0 \text{ as } l \to \infty. \tag{7.54}$$

To see this, suppose that (7.54) holds, and choose A to be the event $\mathbf{y}_1^k = \mathbf{u}$, and similarly, choose B to be the event $y_{l+k+1}^{l+2k} = \mathbf{v}$, for some $\mathbf{u}, \mathbf{v} \in \mathbb{M}^k$.

Then it is clear that $A \cap B$ is the event that a string of length $l + 2k$ begins with $\mathbf{u}$ and ends with $\mathbf{v}$. Thus

$$P(A) = f_{\mathbf{u}},\ P(B) = f_{\mathbf{v}},\ P(A \cap B) = \sum_{\mathbf{w} \in \mathbb{M}^l} f_{\mathbf{uwv}}.$$

Hence (7.54) implies (7.53). To show the converse, suppose (7.53) holds. Then (7.54) also holds for elementary events A and B. Since k is a *fixed* number and the alphabet of the process is finite, both of the σ-algebras Σ_1^k, Σ_{l+k+1}^{2k} are *finite* unions of elementary events. Hence (7.53) is enough to imply (7.54). It is not known whether (7.54) is *strictly weaker* than α-mixing for processes assuming values over a finite alphabet.

Now we state the main result of this section.

Theorem 7.10 *Suppose the process $\{Y_t\}$ is α-mixing and has finite Hankel rank r. Let $\{r, \mathbf{x}, \mathbf{y}, C^{(\mathbf{u})}, u \in \mathbb{M}\}$ be any regular quasi-realization of the process, and define*

$$S := \sum_{u \in \mathbb{M}} C^{(\mathbf{u})}.$$

Then $S^l \to \mathbf{yx}$ as $l \to \infty$, $\rho(S) = 1$, $\rho(S)$ is a simple eigenvalue of S, and all other eigenvalues of S have magnitude strictly less than one.

This theorem can be compared with [6, Theorem 4].

Proof. It is enough to prove the theorem for the *particular* quasi-realization $\{r, \boldsymbol{\theta}, \boldsymbol{\phi}, D^{(u)}, u \in \mathbb{M}\}$ defined in (7.35). This is because there exists a non-singular matrix T such that $C^{(u)} = T^{-1} D^{(u)} T$ for all u, and as a result the matrices $\sum_{u \in \mathbb{M}} C^{(u)}$ and $\sum_{u \in \mathbb{M}} D^{(u)}$ have the same spectrum. The α-mixing property implies that, for each $\mathbf{i} \in I, \mathbf{j} \in J$, we have

$$\sum_{\mathbf{w} \in \mathbb{M}^l} f_{\mathbf{iwj}} \to f_{\mathbf{i}} f_{\mathbf{j}} \text{ as } l \to \infty. \tag{7.55}$$

This is a consequence of (7.53) since both I and J are subsets of $\mathbb{M}^k$. Now note that, for each fixed $\mathbf{w} \in \mathbb{M}^l$, we have from (7.34) that

$$[f_{\mathbf{iwj}}, \mathbf{i} \in I, \mathbf{j} \in J] = F_{I,J}^{(\mathbf{w})} = D^{(\mathbf{w})} F_{I,J}, \tag{7.56}$$

where, as per earlier convention, we write

$$D^{(\mathbf{w})} := D^{(w_1)} \ldots D^{(w_l)}.$$

It is clear that

$$\sum_{\mathbf{w} \in \mathbb{M}^l} D^{(\mathbf{w})} = \left[\sum_{u \in \mathbb{M}} D^{(u)} \right]^l = S^l. \tag{7.57}$$

Now (7.55) implies that

$$\sum_{\mathbf{w} \in \mathbb{M}^l} [f_{\mathbf{iwj}}, \mathbf{i} \in I, \mathbf{j} \in J] \to [f_{\mathbf{i}}, i \in I][f_{\mathbf{j}}, j \in J] =: F_{I,0} F_{0,J},$$

where $F_{I,0}$ is an r-dimensional column vector and $F_{0,J}$ is an r-dimensional row vector. Moreover, combining (7.56) and (7.57) shows that

$$S^l F_{I,J} \to F_{I,0} F_{0,J},$$

and since $F_{I,J}$ is nonsingular, that

$$S^l \to F_{I,0} F_{0,J} F_{I,J}^{-1} = \boldsymbol{\phi\theta} \text{ as } l \to \infty.$$

So the conclusion is that S^l approaches $\boldsymbol{\phi\theta}$, which is a rank one matrix, as $l \to \infty$. Moreover, this rank one matrix has one eigenvalue at one and the rest at zero. To establish this, we show that

$$F_{0,J} F_{I,J}^{-1} F_{I,0} = 1.$$

This is fairly straightforward. Note that $F_{0,J} F_{I,J}^{-1} = \boldsymbol{\theta}$ and $F_{I,0} = \boldsymbol{\phi}$ as defined in (7.35). Then taking $\mathbf{u}$ to be the empty string in (7.34) (and of course, substituting $\mathbf{x} = \boldsymbol{\theta}, \mathbf{y} = \boldsymbol{\phi}$) shows that $\boldsymbol{\theta\phi} = 1$, which is the desired conclusion. Let A denote the rank one matrix

$$A := F_{I,0} F_{0,J} F_{I,J}^{-1}.$$

Then $S^l \to A$ as $l \to \infty$. Suppose the spectrum of the matrix S is $\{\lambda_1, \ldots, \lambda_n\}$, where $n = m^k$, and $|\lambda_1| = \rho(S)$. Then, since the spectrum of S^l is precisely $\{\lambda_1^l, \ldots, \lambda_n^l\}$, it follows that

$$\{\lambda_1^l, \ldots, \lambda_n^l\} \to \{1, 0, \ldots, 0\} \text{ as } l \to \infty.$$

Here we make use of the facts that A is a rank one matrix, and that its spectrum consists of $n-1$ zeros plus one. This shows that S has exactly one eigenvalue on the unit circle, namely at $\lambda = 1$, and the remaining eigenvalues all have magnitude strictly less than one. □

Corollary 7.11 *Suppose a stationary process $\{Y_t\}$ is α-mixing and has a regular realization. Then the underlying Markov chain is aperiodic and irreducible.*

Proof. Suppose that the process under study has a regular realization (and not just a regular quasi-realization). Let A denote the state transition matrix of the corresponding Markov process $\{X_t\}$. From Theorem 7.9, it follows that A is similar to the matrix S defined in Theorem 7.10. Moreover, if the process $\{Y_t\}$ is α-mixing, then the matrix A (which is similar to S) satisfies the strong Perron property. In other words, it has only one eigenvalue on the unit circle, namely a simple eigenvalue at one. Hence the Markov chain $\{X_t\}$ is irreducible and aperiodic. □

7.6 ULTRA-MIXING PROCESSES

In the previous two sections, we studied the existence of quasi-realizations. In this section, we study the existence of "true" (as opposed to quasi) realizations. We introduce a new property known as "ultra-mixing" and show that if a process has finite Hankel rank, and is both α-mixing as well as ultra-mixing, then modulo a technical condition it has a HMM where the underlying Markov chain is itself α-mixing (and hence aperiodic and irreducible) or else satisfies a "consistency condition." The converse is also true, modulo another technical condition.

The material in this section is strongly influenced by [6]. In that paper, the author *begins* with the assumption that the stochastic process under study is generated by an irreducible HMM (together with a few other assumptions), and then gives a constructive procedure for constructing an irreducible HMM for the process. Thus the paper does not give a set of conditions for the existence of a HMM *in terms of the properties of the process under study*. Moreover, even with the assumptions in [6], the order of the HMM constructed using the given procedure can in general be much larger than the order of the HMM that generates the process in the first place. In contrast, in the present paper we give conditions explicitly in terms of the process under study, that are sufficient to guarantee the existence of an irreducible HMM. However, the proof techniques used here borrow heavily from [6].

We begin with a rather "obvious" result that sets the foundation for the material to follow.

Lemma 7.12 *Suppose $\{Y_t\}$ is a stationary process over a finite alphabet $\mathcal{M}$. Then the process $\{Y_t\}$ has a "joint Markov process" HMM if and only if there exist an integer n, a stochastic row vector $\mathbf{h}$, and $n \times n$ nonnegative matrices $G^{(1)}, \dots, G^{(m)}$ such that the following statements are true.*

1. *The matrix $Q := \sum_{u \in \mathcal{M}} G^{(u)}$ is stochastic, in that $Q\mathbf{e}_n = \mathbf{e}_n$.*

2. *$\mathbf{h}$ is a row eigenvector $\mathbf{h}$ of Q corresponding to the eigenvalue one, i.e., $\mathbf{h}Q = \mathbf{h}$.*

3. *For every $\mathbf{u} \in \mathcal{M}^*$, we have*

$$f_{\mathbf{u}} = \mathbf{h}G^{(u_1)} \cdots G^{(u_l)}\mathbf{e}_n,$$

where $l = |\mathbf{u}|$.

In this case there exists a Markov process $\{X_t\}$ with the state space $\mathbb{N} := \{1, \dots, n\}$ such that the joint process $\{(X_t, Y_t)\}$ is a Type 3 HMM.

Proof. One half of this lemma has already been proven in the course of proving Theorem 7.1. Suppose $\{Y_t\}$ has a "joint Markov process" HMM model. Let $\{X_t\}$ denote the associated Markov process. Define the matrices $M^{(1)}, \dots, M^{(m)}$ as in (6.6), and let $\boldsymbol{\pi}$ denote the stationary distribution of the process $\{X_t\}$. Then it is clear that the conditions of the lemma are satisfied with $\mathbf{h} = \boldsymbol{\pi}$ and $G^{(u)} = M^{(u)}$ for each $u \in \mathcal{M}$.

To prove the converse, suppose $\mathbf{h}, G^{(1)}, \dots, G^{(m)}$ exist that satisfy the stated conditions. Let $\{Z_t\}$ be a stationary Markov process with the state transition matrix

$$A_Z := \begin{bmatrix} G^{(1)} & G^{(2)} & \ldots & G^{(m)} \\ \vdots & \vdots & \vdots & \vdots \\ G^{(1)} & G^{(2)} & \ldots & G^{(m)} \end{bmatrix},$$

and the stationary distribution

$$\boldsymbol{\pi}_Z = [\mathbf{h}G^{(1)}|\ldots|\mathbf{h}G^{(m)}].$$

To show that $\boldsymbol{\pi}_Z$ is indeed a stationary distribution of A_Z, partition $\boldsymbol{\pi}_Z$ in the obvious fashion as $[\boldsymbol{\pi}_1 \ldots \boldsymbol{\pi}_m]$, and observe that $\boldsymbol{\pi}_v = \mathbf{h}G^{(v)}$. Then, because of the special structure of the matrix A_Z, in order to be a stationary distribution of the Markov chain, the vector $\boldsymbol{\pi}_Z$ needs to satisfy the relationship

$$\left[\sum_{v \in \mathcal{M}} \boldsymbol{\pi}_v\right] \cdot G^{(u)} = \boldsymbol{\pi}_u. \tag{7.58}$$

Now observe that

$$\sum_{v \in \mathcal{M}} \boldsymbol{\pi}_v = \mathbf{h} \sum_{v \in \mathcal{M}} G^{(v)} = \mathbf{h}Q = \mathbf{h}.$$

Hence the desired relationship (7.58) follows readily. Now the stationary distribution of the X_t process is clearly $\sum_{v \in \mathcal{M}} \mathbf{h}G^{(v)} = \mathbf{h}$. Hence, by the formula (7.3), it follows that the frequencies of the Y_t process are given by

$$f_{\mathbf{u}} = \mathbf{h}G^{(u_1)} \cdots G^{(u_l)}\mathbf{e}_n.$$

This is the desired conclusion. □

Before presenting the sufficient condition for the existence of a HMM, we recall a very important result from [6]. Consider a "joint Markov process" HMM where the associated matrix A (the transition matrix of the $\{X_t\}$ process) is irreducible. In this case, it is well known and anyway rather easy to show that the state process $\{X_t\}$ is α-mixing if and only if the matrix A is aperiodic in addition to being irreducible. If A is aperiodic (so that the state process is α-mixing), then the output process $\{Y_t\}$ is also α-mixing. However, the converse is not always true. It is possible for the output process to be α-mixing even if the state process is not. Theorem 5 of [6] gives necessary and sufficient conditions for this to happen. We reproduce this important result below.

Suppose a "joint Markov process" HMM has n states and that the state transition matrix A is irreducible. Let $\boldsymbol{\pi}$ denote the unique positive stationary probability distribution of the X_t process. As in (6.6), define the matrices $M^{(u)}, u \in \mathcal{M}$ by

$$m_{ij}^{(u)} = \Pr\{X_1 = j \wedge Y_1 = u | X_0 = i\}, 1 \le i, j \le n, u \in \mathcal{M}.$$

Let p denote the number of eigenvalues of A on the unit circle (i.e., the period of the Markov chain). By renumbering the states if necessary, rearrange A

so that it has the following cyclic form:

$$A = \begin{bmatrix} \mathbf{0} & \mathbf{0} & \dots & \mathbf{0} & A_1 \\ A_p & \mathbf{0} & \dots & \mathbf{0} & \mathbf{0} \\ \mathbf{0} & A_{p-1} & \dots & \mathbf{0} & \mathbf{0} \\ \vdots & \vdots & \vdots & \vdots & \vdots \\ \mathbf{0} & \mathbf{0} & \dots & A_2 & \mathbf{0} \end{bmatrix}. \tag{7.59}$$

The matrices $M^{(u)}$ inherit the same zero block structure as A; so the notation $M_i^{(u)}$ is unambiguous. For a string $\mathbf{u} \in \mathcal{M}^l$, define

$$M_i^{(\mathbf{u})} := M_i^{(u_1)} M_{i+1}^{(u_2)} \dots M_{i+l-1}^{(u_l)},$$

where the subscripts on M are taken modulo p. Partition $\boldsymbol{\pi}$ into p equal blocks, and label them as $\boldsymbol{\pi}_1$ through $\boldsymbol{\pi}_p$.

Theorem 7.13 *The output process $\{Y_t\}$ is α-mixing if and only if, for every string $\mathbf{u} \in \mathcal{M}^*$, the following "consistency conditions" hold:*

$$\begin{aligned} \boldsymbol{\pi}_1 M_1^{(\mathbf{u})} \mathbf{e}_{(n/p)} = \boldsymbol{\pi}_2 M_p^{(\mathbf{u})} \mathbf{e}_{(n/p)} &= \boldsymbol{\pi}_3 M_{p-1}^{(\mathbf{u})} \mathbf{e}_{(n/p)} = \dots \\ = \boldsymbol{\pi}_p M_2^{(\mathbf{u})} \mathbf{e}_{(n/p)} &= \frac{1}{p} \boldsymbol{\pi} M^{(\mathbf{u})} \mathbf{e}_n. \end{aligned} \tag{7.60}$$

For a proof, see [6, Theorem 5].

In earlier sections, we studied the spectrum of various matrices under the assumption that the process under study is α-mixing. For present purposes, we introduce a different kind of mixing property.

Definition 7.14 *Given the process $\{Y_t\}$, suppose it has finite Hankel rank, and let k denote the unique integer defined in Lemma 7.5. Then the process $\{Y_t\}$ is said to be* **ultra-mixing** *if there exists a sequence $\{\delta_l\} \downarrow 0$ such that*

$$\left| \frac{f_{\mathbf{iu}}}{f_{\mathbf{u}}} - \frac{f_{\mathbf{iuv}}}{f_{\mathbf{uv}}} \right| \leq \delta_l, \ \forall \mathbf{i} \in \mathcal{M}^k, \mathbf{u} \in \mathcal{M}^l, \mathbf{v} \in \mathcal{M}^*. \tag{7.61}$$

Note that, the way we have defined it here, the notion of ultra-mixing is defined only for processes with finite Hankel rank.

In [73], Kalikow defines a notion that he calls a "uniform martingale," which is the same as an ultra-mixing stochastic process. He shows that a stationary stochastic process over a finite alphabet is a uniform martingale if and only if it is also a "random Markov process," which is defined as follows: A process $\{(Y_t, N_t)\}$ where $Y_t \in \mathcal{M}$ and N_t is a positive integer (natural number) for each t is said to be a "random Markov process" if (i) the process $\{N_t\}$ is independent of the $\{Y_t\}$ process, and (ii) for each t, the conditional distribution of $Y_t | Y_{-\infty}^{t-1}$ is the same as that of $Y_t | Y_{t-N_t}^{t-1}$. Observe that if N_t equals a fixed integer N for all t, then the above condition says that $\{Y_t\}$ is an N-step Markov process. Hence a "random Markov process" is an N_t-step Markov process where the length of the "memory" N_t is itself random and independent of Y_t. One of the main results of [73] is that the ultra-mixing property is equivalent to the process being random Markov. However, the random Markov property seems to be quite different in spirit from a process having a HMM.

The ultra-mixing property can be interpreted as a kind of long-term independence. It says that the conditional probability that a string begins with $\mathbf{i}$, given the next l entries, is just about the same whether we are given just the next l entries, or the next l entries as well as the still later entries. This property is also used in [6]. It does not appear straightforward to relate ultra-mixing to other notions of mixing such as α-mixing. This can be seen from the treatment of [6, Section 11], where the author assumes (in effect) that the process under study is *both* ultra-mixing as well as α-mixing.

7.7 A SUFFICIENT CONDITION FOR THE EXISTENCE OF HMMS

Starting with the original work of Dharmadhikari [44], "cones" have played a central role in the construction of HMMs. Moreover, cones also play an important role in the so-called positive realization problem. Hence it is not surprising that the conditions given here also borrow a little bit from positive realization theory. See [15] for a survey of the current status of this problem.

Recall that a set $\mathcal{S} \subseteq \mathbb{R}^r$ is said to be a "cone" if $\mathbf{x}, \mathbf{y} \in \mathcal{S} \Rightarrow \alpha\mathbf{x} + \beta\mathbf{y} \in \mathcal{S}\ \forall \alpha, \beta \geq 0$. The term "convex cone" is also used to describe such an object. Given a (possibly infinite) set $\mathcal{V} \subseteq \mathbb{R}^r$, the symbol Cone($\mathcal{V}$) denotes the smallest cone containing $\mathcal{V}$, or equivalently, the intersection of all cones containing $\mathcal{V}$. If $\mathcal{V} = \{\mathbf{v}_1, \ldots, \mathbf{v}_n\}$ is a finite set, then it is clear that

$$\text{Cone}(\mathcal{V}) = \{\sum_{i=1}^{n} \alpha_i \mathbf{v}_i : \ \alpha_i \geq 0\ \forall i\}.$$

In such a case, Cone($\mathcal{V}$) is said to be "polyhedral" and $\mathbf{v}_1, \ldots, \mathbf{v}_n$ are said to be "generators" of the cone. Note that, in the way we have defined the concept here, the generators of a polyhedral cone are not uniquely defined. It is possible to refine the definition; however, the above definition is sufficient for the present purposes. Finally, given a cone $\mathcal{C}$ (polyhedral or otherwise), the "polar cone" $\mathcal{C}^p$ is defined by

$$\mathcal{C}^p := \{\mathbf{y} \in \mathbb{R}^r : \mathbf{y}^t\mathbf{x} \geq 0\ \forall \mathbf{x} \in \mathcal{C}\}.$$

It is easy to see that $\mathcal{C}^p$ is also a cone, and that $\mathcal{C} \subseteq (\mathcal{C}^p)^p$.

Next, we introduce two cones that play a special role in the proof. Suppose as always that the process under study has finite Hankel rank, and define the integer k as in Lemma 7.5. We begin with the quasi-realization $\{r, \boldsymbol{\theta}, \boldsymbol{\phi}, D^{(u)}\}$ defined in (7.51). Now define

$$\mathcal{C}_c := \text{Cone}\{D^{(\mathbf{u})}\boldsymbol{\phi} : \mathbf{u} \in \mathcal{M}^*\},$$

$$\mathcal{C}_o := \{\mathbf{y} \in \mathbb{R}^r : \boldsymbol{\theta} D^{(\mathbf{v})}\mathbf{y} \geq 0,\ \forall \mathbf{v} \in \mathcal{M}^*\}.$$

The subscripts o and c have their legacy from positive realization theory, where $\mathcal{C}_c$ is called the "controllability cone" and $\mathcal{C}_o$ is called the "observability cone." See for example [15]. However, in the present context, we could have used any other symbols. Note that from (7.34) and (7.35) we have

$$\boldsymbol{\theta} D^{(\mathbf{v})} D^{(\mathbf{u})} \boldsymbol{\phi} = f_{\mathbf{uv}} \geq 0, \ \forall \mathbf{u}, \mathbf{v} \in \mathcal{M}^*.$$

Hence $D^{(\mathbf{u})}\boldsymbol{\phi} \in \mathcal{C}_o \ \forall \mathbf{u} \in \mathcal{M}^*$, and as a result $\mathcal{C}_c \subseteq \mathcal{C}_o$. Moreover, both $\mathcal{C}_c$ and $\mathcal{C}_o$ are invariant under $D^{(w)}$ for each $w \in \mathcal{M}$. To see this, let $w \in \mathcal{M}$ be arbitrary. Then $D^{(w)} D^{(\mathbf{u})} \boldsymbol{\phi} = D^{(w\mathbf{u})}\boldsymbol{\phi}$, for all $\mathbf{u} \in \mathcal{M}^*$. Hence

$$D^{(w)} \mathcal{C}_c = \text{Cone}\{D^{(w\mathbf{u})}\boldsymbol{\phi} : \mathbf{u} \in \mathcal{M}^*\} \subseteq \mathcal{C}_c.$$

Similarly, suppose $\mathbf{y} \in \mathcal{C}_o$. Then the definition of $\mathcal{C}_o$ implies that $\boldsymbol{\theta} D^{(\mathbf{v})}\mathbf{y} \geq 0$ for all $\mathbf{v} \in \mathcal{M}^*$. Therefore

$$\boldsymbol{\theta} D^{(\mathbf{v})} D^{(w)} \mathbf{y} = \boldsymbol{\theta} D^{(\mathbf{v}w)} \mathbf{y} \geq 0 \ \forall \mathbf{v} \in \mathcal{M}^*.$$

Hence $D^{(w)}\mathbf{y} \in \mathcal{C}_c$. The key difference between $\mathcal{C}_c$ and $\mathcal{C}_o$ is that the former cone need not be closed, whereas the latter cone is always closed as is easy to show.

In order to state the sufficient condition for the existence of a HMM, a few other bits of notation are introduced. Suppose the process under study has finite Hankel rank, and let k be the unique integer defined in Lemma 7.5. Let r denote the rank of the Hankel matrix, and choose subsets $I, J \subseteq \mathcal{M}^k$ such that $|I| = |J| = r$ and $F_{I,J}$ has rank r. For each finite string $\mathbf{u} \in \mathcal{M}^*$, define the vectors

$$\mathbf{p_u} := \frac{1}{f_{\mathbf{u}}} F_{I,0}^{(\mathbf{u})} = [f_{\mathbf{iu}}/f_{\mathbf{u}}, i \in I] \in [0,1]^{r \times 1},$$

$$\mathbf{q_u} := \frac{1}{f_{\mathbf{u}}} F_{0,J}^{(\mathbf{u})} = [f_{\mathbf{uj}}/f_{\mathbf{u}}, j \in J] \in [0,1]^{1 \times r}.$$

The interpretation of $\mathbf{p_u}$ is that the $\mathbf{i}$-th component of this vector is the conditional probability, given that the last part of a sample path consists of the string $\mathbf{u}$, that the immediately preceding k symbols are $\mathbf{i}$. The vector $\mathbf{q_u}$ is interpreted similarly. The $\mathbf{j}$-th component of this vector is the conditional probability, given that the first part of a sample path consists of the string $\mathbf{u}$, that the next k symbols are $\mathbf{j}$.

Lemma 7.15 *Let $\|\cdot\|_1$ denote the ℓ_1-norm on $\mathbb{R}^r$. Then there exists a constant $\gamma > 0$ such that*

$$\gamma \leq \|\mathbf{p_u}\|_1 \leq 1, \gamma \leq \|\mathbf{q_u}\|_1 \leq 1, \ \forall \mathbf{u} \in \mathcal{M}^*.$$

Proof. Note that the vector $[f_{\mathbf{iu}}/f_{\mathbf{u}}, \mathbf{i} \in \mathcal{M}^k]$ is a probability vector, in the sense that its components are nonnegative and add up to one. Hence this vector has ℓ_1-norm of one. Since $\mathbf{p_u}$ is a subvector of it, it follows that $\|\mathbf{p_u}\|_1 \leq 1$. On the other hand, we have

$$[f_{\mathbf{iu}}/f_{\mathbf{u}}, \mathbf{i} \in \mathcal{M}^k] = U\mathbf{p_u}, \ \forall \mathbf{u} \in \mathcal{M}^*,$$

and U has full column rank. Hence $\|\mathbf{p_u}\|_1$ is bounded away from zero independently of $\mathbf{u}$. Similar arguments apply to $\mathbf{q_u}$. □

The vectors $\mathbf{p_u}$ and $\mathbf{q_u}$ satisfy some simple recurrence relationships.

Lemma 7.16 *Suppose* $\mathbf{u}, \mathbf{v} \in \mathcal{M}^*$. *Then*

$$D^{(\mathbf{u})}\mathbf{p_v} = \frac{f_{\mathbf{uv}}}{f_{\mathbf{v}}}\mathbf{p_{uv}}, \ \mathbf{q_u}C^{(\mathbf{v})} = \frac{f_{\mathbf{uv}}}{f_{\mathbf{u}}}\mathbf{q_{uv}}.$$

Proof. From Lemmas 7.6 and 7.7, it follows that

$$F_{I,0}^{(\mathbf{v})} = F_{I,k}^{(\mathbf{v})}\mathbf{e}_{m^k} = D^{(\mathbf{v})}F_{I,J}V\mathbf{e}_{m^k} = D^{(\mathbf{v})}\phi, \ \forall \mathbf{v} \in \mathcal{M}^*.$$

This shows that

$$\mathbf{p_v} = \frac{1}{f_{\mathbf{v}}}F_{I,0}^{(\mathbf{v})} = \frac{1}{f_{\mathbf{v}}}D^{(\mathbf{v})}\phi.$$

Hence, for arbitrary $\mathbf{u}, \mathbf{v} \in \mathcal{M}^*$, we have

$$D^{(\mathbf{u})}\mathbf{p_v} = \frac{1}{f_{\mathbf{v}}}D^{(\mathbf{u})}D^{(\mathbf{v})}\phi = \frac{1}{f_{\mathbf{v}}}D^{(\mathbf{uv})}\phi = \frac{f_{\mathbf{uv}}}{f_{\mathbf{v}}}\mathbf{p_{uv}}.$$

The proof in the case of $\mathbf{q_v}$ is entirely similar. □

Now let us consider the countable collection of probability vectors $\mathcal{A} := \{\mathbf{p_u} : \mathbf{u} \in \mathcal{M}^*\}$. Since $\mathbf{p_u}$ equals $D^{(\mathbf{u})}\phi$ within a scale factor, it follows that $\mathcal{C}_c = \text{Cone}(\mathcal{A})$. Moreover, since $\mathcal{A} \subseteq \mathcal{C}_c \subseteq \mathcal{C}_o$ and $\mathcal{C}_o$ is a closed set, it follows that the set of cluster points of $\mathcal{A}$ is also a subset of $\mathcal{C}_o$.[4] Finally, it follows from Lemma 7.15 that every cluster point of $\mathcal{A}$ has norm no smaller than γ.

Now we state the main result of this section.

Theorem 7.17 *Suppose the process* $\{Y_t\}$ *satisfies the following conditions:*

1. *It has finite Hankel rank.*
2. *It is ultra-mixing.*
3. *It is* α*-mixing.*
4. *The cluster points of the set* $\mathcal{A}$ *of probability vectors are finite in number and lie in the interior of the cone* $\mathcal{C}_o$.

Under these conditions, the process has an irreducible "joint Markov process" hidden Markov model. Moreover the HMM satisfies the consistency conditions (7.60).

Remark: Among the hypotheses of Theorem 7.17, Conditions 1 through 3 are "real" conditions, whereas Condition 4 is a "technical" condition.

The proof proceeds via two lemmas. The first lemma gives insight into the behavior of the matrix $D^{(\mathbf{u})}$ as $|\mathbf{u}| \to \infty$. To put these lemmas in context, define the matrix $S = \sum_{u \in \mathcal{M}} D^{(u)}$. Then by Theorem 7.10, we know that if the process $\{Y_t\}$ is α-mixing, then S^l approaches a rank one matrix as $l \to \infty$. In the present case it is shown that, if the process is ultra-mixing, then *each individual* matrix $D^{(\mathbf{u})}$ approaches a rank one matrix as $|\mathbf{u}| \to \infty$. This result has no counterpart in earlier literature and may be of independent interest.

[4]Recall that a vector $\mathbf{y}$ is said to be a "cluster point" of $\mathcal{A}$ if there exists a sequence in $\mathcal{A}$, no entry of which equals $\mathbf{y}$, converging to $\mathbf{y}$. Equivalently, $\mathbf{y}$ is a cluster point if $\mathcal{A}$ if every neighborhood of $\mathbf{y}$ contains a point of $\mathcal{A}$ not equal to $\mathbf{y}$.

Lemma 7.18 *Let $\| \cdot \|_{i1}$ denote the norm on the set of $m^k \times m^k$ matrices induced by the ℓ_1-norm on $\mathbb{R}^{m^k}$. Suppose the process $\{Y_t\}$ is ultra-mixing. Define*

$$\mathbf{b}_U := \mathbf{e}_{m^k}^t U \in \mathbb{R}^{1 \times r}.$$

Then

$$\| \frac{1}{f_{\mathbf{u}}} D^{(\mathbf{u})} - \frac{1}{f_{\mathbf{u}}} \mathbf{p}_{\mathbf{u}} \mathbf{b}_U D^{(\mathbf{u})} \|_{i1} \leq r \delta_{|\mathbf{u}|} \| F_{I,J}^{-1} \|_{i1}, \tag{7.62}$$

where $\{\delta_l\}$ is the sequence in the definition of the ultra-mixing property, and $|\mathbf{u}|$ denotes the length of the string u.

Proof. (Of Lemma 7.18): If we substitute $\mathbf{j}$ for $\mathbf{v}$ in (7.61), we get

$$\left| \frac{f_{\mathbf{iu}}}{f_{\mathbf{u}}} - \frac{f_{\mathbf{iuj}}}{f_{\mathbf{uj}}} \right| \leq \delta_{|\mathbf{u}|}.$$

For each $\mathbf{j} \in J$, we have that $f_{\mathbf{uj}}/f_{\mathbf{u}} \leq 1$. Hence we can multiply both sides of the above equation by $f_{\mathbf{uj}}/f_{\mathbf{u}} \leq 1$, which gives

$$\left| \frac{f_{\mathbf{iu}}}{f_{\mathbf{u}}} \cdot \frac{f_{\mathbf{uj}}}{f_{\mathbf{u}}} - \frac{f_{\mathbf{iuj}}}{f_{\mathbf{uj}}} \cdot \frac{f_{\mathbf{uj}}}{f_{\mathbf{u}}} \right| = \left| \frac{f_{\mathbf{iu}}}{f_{\mathbf{u}}} \cdot \frac{f_{\mathbf{uj}}}{f_{\mathbf{u}}} - \frac{f_{\mathbf{iuj}}}{f_{\mathbf{u}}} \right| \leq \delta_{|\mathbf{u}|} \cdot \frac{f_{\mathbf{uj}}}{f_{\mathbf{u}}} \leq \delta_{|\mathbf{u}|}.$$

Now define the $r \times r$ matrix $R^{(\mathbf{u})}$ by

$$(R^{(\mathbf{u})})_{\mathbf{ij}} := \frac{f_{\mathbf{iu}}}{f_{\mathbf{u}}} \cdot \frac{f_{\mathbf{uj}}}{f_{\mathbf{u}}} - \frac{f_{\mathbf{iuj}}}{f_{\mathbf{u}}}.$$

Then (see for example [128])

$$\|R^{(\mathbf{u})}\|_{i1} = \max_{\mathbf{j} \in \mathcal{M}^k} \sum_{\mathbf{i} \in \mathcal{M}^k} |(R^{(\mathbf{u})})_{\mathbf{ij}}| \leq r \delta_{|\mathbf{u}|}.$$

Next, note that

$$R^{(\mathbf{u})} = \mathbf{p}_{\mathbf{u}} \mathbf{q}_{\mathbf{u}} - \frac{1}{f_{\mathbf{u}}} D^{(\mathbf{u})} F_{I,J}.$$

Hence we have established that

$$\| \frac{1}{f_{\mathbf{u}}} D^{(\mathbf{u})} F_{I,J} - \mathbf{p}_{\mathbf{u}} \mathbf{q}_{\mathbf{u}} \|_{i1} \leq r \delta_{|\mathbf{u}|}. \tag{7.63}$$

Therefore

$$\| \frac{1}{f_{\mathbf{u}}} D^{(\mathbf{u})} - \mathbf{p}_{\mathbf{u}} \mathbf{q}_{\mathbf{u}} F_{I,J}^{-1} \|_{i1} \leq r \delta_{|\mathbf{u}|} \| F_{I,J}^{-1} \|_{i1}.$$

Thus the proof is complete once it is shown that

$$\mathbf{q}_{\mathbf{u}} F_{I,J}^{-1} = \frac{1}{f_{\mathbf{u}}} \mathbf{b}_U D^{(\mathbf{u})}.$$

But this last step is immediate, because

$$f_{\mathbf{u}} \mathbf{q}_{\mathbf{u}} F_{I,J}^{-1} = F_{0,J}^{(\mathbf{u})} F_{I,J}^{-1} = \mathbf{e}_{m^k}^t U F_{I,J}^{(\mathbf{u})} F_{I,J}^{-1} = \mathbf{e}_{m^k}^t U D^{(\mathbf{u})} F_{I,J} F_{I,J}^{-1} = \mathbf{b}_U D^{(\mathbf{u})}.$$

This completes the proof. □

The reader may wonder about the presence of the factor $1/f_{\mathbf{u}}$ in (7.62). Obviously, in any reasonable stochastic process, the probability $f_{\mathbf{u}}$ approaches zero as $|\mathbf{u}| \to \infty$. Hence, unless we divide by this quantity, we would get an inequality that is trivially true because both quantities individually approach zero. In contrast, (7.63) shows that the matrix $(1/f_{\mathbf{u}})D^{(\mathbf{u})}$ is both bounded and bounded away from zero for all $\mathbf{u} \in \mathcal{M}^*$.

Thus Lemma 7.18 serves to establish the behavior of the matrix $D^{(\mathbf{u})}$ as $|\mathbf{u}| \to \infty$. Whatever be the vector $\mathbf{x} \in \mathbb{R}^r$, the vector $(1/f_{\mathbf{u}})D^{(\mathbf{u})}\mathbf{x}$ approaches $(1/f_{\mathbf{u}})\mathbf{p}_{\mathbf{u}}\mathbf{b}_U D^{(\mathbf{u})}\mathbf{x}$ and thus eventually gets "aligned" with the vector $\mathbf{p}_{\mathbf{u}}$ as $|\mathbf{u}| \to \infty$.

Lemma 7.19 *Suppose the process under study is ultra-mixing, and that the cluster points of the probability vector set $\mathcal{A}$ are finite in number and belong to the interior of the cone $\mathcal{C}_c$. Then there exists a polyhedral cone $\mathcal{P}$ such that*

1. *$\mathcal{P}$ is invariant under each $D^{(u)}, \mathbf{u} \in \mathcal{M}$.*
2. *$\mathcal{C}_c \subseteq \mathcal{P} \subseteq \mathcal{C}_o$.*
3. *$\phi \in \mathcal{P}$.*
4. *$\boldsymbol{\theta}^t \in \mathcal{P}^p$.*

Remark: In some sense this is the key lemma in the proof of the main theorem. It is noteworthy that the hypotheses do *not* include the assumption that the process under study is α-mixing.

Proof. First, note that, given any $\epsilon > 0$, there exists an $L = L(\epsilon)$ such that the following is true: For each $\mathbf{w} \in \mathcal{M}^*$ with $|\mathbf{w}| > L$, write $\mathbf{w} = \mathbf{uv}$ with $|\mathbf{u}| = L$. Then $\|\mathbf{p_w} - \mathbf{p_u}\|_1 \leq \epsilon$. To see this, given $\epsilon > 0$, choose L such that $\delta_L \leq \epsilon/m^k$. Then (7.61) implies that $\|\mathbf{p_u} - \mathbf{p_w}\|_1 \leq \epsilon$.

By assumption, the set of probability vectors $\mathcal{A} := \{\mathbf{p}_u : \mathbf{u} \in \mathcal{M}^k\}$ has only finitely many cluster points. Let us denote them as $\mathbf{x}_1, \ldots, \mathbf{x}_n$. By assumption again, each of these vectors lies in the interior of $\mathcal{C}_o$. Hence there exists an $\epsilon > 0$ such that the sphere (in the ℓ_1-norm) centered at each $\mathbf{x}_i$ of radius 2ϵ is also contained in $\mathcal{C}_o$.

Next, note that there exists an integer L such that *every* vector $\mathbf{p_u}$ with $|\mathbf{u}| \geq L$ lies within a distance of ϵ (in the ℓ_1-norm) from at least one of the $\mathbf{x}_i$. In other words, there exists an integer L such that

$$\min_{1 \leq i \leq n} \|\mathbf{p_u} - \mathbf{x}_i\|_1 \leq \epsilon, \ \forall \mathbf{u} \in \mathcal{M}^l \text{ with } l > L.$$

To see why this must be so, assume the contrary. Thus there exists a sequence $\mathbf{p}_{\mathbf{u}_j}$ such that $\|\mathbf{p}_{\mathbf{u}_j} - \mathbf{x}_i\|_1 > \epsilon$ for all i, j. Now the sequence $\{\mathbf{p}_{\mathbf{u}_j}\}$ is bounded and therefore has a convergent subsequence. The limit of this convergent subsequence cannot be any of the $\mathbf{x}_i$ by the assumption that $\|\mathbf{p}_{\mathbf{u}_j} - \mathbf{x}_i\|_1 > \epsilon$ for all i, j. This violates the earlier assumption that $\mathbf{x}_1, \ldots, \mathbf{x}_n$ are *all* the cluster points of the set $\mathcal{A}$.

Now choose a set $\mathbf{z}_1, \ldots, \mathbf{z}_r$ of basis vectors for $\mathbb{R}^r$ such that each $\mathbf{z}_j$ has unit norm. For instance, we can take $\mathbf{z}_j$ to be the unit vector with a 1 in position j and zeros elsewhere. With ϵ already defined above, define the unit vectors

$$\mathbf{y}_{i,j}^+ := \frac{\mathbf{x}_i + 2\epsilon \mathbf{z}_j}{\|\mathbf{x}_i + 2\epsilon \mathbf{z}_j\|_1}, \; \mathbf{y}_{i,j}^- := \frac{\mathbf{x}_i - 2\epsilon \mathbf{z}_j}{\|\mathbf{x}_i - 2\epsilon \mathbf{z}_j\|_1}, \; 1 \le i \le n, \; 1 \le j \le s.$$

With this definition, it is clear that *every* vector in the ball of radius 2ϵ centered at each $\mathbf{x}_i$ can be written as a nonnegative combination of the set of vectors $\{\mathbf{y}_{i,j}^+, \mathbf{y}_{i,j}^-\}$.

Now define the cone

$$\mathcal{B} := \text{Cone}\{\mathbf{y}_{i,j}^+, \mathbf{y}_{i,j}^-\}.$$

We begin by observing that $\mathbf{p_u} \in \mathcal{B}$ whenever $|\mathbf{u}| \ge L$. This is because each such $\mathbf{p_u}$ lies within a distance of ϵ from one of the $\mathbf{x}_i$ whenever $|\mathbf{u}| \ge L$. In particular, $\mathbf{p_u} \in \mathcal{B}$ whenever $|\mathbf{u}| = L$. Moreover, by (7.1) and (7.2), every $\mathbf{p_v}$ with $|\mathbf{v}| < L$ is a nonnegative combination of $\mathbf{p_u}$ with $|\mathbf{u}| = L$. To see this, let $s := L - |\mathbf{v}|$, and note that

$$f_{\mathbf{v}} \mathbf{p_v} = F_{I,0}^{(\mathbf{v})} = \sum_{\mathbf{w} \in \mathcal{M}^s} F_{I,0}^{(\mathbf{vw})},$$

and each vector $F_{I,0}^{(\mathbf{vw})}$ belongs to $\mathcal{B}$. Hence $\mathbf{p_u} \in \mathcal{B}$ whenever $|\mathbf{u}| < L$. Combining all this shows that $\mathbf{p_u} \in \mathcal{B}$ for all $\mathbf{u} \in \mathcal{M}^*$. As a result, it follows that $\mathcal{C}_c \subseteq \mathcal{B}$.

While the cone $\mathcal{B}$ is polyhedral, it is not necessarily invariant under each $D^{(u)}$. For the purpose of constructing such an invariant cone, it is now shown that $\mathcal{B}$ is invariant under each $D^{(\mathbf{u})}$ whenever $|\mathbf{u}|$ is sufficiently long. By Lemma 7.18, it follows that for every vector $\mathbf{y}$, the vector $(1/f_{\mathbf{u}})D^{(\mathbf{u})}\mathbf{y}$ gets "aligned" with $p_{\mathbf{u}}$ as $|\mathbf{u}|$ becomes large. Therefore it is possible to choose an integer s such that

$$\| \frac{\|\mathbf{p_u}\|_1}{\|D^{(\mathbf{u})}\mathbf{y}\|_1} D^{(\mathbf{u})}\mathbf{y} - \mathbf{p_u}\|_1 \le \epsilon \text{ whenever } |\mathbf{u}| \ge s,$$

whenever $\mathbf{y}$ equals one of the $2nr$ vectors $\mathbf{y}_{i,j}^+, \mathbf{y}_{i,j}^-$. Without loss of generality it may be assumed that $s \ge L$. In particular, the vectors $D^{(\mathbf{u})}\mathbf{y}_{i,j}^+$ and $D^{(\mathbf{u})}\mathbf{y}_{i,j}^-$, after normalization, are all within a distance of ϵ from $\mathbf{p_u}$, which in turn is within a distance of ϵ from some $\mathbf{x}_t$. By the triangle inequality, this implies that the normalized vectors corresponding to $D^{(\mathbf{u})}\mathbf{y}_{i,j}^+$ and $D^{(\mathbf{u})}\mathbf{y}_{i,j}^-$ are all within a distance of 2ϵ from some $\mathbf{x}_t$, and hence belong to $\mathcal{B}$. In other words, we have shown that

$$D^{(\mathbf{u})}\mathcal{B} \subseteq \mathcal{B} \; \forall \mathbf{u} \text{ with } |\mathbf{u}| \ge s.$$

Now we are in a position to construct the desired polyhedral cone $\mathcal{P}$. Define

$$\mathcal{B}_i := \{D^{(\mathbf{u})}\mathcal{B} : |\mathbf{u}| = i\}, 1 \le i \le s - 1.$$

Thus $\mathcal{B}_i$ is the set obtained by multiplying each vector in $\mathcal{B}$ by a matrix of the form $D^{(\mathbf{u})}$ where $\mathbf{u}$ has length precisely i. It is easy to see that, since $\mathcal{B}$ is polyhedral, so is each $\mathcal{B}_i$. Now define

$$\mathcal{P} := \text{Cone}\{\mathcal{B}, \mathcal{B}_1, \ldots, \mathcal{B}_{s-1}\}.$$

For this cone, we establish in turn each of the four claimed properties.

Property 1: By definition we have that $D^{(u)}\mathcal{B}_i \subseteq \mathcal{B}_{i+1}\ \forall u \in \mathcal{M}$, whenever $0 \leq i \leq s-2$, and we take $\mathcal{B}_0 = \mathcal{B}$. On the other hand, $D^{(u)}\mathcal{B}_{s-1} \subseteq \mathcal{B}$ as has already been shown. Hence $\mathcal{P}$ is invariant under $D^{(u)}$ for each $u \in \mathcal{M}$.

Property 2: We have already seen that $\mathbf{p_u} \in \mathcal{B}$ for all $\mathbf{u} \in \mathcal{M}^*$. Hence $\mathcal{C}_c = \text{Cone}\{\mathbf{p_u} : \mathbf{u} \in \mathcal{M}^*\} \subseteq \mathcal{B} \subseteq \mathcal{P}$. To prove the other containment, note that by assumption, the sphere of radius 2ϵ centered at each cluster point $\mathbf{x}_i$ is contained in $\mathcal{C}_o$. Hence $\mathcal{B} \subseteq \mathcal{C}_o$. Moreover, $\mathcal{C}_o$ is invariant under $D^{(u)}$ for each $u \in \mathcal{M}$. Hence $\mathcal{B}_i \subseteq \mathcal{C}_o$ for each $i \in \{1, \ldots, s-1\}$. Finally $\mathcal{P} \in \mathcal{C}_o$.

Property 3: Note that each $\mathbf{p_u}$ belongs to $\mathcal{B}$, which is in turn a subset of $\mathcal{P}$. In particular, $\boldsymbol{\phi} = \mathbf{p}_{\emptyset} \in \mathcal{P}$.

Property 4: Since $\mathcal{P} \subseteq \mathcal{C}_o$, it follows that $(\mathcal{P})^p \supseteq (\mathcal{C}_o)^p$. Hence it is enough to show that $\boldsymbol{\theta}^t \in (\mathcal{C}_o)^p$. But this is easy to establish. Let $\mathbf{y} \in \mathcal{C}_o$ be arbitrary. Then by the definition of $\mathcal{C}_o$ we have that

$$\boldsymbol{\theta} D^{(\mathbf{u})}\mathbf{y} \geq 0\ \forall \mathbf{u} \in \mathcal{M}^*\ \forall \mathbf{y} \in \mathcal{C}_o.$$

In particular, by taking $\mathbf{u}$ to be the empty string (leading to $D^{(\mathbf{u})} = I$), it follows that $\boldsymbol{\theta}\mathbf{y} \geq 0\ \forall \mathbf{y} \in \mathcal{C}_o$. Since $\mathbf{y}$ is arbitrary, this shows that $\boldsymbol{\theta}^t \in (\mathcal{C}_o)^p$. □

Proof. (Of Theorem 7.17): The proof of the main theorem closely follows the material in [6, pp. 117–119]. Let us "recycle" the notation and let $\mathbf{y}_1, \ldots, \mathbf{y}_s$ denote generators of the polyhedral cone $\mathcal{P}$. In other words, $\mathcal{P}$ consists of all nonnegative combinations of the vectors $\mathbf{y}_1, \ldots, \mathbf{y}_s$. Note that neither the integer s nor the generators need be uniquely defined, but this does not matter. Define the matrix

$$Y := [\mathbf{y}_1 | \ldots | \mathbf{y}_s] \in \mathbb{R}^{m^k \times s}.$$

Then it is easy to see that

$$\mathcal{P} = \{Y\mathbf{x} : \mathbf{x} \in \mathbb{R}^s_+\}.$$

Now we can reinterpret the four properties of Lemma 7.19 in terms of this matrix. Actually we need not bother about Property 2.

Property 1: Since $\mathcal{P}$ is invariant under $D^{(u)}$ for each $u \in \mathcal{M}$, it follows that each $D^{(u)}\mathbf{y}_i$ is a nonnegative combination of $\mathbf{y}_1, \ldots, \mathbf{y}_s$. Hence there exist nonnegative matrices $G^{(u)} \in \mathbb{R}^{s \times m^k}_+, u \in \mathcal{M}$ such that

$$D^{(u)}Y = YG^{(u)},\ \forall u \in \mathcal{M}.$$

Property 3: Since $\boldsymbol{\phi} \in \mathcal{P}$, there exists a nonnegative vector $\mathbf{z} \in \mathbb{R}^s_+$ such that

$$\boldsymbol{\phi} = Y\mathbf{z}.$$

Property 4: Since $\boldsymbol{\theta} \in \mathcal{P}^p$, we have in particular that $\boldsymbol{\theta}\mathbf{y}_i \geq 0$ for all i. Hence

$$\mathbf{h} := \boldsymbol{\theta} Y \in \mathbb{R}^s_+.$$

Moreover, $\mathbf{h} \neq \mathbf{0}$, because $\boldsymbol{\theta}\boldsymbol{\phi} = \mathbf{h}\mathbf{z} = 1$, the frequency of the empty string.

With these observations, we can rewrite the expression for the frequency of an arbitrary string $\mathbf{u} \in \mathcal{M}^*$. We have

$$\begin{aligned} f_{\mathbf{u}} &= \boldsymbol{\theta} D^{(u_1)} \cdots D^{(u_l)} \boldsymbol{\phi} \\ &= \boldsymbol{\theta} D^{(u_1)} \cdots D^{(u_l)} Y \mathbf{z} \\ &= \boldsymbol{\theta} D^{(u_1)} \cdots D^{(u_{l-1})} Y G^{(u_l)} \mathbf{z} = \cdots \\ &= \boldsymbol{\theta} Y G^{(u_1)} \cdots G^{(u_l)} \mathbf{z} \\ &= \mathbf{h} G^{(u_1)} \cdots G^{(u_l)} \mathbf{z} \end{aligned} \tag{7.64}$$

The formula (7.64) is similar in appearance to (7.34), but with one very important difference: *Every matrix and vector in (7.64) is nonnegative.* Therefore, in order to construct an irreducible HMM from the above formula, we need to ensure that the matrix $Q := \sum_{u \in \mathcal{M}} G^{(u)}$ is irreducible and row stochastic, that $\mathbf{h}$ satisfies $\mathbf{h} = \mathbf{h}Q$, and that $\mathbf{z} = \mathbf{e}_s$. This is achieved through a set of three reductions. Note that these reductions are the same as in [6, pp. 117-119].

Now for the first time we invoke the assumption that the process $\{Y_t\}$ is α-mixing. From Theorem 7.10, this assumption implies that the matrix $S = \sum_{u \in \mathcal{M}} D^{(u)}$ has the "strong Perron property," namely: the spectral radius of S is one, and one is an eigenvalue of S; moreover, if λ is any eigenvalue of S besides one, then $|\lambda| < 1$. We also know that $\boldsymbol{\phi}$ and $\boldsymbol{\theta}$ are respectively a column eigenvector and a row eigenvector of S corresponding to the eigenvalue one.

Now let us return to the formula (7.64). Define $Q := \sum_{u \in \mathcal{M}} G^{(u)}$ as before. Observe that Q is a nonnegative matrix; hence, by [16, Theorem 1.3.2, p. 6], it follows that the spectral radius $\rho(Q)$ is also an eigenvalue. Moreover, $\rho(Q)$ is at least equal to one, because

$$\mathbf{h}Q = \boldsymbol{\theta} \sum_{u \in \mathcal{M}} Y G^{(u)} = \boldsymbol{\theta} \left(\sum_{u \in \mathcal{M}} D^{(u)} \right) Y = \boldsymbol{\theta} Y = \mathbf{h}.$$

Here we make use of the fact that $\boldsymbol{\theta}$ is a row eigenvector of $\sum_{u \in \mathcal{M}} D^{(u)}$ corresponding to the eigenvalue one.

In what follows, we cycle through three steps in order to arrive at a situation where Q is irreducible and row stochastic. In each step we will be replacing the various matrices by other, smaller matrices that play the same role. To avoid notational clutter, the old and new matrices are denoted by the same symbols.

Step 1: If Q is irreducible, go to Step 3. If Q is reducible, permute rows and columns if necessary and partition Q as

$$Q = \begin{bmatrix} Q_{11} & Q_{12} \\ \mathbf{0} & Q_{22} \end{bmatrix},$$

where Q_{11} is irreducible and has dimension $(s-l) \times (s-l)$, and Q_{22} has dimension $l \times l$ for some $l < s$. (It is not assumed that Q_{22} is irreducible, since an irreducible partition of Q may have more than two "blocks.") Since $Q = \sum_{u \in \mathcal{M}} G^{(u)}$ and each $G^{(u)}$ is nonnegative, if we partition each $G^{(u)}$ commensurately, then the block zero structure of Q will be reflected in each $G^{(u)}$. Now there are two possibilities: Either $\rho(Q_{11}) = 1$, or it is not. If $\rho(Q_{11}) = 1$, go to Step 2. If $\rho(Q_{11}) \neq 1$, proceed as follows: Let $\lambda_1 = \rho(Q_{11}) \neq 1$. Note that, by Theorem 3.25, it is possible to choose a strictly positive eigenvector of Q_{11} corresponding to the eigenvalue $\rho(Q_{11})$, since Q_{11} is irreducible. So choose a positive vector $\mathbf{x}_1 \in \mathbb{R}_+^{s-l}$ such that $Q_{11}\mathbf{x}_1 = \lambda_1 \mathbf{x}_1$. Then clearly $Q\mathbf{x} = \lambda_1 \mathbf{x}$, where $\mathbf{x} = [\mathbf{x}_1^t \ \ \mathbf{0}^t]^t$. Since $\lambda_1 \neq 1$, it follows that $\mathbf{hx} = 0$, because a row eigenvector and a column eigenvector corresponding to different eigenvalues are orthogonal. So if we partition $\mathbf{h}$ as $[\mathbf{h}_1 \ \ \mathbf{h}_2]$, then $\mathbf{h}_1 = \mathbf{0}$ since $\mathbf{x}_1$ is a positive vector. Now observe that each $G^{(u)}$ has the same block-triangular structure as Q. Hence, by a slight abuse of notation, let us define, for every string $\mathbf{u} \in \mathcal{M}^*$,

$$G^{(\mathbf{u})} = \begin{bmatrix} G_{11}^{(\mathbf{u})} & G_{12}^{(\mathbf{u})} \\ \mathbf{0} & G_{22}^{(\mathbf{u})} \end{bmatrix}.$$

Let us partition $\mathbf{z}$ commensurately. Because the first block of $\mathbf{h}$ is zero, it is easy to verify that, for every $\mathbf{u} \in \mathcal{M}^*$, we have

$$f_{\mathbf{u}} = \mathbf{h}G^{(\mathbf{u})}\mathbf{z} = \mathbf{h}_2 G_{22}^{(\mathbf{u})} \mathbf{z}_2,$$

where $\mathbf{z}_2$ consists of the last l components of $\mathbf{z}$. Hence we can partition Y as $[Y_1 | Y_2]$ where $Y_2 \in \mathbb{R}^{r \times l}$ and make the following substitutions:

$$s \leftarrow l, Y \leftarrow Y_2, G^{(u)} \leftarrow G_{22}^{(u)} \ \forall u \in \mathcal{M}, \mathbf{h} \leftarrow \mathbf{h}_2, \mathbf{z} \leftarrow \mathbf{z}_2.$$

With these substitutions, we have reduced the number of columns of Y from s to l, and the relationship (7.64) continues to hold. Now go back to Step 1.

Step 2: If we have reached this point, then Q is reducible, and if it is partitioned as above, we have $\rho(Q_{11}) = 1$. Choose a positive vector $\mathbf{x}_1$ such that $Q_{11} = \mathbf{x}_1$. Then $Q\mathbf{x} = \mathbf{x}$, where as before $\mathbf{x} = [\mathbf{x}_1^t \ \ \mathbf{0}^t]^t$. Next, note that

$$SY\mathbf{x} = \left(\sum_{u \in \mathcal{M}} D^{(u)} \right) Y\mathbf{x} = Y \left(\sum_{u \in \mathcal{M}} G^{(u)} \right) \mathbf{x} = YQ\mathbf{x} = Y\mathbf{x}.$$

Hence $Y\mathbf{x}$ is a column eigenvector of S corresponding to the eigenvalue one. However, from Theorem 7.10, the α-mixing property implies that S has a simple eigenvalue at one, with corresponding column eigenvector $\boldsymbol{\phi} = F_{I,0}$. Hence $F_{I,0}$ equals $Y\mathbf{x}$ times some scale factor, which can be taken as one without loss of generality (since both vectors are nonnegative). Partition Y as $[Y_1 \ \ Y_2]$ where $Y_1 \in \mathbb{R}^{r \times (s-l)}$. Then

$$F_{I,0} = [Y_1 \ \ Y_2] \begin{bmatrix} \mathbf{x}_1 \\ \mathbf{0} \end{bmatrix} = Y_1 \mathbf{x}_1.$$

Moreover, since each $G^{(u)}$ inherits the zero structure of Q, we have that

$$D^{(u)}[Y_1 \ \ Y_2] = [Y_1 \ \ Y_2] \left[\begin{array}{cc} G_{11}^{(u)} & G_{12}^{(u)} \\ \mathbf{0} & G_{22}^{(u)} \end{array} \right].$$

In particular, we have that $D^{(u)}Y_1 = Y_1 G_{11}^{(u)}$. This means that $F_{I,0}$ lies in the cone generated by the columns of Y_1, and that this cone is invariant under $D^{(u)}$ for each $u \in \mathcal{M}$. So if we define $\mathbf{h}_1 := \boldsymbol{\theta} Y_1$, then because of the zero block in $\mathbf{x}$ it follows that

$$f_{\mathbf{u}} = \boldsymbol{\theta} D^{(\mathbf{u})} F_{I,0} = \mathbf{h}_1 G_{11}^{(\mathbf{u})} \mathbf{x}_1.$$

So now we can make the substitutions

$$s \leftarrow s - l, Y \leftarrow Y_1, G^{(u)} \leftarrow G_{11}^{(\mathbf{u})}, \mathbf{h} \leftarrow \mathbf{h}_1, \mathbf{z} \leftarrow \mathbf{x}_1.$$

With these substitutions we have reduced the number of columns of Y from s to $s - l$, and the relationship (7.64) continues to hold. Moreover, the resulting matrix Q is the old Q_{11}, which is irreducible. Now go to Step 3.

Step 3: When we reach this stage, (7.64) continues to hold, but with two crucial additional features: Q is irreducible and $\rho(Q) = 1$. As before, let s denote the size of the matrix Q, and write $\mathbf{z} = [z_1 \ldots z_s]^t$, where each z_i is positive. Define $Z = \text{Diag}\{z_1, \ldots, z_s\}$. Now (7.64) can be rewritten as

$$f_{\mathbf{u}} = \mathbf{h} Z Z^{-1} G^{(u_1)} Z \cdot Z^{-1} G^{(u_2)} Z \ldots Z^{-1} G^{(u_l)} Z \cdot Z^{-1} \mathbf{z}.$$

Thus (7.64) holds with the substitutions

$$G^{(u)} \leftarrow Z^{-1} G^{(u)} Z, \mathbf{h} \leftarrow \mathbf{h} Z, \mathbf{z} \leftarrow Z^{-1} \mathbf{z}.$$

In this process, Q gets replaced by $Z^{-1} Q Z$. Now observe that

$$Z^{-1} Q Z \mathbf{e}_s = Z^{-1} Q \mathbf{z} = Z^{-1} \mathbf{z} = \mathbf{e}_s.$$

In other words, the matrix $Z^{-1} Q Z$ is row stochastic. It is obviously nonnegative and irreducible. Moreover, we have that $\mathbf{hz} = 1$ since it is the frequency of the empty string, which by definition equals one. Hence the row vector $\mathbf{h} Z^{-1}$ is row stochastic in that its entries add up to one. Hence, after we make the substitutions, (7.64) holds with the additional properties that (i) $Q := \sum_{u \in \mathcal{M}} G^{(u)}$ is row-stochastic, (ii) $\mathbf{h}$ is row-stochastic and satisfies $\mathbf{h} = \mathbf{h} Q$, and (iii) $\mathbf{z} = \mathbf{e}_s$. Now it follows from Lemma 7.12 that the process $\{Y_t\}$ has a "joint Markov process" HMM. Moreover, the matrix Q is irreducible.

Thus far it has been established that the stochastic process $\{Y_t\}$ has an irreducible HMM. Moreover, this process is assumed to be α-mixing. So from Theorem 7.13, it finally follows that either the corresponding state transition matrix is aperiodic, or else the consistency conditions (7.60) hold. □

Theorem 7.17 gives *sufficient* conditions for the existence of an irreducible HMM that satisfies some consistency conditions in addition. It is therefore natural to ask how close these sufficient conditions are to being necessary. The paper [6, Lemma 2] answers this question.

Theorem 7.20 *Given an irreducible HMM with n states and m outputs, define its period p. Rearrange the state transition matrix A as in Theorem 7.13, permute the matrices $M^{(u)}, u \in \mathcal{M}$ correspondingly, and define the blocks $M_i^{(u)}$ in analogy with the partition of A. Suppose in addition that there exists an index $q \leq s$ such that the following property holds: For every string $\mathbf{u} \in \mathcal{M}^q$ and every integer r between 1 and p, every column of the product $M_r^{(u_1)} M_{r+1}^{(u_2)} \ldots M_{r+q-1}^{(u_q)}$ is either zero or else is strictly positive. In this computation, any subscript M_i is replaced by $i \bmod p$ if $i > p$. With this property, the HMM is α-mixing and also ultra-mixing.*

Thus we see that there is in fact a very small gap between the sufficiency condition presented in Theorem 7.17 and the necessary condition discovered earlier in [6]. If the sufficient conditions of Theorem 7.17 are satisfied, then there exists an irreducible HMM that also satisfies the consistency conditions (7.60). Conversely, if an irreducible HMM satisfies the consistency conditions (7.60) and one other technical condition, then it satisfies three out of the four hypotheses of Theorem 7.17, the only exception being the technical condition about the cluster points lying in the interior of the cone $\mathcal{C}_c$.

We conclude this section by discussing the nature of the "technical" conditions in the hypotheses of Theorems 7.17 and 7.20. The idea is to show that, in a suitably defined topology, each of the conditions is satisfied by an "open dense subset" of stochastic processes. Thus, if the given process satisfies the condition, so does any sufficiently small perturbation of it, whereas if a given process fails to satisfy the condition, an arbitrarily small perturbation will cause the condition to hold.

Let us begin with the fourth hypothesis of Theorem 7.13. We follow [102] and define a topology on the set of all stationary stochastic processes assuming values in $\mathcal{M}$. Suppose we are given two stochastic processes assuming values in a common finite alphabet $\mathcal{M}$. Let $f_{\mathbf{u}}, g_{\mathbf{u}}, \mathbf{u} \in \mathcal{M}^*$ denote the frequency vectors of the two stochastic processes. This is equivalent to specifying the joint distribution of l-tuples of each stochastic process, for every integer l. If we arrange all strings $\mathbf{u} \in \mathcal{M}^*$ in some appropriate lexical ordering (say first lexical), then each of $[f_{\mathbf{u}}, \mathbf{u} \in \mathcal{M}^*], [g_{\mathbf{u}}, \mathbf{u} \in \mathcal{M}^*]$ is a vector with a countable number of components, and each component lies between 0 and 1.[5] Let the symbols $\mathbf{f}, \mathbf{g}$, without any subscript, denote these vectors belonging to ℓ_∞. We might be tempted to compare the two stochastic processes by computing the norm $\|\mathbf{f} - \mathbf{g}\|_\infty$. The difficulty with this approach is that, as the length of the string $\mathbf{u}$ approaches infinity, the likelihood of that sequence will in general approach zero. Thus, in any "reasonable" stochastic process, the difference $f_{\mathbf{u}} - g_{\mathbf{u}}$ will approach zero as $|\mathbf{u}| \to \infty$, but this tells us nothing about how close the two probability laws are. To get around this

[5]Note that there is a lot of redundancy in this description of a stochastic process because, as we have already seen, the joint distribution of l-tuples can be uniquely determined from the joint distribution of s-tuples if $s > l$.

difficulty, for each $\mathbf{u} \in \mathcal{M}^*$, we define the vector $\mathbf{p}_{|\mathbf{u}} \in [0,1]^m$ as follows:

$$\mathbf{p}_{|\mathbf{u}} = \frac{1}{f_{\mathbf{u}}} \mathbf{f}_{\mathbf{u}v, v \in \mathcal{M}} = \left[\frac{f_{\mathbf{u}v}}{f_{\mathbf{u}}}, v \in \mathcal{M} \right].$$

Thus $\mathbf{p}_{|\mathbf{u}}$ is just the conditional distribution of the *next* symbol, given the past history $\mathbf{u}$. The advantage of $\mathbf{p}_{|\mathbf{u}}$ is that, even as $|\mathbf{u}|$ becomes large, the elements of this vector must still add up to one, and as a result they cannot all go to zero. With this convention, let us list all strings $\mathbf{u} \in \mathcal{M}^*$ in some appropriate lexical ordering (say first lexical), and for each $\mathbf{u}$ let us define the conditional distribution vectors $\mathbf{p}_{|\mathbf{u}}$ corresponding to $\{f_{\mathbf{u}}\}$, and the conditional distribution vectors $\mathbf{q}_{|\mathbf{u}}$ corresponding to the vector $\{\mathbf{g}_{\mathbf{u}}\}$. Finally, let us define the vectors

$$\tilde{\mathbf{p}} := [\mathbf{p}_{|\mathbf{u}}, \mathbf{u} \in \mathcal{M}^*], \tilde{\mathbf{q}} := [\mathbf{q}_{|\mathbf{u}}, \mathbf{u} \in \mathcal{M}^*].$$

Thus both $\tilde{\mathbf{p}}, \tilde{\mathbf{q}}$ have a countable number of components, since $\mathcal{M}^*$ is a countable set. Thus the ℓ_∞ norm of the difference $\tilde{\mathbf{p}} - \tilde{\mathbf{q}}$ is a measure of the disparity between the two stochastic processes. This is essentially the distance measure introduced in [102]. With this measure, it is easy to see that the fourth hypothesis of Theorem 7.13 is truly technical: If a given stochastic process satisfies the condition about the cluster points, then so will any sufficiently small perturbation of it, while if a given stochastic process fails to satisfy this condition, any sufficiently small perturbation of it will cause the condition to be satisfied.

Now let us turn to the condition in Theorem 7.20. Given two HMMs over a common state space, a natural metric is

$$\sum_{u \in \mathcal{M}} \|M_1^{(u)} - M_2^{(u)}\|,$$

where $\|\cdot\|$ is any reasonable matrix norm. Again, it is easy to see that the condition in Theorem 7.20 about the various columns being either identically zero or strictly positive is "technical." In fact, if for a HMM some elements of the matrices $M_r^{(u_1)} M_{r+1}^{(u_2)} \ldots M_{r+q-1}^{(u_q)}$ are zero, then by simply making an arbitrarily small perturbation in the matrices we can ensure that every entry is strictly positive.

PART 3
Applications to Biology

Chapter Eight

Some Applications to Computational Biology

In the preceding chapters, we have introduced a great many ideas regarding Markov processes and hidden Markov processes. In the last part of the book, we study ways in which some of these ideas can be applied to problems in computational biology. One of the first observations to be made is that, unlike problems in engineering, problems in computational biology do not always perfectly fit into a nice mathematical framework. Instead, one uses a mathematical framework as a starting point, and makes some modifications to suit the practicalities of biology. Usually, in the process, "rigor" is not always maintained. However, this can be justified on the grounds that the aim of computational biology is not to prove theorems (as we did in the first two parts of this book), but rather to address some very real applications. One of the consequences of this is that algorithms in computational biology evolve very rapidly, and any attempt to catalog them would become dated equally rapidly. In view of that, in the last part of the book we present some historical algorithms that were among the first to apply the methods of hidden Markov models to biological problems. Over time these algorithms have been modified but the core philosophy of the later algorithms is still similar to that of their progenitors. Thus the contents of this part of the book can serve to provide an introduction to some current methods in computational biology.

In the present chapter we introduce three important problems, namely sequence alignment, gene finding, and protein classification. Two relatively encyclopedic, but slightly aged, references are [97] and [11]. We revisit the sequence alignment problem again in Chapter 9 when the BLAST algorithm is discussed.

The chapter is organized as follows. In Section 8.1, we give a very brief introduction to some relevant aspects of biology. In Section 8.2, we discuss the problem of optimal gapped alignment between two sequences. This is a way to detect similarity between two sequences over a common alphabet, such as the four-symbol alphabet of nucleotides, or the 20-symbol alphabet of amino acids. Though an "exact" solution to this problem can be found using dynamic programming, it turns out that an approximate solution is often good enough in many situations. This was one of the motivations for the development of BLAST theory, which is presented in Chapter 9. In Section 8.3, we discuss some widely used algorithms for finding genes from DNA sequences (genomes). Finally, in Section 8.4, we discuss a special type of hidden Markov model, called a profile HMM, which is commonly used to classify proteins into a small number of groups.

8.1 SOME BASIC BIOLOGY

It would be more accurate to name this section "A view of biology from the perspective of computation." The section represents an attempt to put forward a *simplified view* of the basic issues in genomics and proteomics, which are perhaps the two aspects of biology that lend themselves most readily to computational analysis. To persons with a mathematical training, biology appears to be a bewildering array of terminology and conventions. This section therefore represents an attempt to simplify the subject for the benefit of those who wish to understand what the basic issues are, at least from a computational standpoint, and then move on to tackle some of the outstanding problems.

Because of the simplification involved, some of the description below may not always be 100% accurate from a biological standpoint. For instance, it will be said below that the only difference between DNA and RNA is that the letter T gets replaced by the letter U. Biologically speaking, this statement is incorrect: T (thymine) has a different chemical structure and biological function from U (uracil). But for the intended readers of this book, the key point is that if a stretch of DNA is viewed as a long string over the four symbol alphabet $\{A, C, G, T\}$ (as explained below), then the corresponding stretch of RNA is obtained simply by substituting the symbol U for the symbol T wherever the latter occurs. Thus, *in this limited sense*, there is no difference between T and U. Thus readers are advised to keep this point in mind, and not treat everything said here as being "biologically true." However, all of the specific problem statements given here are "computationally true"–there is no lack of precision either in the problem formulations or in the solutions put forward.

8.1.1 The Genome

The genetic material in all living things is DNA, or deoxyribonucleic acid, which is an enormously complex molecule built up out of just four building blocks, known as nucleotides. The discovery of the nucleotides is generally credited to Phoebus Levene during the 1920s. The four nucleotides share a common phosphate backbone, and contain one of four nucleic acids, or bases, namely: adenine, cytosine, guanine and thymine. It is customary to denote the four nucleotides by the initial letters of the bases they contain, as A, C, G, T. There is no particular preferred order for listing the nucleotides, so they are listed here in the English alphabetical order. Figure 8.1 shows the four nucleic acids, while Figure 8.2 shows the corresponding four nucleotides including both the phosphate backbone and the base.

Because all four nucleotides share a common phosphate backbone, any nucleotide can "plug into" any other nucleotide, like a Lego toy. This is

adenine (A) guanine (G) cytosine (C) thymine (T)

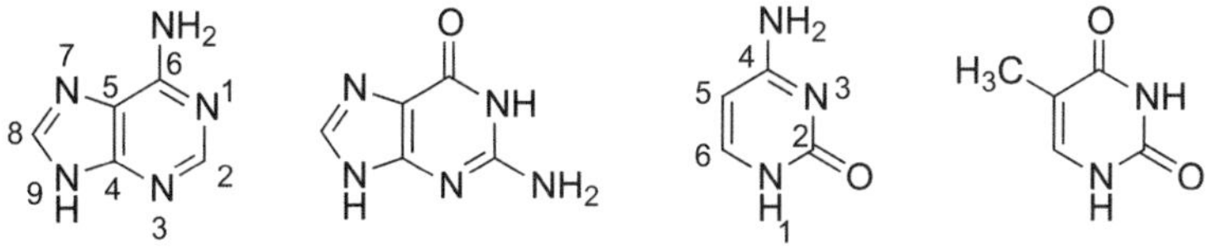

Figure 8.1 The Four Nucleic Acids. Reproduced with permission from *Nucleic Acids Book* by T. Brown Jr.

depicted in Figure 8.2. The so-called $3'$ end of one nucleotide forms a very strong covalent chemical bond with the $5'$ end of the next nucleotide. Thus it is possible to assemble the nucleotides in any order we wish. Such strings of nucleotides, assembled as per our specifications, are referred to as a strand of "oligonucleotides." While it has been known for decades that it is possible to produce "arbitrary" sequences of nucleotides, it is only within the past few years that it has become commercially feasible to produce oligonucleotide sequences on demand, thus leading to an entirely new field known as "synthetic biology."

The DNA molecule consists of two strands of oligonucleotides that run in opposite directions. In addition to the very strong covalent bond between successive nucleotides on the same strand, the DNA molecule also has much weaker hydrogen bonds between nucleotides on opposite strands, which is a property known as "reverse complementarity." Thus if one strand contains A, then its counterpart on the opposite side must be T. Similarly, if one strand contains C, then its counterpart on the opposite must be G. Figure 8.3 depicts the reverse complementarity property. It can be seen from this figure that the $A \leftrightarrow T$ bond is somewhat weaker than the $C \leftrightarrow G$ bond, because the former consists of two hydrogen atoms bonding, while the latter consists of three hydrogen atoms bonding.

The DNA molecule is shaped like a double helix as shown in Figure 8.4. The discovery of the shape of the DNA molecule and the chemical bonds that hold it together (the horizontal beads in Figure 8.4) was made by James Watson and Francis Crick in 1953. Prior to that, Edwin Chargaff had discovered that in DNA, adenine and thymine occur in equal proportions, and that cytosine and guanine also occur in equal proportions. The reverse complementarity property in the double helix structure explained why this must be so.

Determining the structure of DNA was considered to be a very important problem at that time, and several persons were working on it, including the famous chemist Linus Pauling. Experimental evidence for the helical structure of DNA was obtained by Rosalind Franklin based on X-ray diffraction techniques, though she (according to some reports) did not readily believe in the helical structure of DNA. Maurice Wilkins was a colleague of hers at King's College in London (and according to some reports, her boss) who

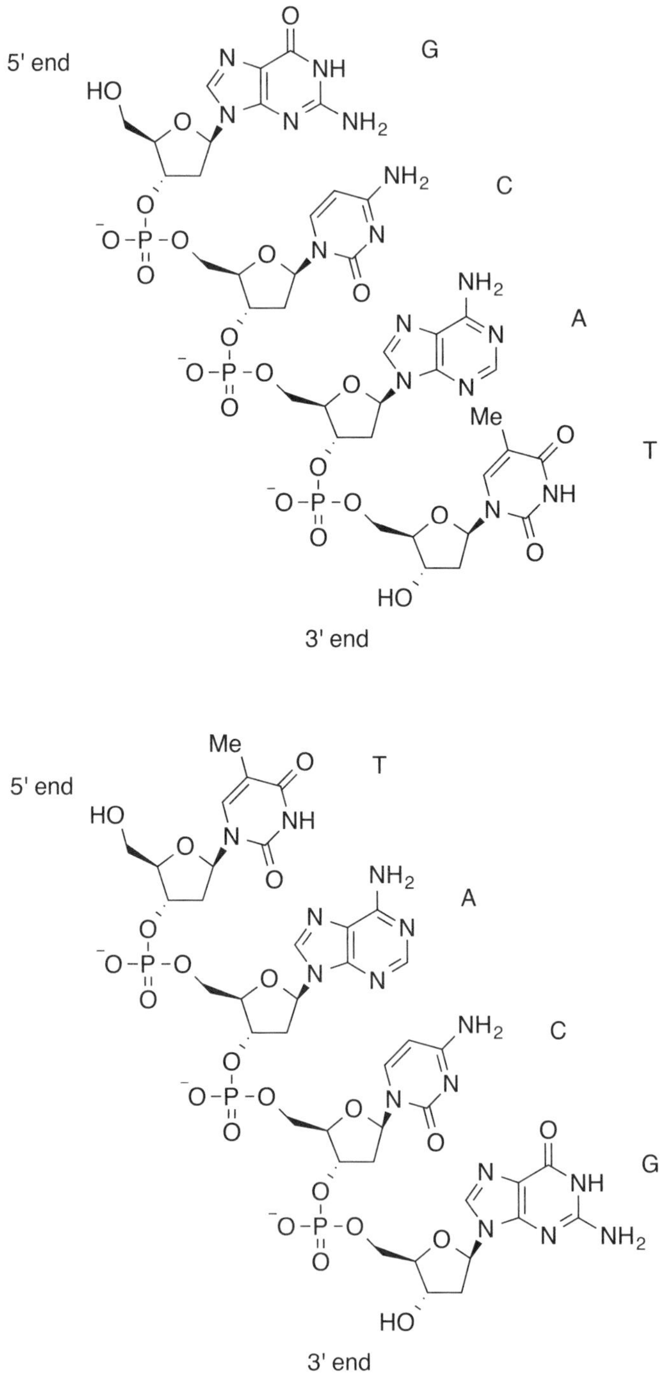

Figure 8.2 The Four Nucleotides with Bases and Phosphate Backbones. Reproduced with permission from *Nucleic Acids Book* by T. Brown Jr.

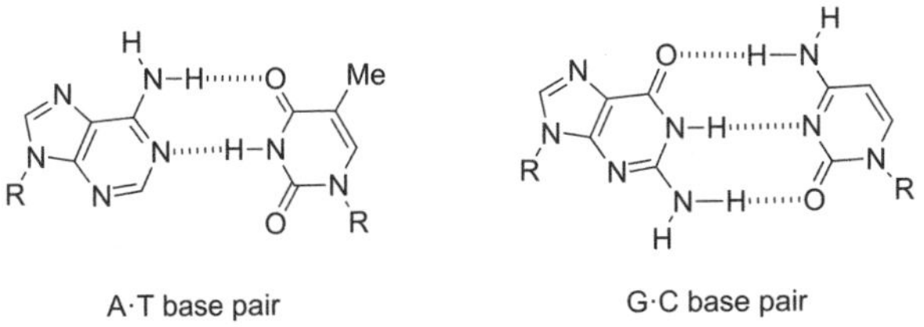

A·T base pair G·C base pair

Figure 8.3 Reverse Complementarity Including Hydrogen Bonds between A and T, and between C and G. Reproduced with permission from *Nucleic Acids Book* by T. Brown Jr.

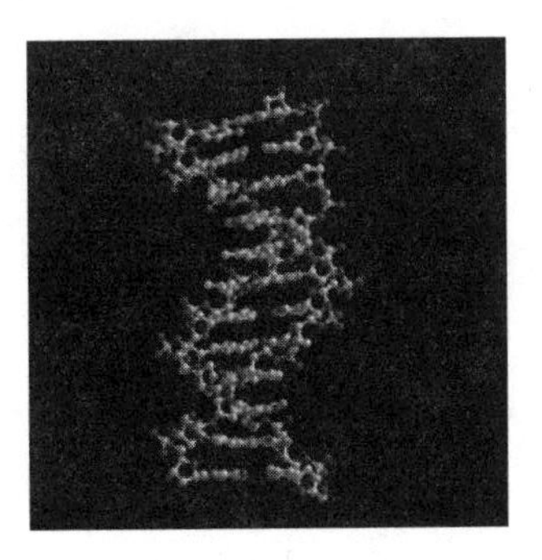

Figure 8.4 Double Helix Structure of DNA

was at the time trying to construct models based on available experimental evidence. Watson, Crick, and Wilkins shared the Nobel Prize in medicine in 1962 for their discoveries. Unfortunately Rosalind Franklin had passed away by that time, so one can only speculate as to whether she too would have received recognition in some form for the discovery. For Watson's own account of the discovery of the double helix, see [136]. The other two winners have also written their own version of events [31, 140], in the case of Wilkins 50 years after the event. For a counter-viewpoint that claims that Rosalind Franklin deserves far more credit than she has received, see [115]. In some "lesser" organisms such as viruses and bacterial phages, the DNA molecule folds back on itself, but still retains the double helix structure.

Each cell within a living organism contains a copy of the DNA. Human DNA consists of roughly *3.3 billion* nucleotides in each strand of the double helix. If the two strands of the double helix were to be separated and one strand were to be stretched out, the total length would be about *two meters*, though there are several estimates available in the literature. Similarly, estimates for the number of cells in the human body also vary widely, but a recent study [17] suggests an average figure of 37.2 trillion. Thus the total length of the DNA in the human body can be estimated at 75 trillion meters, roughly enough to make 250 round trips from the earth to the sun. And yet the chemical interactions are so strong that the DNA is very tightly wound up within itself, and each of the fifty trillion or so cells within the human body contains a copy of this two meter-long molecule. It is clear that, while one dimension of the DNA is very high, the other two dimensions are very tiny, which is why this compactification is possible.

It is important to understand that each strand of the double-helix DNA has a definite spatial direction. The starting point for each strand is called the $5'$ end, while the ending point is called the $3'$ end. If we think of a strand of DNA as a "tape," then there is only one way to read the tape–it cannot be "read backwards." Because of the definite spatial direction, expressions such as "the previous nucleotide," "the next nucleotide," "upstream," and "downstream" are completely unambiguous. This feature permits us to model the *spatial orientation* by a *temporal orientation* and use modeling methods based on stochastic processes. If the spatial orientation were to be arbitrary, we could not do this.

The "genome" of an organism is just a listing out, symbol by symbol, of the sequence of nucleotides that makes up one strand of the DNA. Because of reverse complementarity, if we know the listing of one strand, we know unambiguously the listing of the other strand. Since DNA occurs in two strands and the bases in each strand must "pair up" according to reverse complementarity, the length of a genome is specified in "base pairs." The typical length of the genome varies depending upon the nature and complexity of the organism. Viruses, which cannot survive on their own but need a host in order to replicate, typically have genomes that are several thousand base pairs long. Bacteria, which are the simplest self-sustaining life forms (meaning that they can reproduce on their own without a host, in contrast to viruses), have genomes that are a few million base pairs long. The mosquito has a genome that is about 300 million base pairs, the mouse genome is about 2.4 billion base pairs, while the human genome is about 3.3 billion base pairs long. The genome of wheat is still being assembled, but it appears that the genome will be much longer than that of humans. That last statistic ought to dispel the idea that the length of the genome is somehow monotonically related to the "intelligence" of the organism.

The determination of the genome of organisms is one of the great triumphs of experimental biology, because the genome is one of the most "unambiguous" representations of a life-form. One of the first to devise biochemical methods for determining the genome was Frederick Sanger, one of only three persons to receive two scientific Nobel Prizes.[1] A draft of the human genome was determined and published simultaneously in February 2001 by two groups: the International Human Genome Research Consortium (IHGRC) [69] and Celera Genomics [125], a private company that subsequently went out of business. As technology improves, we can aspire to a situation whereby it will be both quick and inexpensive to determine the genome of a large number of humans. See the first chapter of [105] for an excellent summary of the experimental methods and computational algorithms involved in "sequencing" and "assembling" a genome, that is, determining the string of symbols that comprise the genome. Moreover, it is noteworthy that the genome is a "digital" representation of life, in the sense that the symbols at each location in the genome can have only a finite number of possible values (four).

In spite of its enormous length, it appears that there is a great deal of redundancy in the human genome. The overlap between the human genome and the mouse genome is about 80%, whereas between a human and a chimpanzee (our nearest neighbor in the animal kingdom) the overlap is about 98%. Between two humans the overlap is still more striking. It is estimated that the genome sequences of two humans will agree in about 99.9% of the locations, and differ only in about 0.1%, or about 3 million base pairs. All that distinguishes one human from another, be it height, weight, color of eyes, color of hair, etc. can presumably be attributed to this tiny variation in the genome, aside from environmental factors. These variations from the "consensus" human genome are called Single Nucleotide Polymorphisms (SNPs), often pronounced as "snips." Even "identical" twins will not have identical genomes; rather, the overlap in such a case will be about 99.99%, as opposed to 99.9% in the case of two unrelated humans. If the genome of an organism is 100% reproduced to create another organism, the second one is called a "clone" of the first. The cloning of life forms is both a fascinating

[1]Sanger received two Nobel Prizes in chemistry, in 1958 and in 1980, the latter being for the method known as "Sanger sequencing." The other persons to receive two scientific Nobel Prizes were John Bardeen (two prizes in physics) and Marie Curie (one prize each in physics and in chemistry). Linus Pauling won one Nobel Prize in chemistry and the Nobel Peace Prize. Unfortunately Sanger passed away just as the present book was being completed.

as well as a controversial subject.

One of the most exciting challenges in computational biology is correlating an individual's genotype with his/her "phenotype," for example, the person's propensity to disease, responsiveness to a drug or treatment regime, or even potential adverse reactions to a drug. The current status is that in *some very specific situations*, we know that a particular SNP causes a specific disorder. Among the very first disorders to be tied unambiguously to a specific SNP is sickle cell anemia, which causes red blood cells to be shaped like a crescent (or a sickle), as opposed to the round shape of a normal red blood cell. It was discovered that sickle cell anemia is caused by just one substitution: the codon GAG that codes for glutamine gets replaced by GTG, which codes for valine.[2] Thus *exactly one SNP* at just the right place in the genome can cause sickle cell anemia. Other examples of genetic disorders are Huntington chorea and cystic fibrosis. Cystic fibrosis is in many ways well-suited for study using computational techniques, because while the *location* of the mutations that causes the disorder is well known, there are literally hundreds of mutations that have been discovered thus far among patients afflicted by this disorder. It would therefore be very interesting to correlate, using computational techniques, the particular mutation with the particular manifestation of cystic fibrosis. Unfortunately, most disorders are far more complex, and cannot be related to the malfunction of one specific gene. Instead, most disorders are caused by the malfunctioning of several genes.

8.1.2 The Genetic Code

As we have already seen, the genome of an organism is just an enormously long string over the four-symbol nucleotide alphabet $\{A, C, G, T\}$. The genome of an organism is the most "low level" description of an organism. One can think of the genome as a kind of "raw data" that needs to be turned into "information." This is one of the classic challenges of computational biology.

The next level of complexity in the genome arises from genes and proteins. The DNA of an organism consists in effect of two parts: (i) the genes whose function is to produce proteins, and (ii) the intergenic regions, sometimes referred to "junk" DNA. Proteins are the sustenance of life, and DNA must continually replicate itself so that the production of proteins can go on uninterrupted. Proteins were discovered as early as the beginning of the nineteenth century, in fact much earlier than genes. Practically all the proteins discovered at that time were essential dietary ingredients; for example, vitamins are proteins or combinations of proteins. By the 1840s it was already known that every protein consists of a sequence of amino acids, which are 20 in number. These 20 amino acids are denoted either by a single letter, or by three letters, as shown in Table 8.1. Just as the four nucleotides are

---

[2]Codons are introduced later in this section.

Amino Acid	**Three-Letter Code**	**Single-Letter Code**
glycine	Gly	G
alanine	Ala	A
valine	Val	V
leucine	Leu	L
isoleucine	Ile	I
methionine	Met	M
phenylalanine	Phe	F
tryptophan	Trp	W
proline	Pro	P
serine	Ser	S
threonine	Thr	T
cysteine	Cys	C
tyrosine	Tyr	Y
asparagine	Asn	N
glutamine	Gln	Q
aspartic acid	Asp	D
glutamic acid	Glu	E
lysine	Lys	K
arginine	Arg	R
histidine	His	H

Table 8.1 List of Amino Acids and Their Three-Letter and Single-Letter Codes

the building blocks of DNA, the 20 amino acids are the building blocks of proteins. Thus, just we can think of the genome as a string over the four-symbol alphabet of nucleotides, we can think of a protein as a string over the 20-symbol alphabet of amino acids. The listing out of a protein in terms of its sequence of amino acids is called the **primary structure**. Thus we can think of describing a protein in terms of its primary structure as being analogous to describing an organism in terms of its genome. Both are the "lowest level" descriptions, and additional work is needed to extract useful information in either case. Typically a protein consists of several hundred, or perhaps a few thousand, amino acids. At the other end, proteins consisting of as few as 50 amino acids are also known.

Once the double helix structure of DNA was discovered in 1953, the scientific community attempted to understand how DNA gets converted to proteins. The working hypothesis, which is by now quite universally accepted, states that first the double-stranded DNA molecule is separated into its two individual strands. Then particular stretches of DNA form the template for conversion to RNA. In the process, thymine (T) gets replaced by uracil (U). RNA is a single-stranded molecule and is thus somewhat unstable chemically. (However, double-stranded RNA has been discovered recently.) This process is known as "transcription." Then triplets of RNA nucleotides A, C, G, U get converted into amino acids through a process known as "translation." The entire hypothesis is labeled as "the central dogma" of biology. This much was understood by 1960, but what happened next was still not clear.

Recall that the four nucleotides that make up DNA were discovered during the 1920s. The basis of the conversion of DNA to proteins, called the "genetic code," was discovered in full only in the 1960s. In the early 1960s, Marshall Nirenberg succeeded in showing that the triplet UUU produced the amino acid phenylaline. In quick succession he and his colleagues succeeded in showing that several amino acids were produced by various triplets of nucleotides. See [99] for a summary of their work at the time. Subsequently Khorana discovered another method for synthesizing all the 64 possible combinations of nucleotides [91], and used this to decipher which triplet of nucleotides produced which amino acid, that is, to discover the complete genetic code [92, 80]. Since there are $4^3 = 64$ triplets (called "codons") and only 20 amino acids, there had to be some redundancy. Khorana further showed that each protein-coding RNA ended with one of three sequences, called the **stop codons**, namely UAA, UAG, UGA. Finally, he also showed that every protein-coding RNA began with the "start" codons AUG or GUG. However, while a stop codon cannot occur in the middle of a protein sequence, a "start" codon can also occur in the middle of a protein, where it codes for the amino acid methionine. In recognition of this seminal work, Nirenberg and Khorana shared the Nobel Prize in medicine in 1968, along with Robert Holley, who discovered tRNA (translation RNA).

Figure 8.5 depicts the genetic code in compact form, showing the three-letter code of the amino acid produced by each codon. Strictly speaking, RNA codons consist of triplets from the RNA alphabet $\{A, C, G, U\}$. However, since the RNA sequence is obtained simply by replacing the symbol T by the symbol U (ignoring the chemical significance of such a substitution), one can think of "codons" as either triplets over the alphabet $\{A, C, G, T\}$ or over the alphabet $\{A, C, G, U\}$. Both conventions are used in the biology community, depending on convenience. Thus one can think of TAA, TAG, TGA as stop codons, and of ATG, GTG as the start codons.

From the above table, it is clear that there is a great deal of subtlety in the genetic code. If we think of the genetic code as a map from the 64-symbol set $\{A, C, G, U\}^3$ into the 21 symbol set consisting of the 20 amino acids plus the stop codon, then the structure of the map is not at all clear. The size of the preimage of the various "output symbols" (amino acids or stop codon) ranges from a high of six for Leucine and Serine to a low of one for many amino acids. Several persons have proposed various speculative explanations for the structure of the genetic code, but until now there is no universally accepted explanation.

An interesting aspect of the above chain of discoveries is that some *physicists* played a central role in motivating much of this work. After the discovery of individual nucleotides, the famous physicist Erwin Schrödinger suggested very strongly that *there must be a genetic code*, that is, a way

First letter	U	C	A	G	Third letter
			Second letter		
U	UUU Phe	UCU Ser	UAU Tyr	UGU Cys	U
	UUC Phe	UCC Ser	UAC Tyr	UGC Cys	C
	UUA Leu	UCA Ser	UAA Stop	UGA Stop	A
	UUG Leu	UCG Ser	UAG Stop	UGG Trp	G
C	CUU Leu	CCU Pro	CAU His	CGU Arg	U
	CUC Leu	CCC Pro	CAC His	CGC Arg	C
	CUA Leu	CCA Pro	CAA Gln	CGA Arg	A
	CUG Leu	CCG Pro	CAG Gln	CGG Arg	G
A	AUU Ile	ACU Thr	AAU Asn	AGU Ser	U
	AUC Ile	ACC Thr	AAC Asn	AGC Ser	C
	AUA Ile	ACA Thr	AAA Lys	AGA Arg	A
	AUG Met	ACG Thr	AAG Lys	AGG Arg	G
G	GUU Val	GCU Ala	GAU Asp	GGU Gly	U
	GUC Val	GCC Ala	GAC Asp	GGC Gly	C
	GUA Val	GCA Ala	GAA Glu	GGA Gly	A
	GUG Val	GCG Ala	GAG Glu	GGG Gly	G

Figure 8.5 The Genetic Code in Tabular Form. Reproduced with permission from Stephen Carr.

of associating strings of nucleotides with amino acids. Schrödinger's best known contribution is of course the "wave function" formulation of quantum mechanics, which eventually supplanted the earlier "matrix mechanics" formulation put forth by Werner Heisenberg. Another physicist, George Gamow, suggested on the basis of numerical arguments ($4^3 > 20$) that the genetic code consisted of a map from triplets of nucleotides, that is codons, into amino acids. A very readable description of the entire discovery process can be found in the web site of the Nobel Prize under either Khorana or Nirenberg.

8.2 OPTIMAL GAPPED SEQUENCE ALIGNMENT

The gapped sequence alignment problem arises in connection with detecting similarity between two sequences. Though the problem formulation given below is quite abstract, the most common applications of this problem are to detecting similarities between two different stretches of DNA, or between the primary sequences of two different proteins.

When comparing two sequences over a common alphabet, two approaches are possible, namely: similarity at a symbol by symbol level, or similarity of the statistical properties of the two sequences. One of the important observations in genome analysis is that genes with similar functions in different organisms have very similar DNA sequences; that is to say, they are "conserved." For this reason, a symbol by symbol comparison between two sequences is often employed when one finds a gene in one organism, and wants to deduce its function within the organism. If the new gene is very similar at a symbol by symbol level to another gene in another organism whose function is known, then it is reasonable to guess that the new gene also has a functionality similar to that of the known gene. Another example arises from protein structure prediction. Suppose a protein has been crystallized and its 3-D structure has been determined using X-ray crystallography. Suppose now that it is desired to determine the 3-D structure of another protein with known primary structure (amino acid sequence). It may happen that this new protein is impossible to crystallize; therefore X-ray crystallography is not applicable. However, if there is a substantial similarity between the primary structure of the new protein and another protein with known 3-D structure, one can reasonably guess that the structure of the new protein is also somewhat similar to that of the protein whose structure is known.

To illustrate the difference between deterministic and stochastic algorithms, let us consider the following hypothetical problem. Suppose one is given two different sequences of heads (H) and tails (T) and is asked whether the same coin could have produced both sequences. Clearly, even if both sequences were produced by the same coin, it is extremely unlikely that heads and tails would appear in exactly the same sequence. Thus a deterministic algorithm based on aligning the two sequences would not yield good results. On the other hand, if we compute *the fraction of heads (or tails)* in the two sequences, and the fractions are quite close, then we can state with some confidence that the same coin produced the two sequences. Such an approach, which typifies a statistical approach to detecting sequence similarity, is often used for "ab initio" annotation of a genome, when only a few genes are known and the rest are to be inferred. This approach is discussed in greater detail in Section 8.3.

8.2.1 Problem Formulation

Suppose we are given two strings $\mathbf{x} = x_1 \ldots x_k$ and $\mathbf{y} = y_1 \ldots y_l$ over a common finite alphabet $\mathbb{N}$. Often, though not always, it is the case that one of the strings is much shorter than the other, say $k \ll l$. In such a case, determining whether $\mathbf{x}$ is a *perfect* substring of $\mathbf{y}$ is computationally straightforward. Indeed, text editors address precisely this problem. Thus it is easy to determine whether or not there exists an index j such that $y_{j+i} = x_i$ for $i = 1, \ldots, k$. The problem remains tractable even if we introduce some "wild card" entries. Thus, a text editor that searches for the string $x_1 \ldots x_s * x_{s+1} \ldots x_k$ within $\mathbf{y}$ looks for indices j_1 and $j_2 \geq j_1$ such that

$$y_{j_1+i} = x_i \text{ for } i = 1, \ldots, s, \text{ and } y_{j_2+i} = x_i \text{ for } i = s+1, \ldots, k.$$

The string $\mathbf{x}$ can in fact be divided into any finite number of segments and the wild cards introduced in-between, and the problem remains tractable. The tractability arises from two factors: First, the locations within the string $\mathbf{x}$ where one or more wild card entries are to be introduced are specified ahead of time. Second, we insist on a *perfect* match between the symbols of the two strings. If we were to change either of these requirements then the problem becomes more difficult.

Now we state the "optimal gapped alignment" problem. Suppose $\mathbf{x} = x_1 \ldots x_k$ is a string over some finite alphabet $\mathbb{N}$ and $\mathbf{y} = y_1 \ldots y_l$ is a string over another finite alphabet $\mathbb{M}$, which may or may not be the same as $\mathbb{N}$. In practice, both strings are usually over the same alphabet; however, the theory does not become any more complicated by permitting two distinct alphabets, possibly of different cardinalities. The problem is to determine an optimal "gapped" alignment between the two strings. Before stating the problem formally, we motivate it through a simple example. Suppose $\mathbb{N} = \mathbb{M} = \{A, C, G, T\}$, the set of nucleotides, and let

$$\mathbf{x} = ACACTGT, \mathbf{y} = TAGACGGAGCTTCAC.$$

Then these two strings can be imperfectly aligned as shown below, with the dash indicating a "gap":

$$\begin{array}{ccccccccccccccc} & & & A & C & - & - & A & C & - & T & G & T & & \\ T & A & G & A & C & G & G & A & G & C & T & - & T & C & A & C \end{array}$$

By judiciously introducing gaps into the two sequences, we are able to achieve "perfect" matches between those symbols within each string that do not lie opposite a gap, with just one mismatch. To measure the quality of the gapped alignment, we introduce a "scoring" function $F : \mathbb{N} \times \mathbb{M} \to \mathbb{R}$ that assigns a real number score $F(i, j)$ to each pair $(i, j) \in \mathbb{N} \times \mathbb{M}$. We can also think of $F(i, j)$ as representing the "similarity" between the symbols i and j instead of "score." Thus in the case where $\mathbb{M} = \mathbb{N}$ meaning that both strings $\mathbf{x}$ and $\mathbf{y}$ are over the same alphabet, we would expect $F(i, j)$ to be large and positive whenever $i = j$, and to be much smaller and possibly negative if $i \neq j$. Note that it is *not* assumed that the matrix $F(i, j)$ is symmetric. We also need to define "gap scores" $F(i, -)$ and $F(-, j)$. Note that in this problem it makes no sense to put one gap opposite another. We can avoid the situation by defining $F(-, -) = -\infty$. In this way, the scoring function F can be extended to $(\mathbb{N} \cup \{-\}) \times (\mathbb{M} \cup \{-\})$.

At this point we need to distinguish between "global" alignment and "local" alignment. In global alignment, we would insist that the end points of the two strings must match, after being augmented by gaps if necessary. Thus, in the example above, we would be forced to place gaps above TAG to the left of the $\mathbf{x}$ string, and above CAC to the right of the $\mathbf{x}$ string. This makes no sense if, as is often the case in practice, one of the strings is significantly longer than the other. In such a case one would study *local* alignment, where the scoring function is counted only between the end points of the shorter string. In either case, the total score of a gapped alignment is the sum of the individual pairwise scores $F(i, j), F(i, -)$, or $F(-, j)$ as we traverse from one end to the other. The optimal gapped alignment problem is to determine the alignment that results in the highest score.

8.2.2 Solution via Dynamic Programming

The problem of optimal gapped alignment can be solved using dynamic programming. The principle of optimality, which we have encountered in

Section 6.2.2, applies here too: If any alignment is optimal, then any subset thereof must also be optimal for the appropriate substrings. Otherwise we could take out that particular part of the alignment, replace it with a better alignment, and improve the overall score. Clearly this property is a consequence of the *additive nature* of the total score.

Using the principle of optimality, we can give a simple recursive scheme for solving the problem of optimal gapped alignment. Let $\mathbf{x}, \mathbf{y}$ be the two strings to be aligned (not necessarily over the same alphabet), using the scoring function $F : \mathbb{N} \times \mathbb{M} \to \mathbb{R}$. For simplicity let us suppose that $F(-, j) = F(i, -) = -\gamma$, the "gap penalty." Let $\mathbf{x}$ have length k and let $\mathbf{y}$ have length l. Suppose we begin aligning from the ends of the two strings, and let $P^*(i, j)$ denote the highest possible score that can be achieved by optimally aligning from the end of $\mathbf{x}$ until position i, and from the end of $\mathbf{y}$ until position j. Think of $P^*(i, j)$ as the optimal payoff until positions i, j. Now the optimal payoff function satisfies the following recursion:

$$P^*(i,j) = \max \begin{cases} P^*(i, j+1) - \gamma \\ P^*(i+1, j) - \gamma \\ P^*(i+1, j+1) + F(x_i, y_j) \end{cases} \tag{8.1}$$

This is because, at position i, j there are only three things we can do:

1. We can introduce a gap above the symbol y_j.
2. We can introduce a gap below the symbol x_i.
3. We can match x_i against y_j.

Let us suppose that we have aligned the two sequences optimally before i, j. In the first alternative, the resulting score would be $P^*(i, j+1) - \gamma$, because $P^*(i, j+1)$ is the optimal score of aligning $x_i, \ldots, x_k$ against $y_{j+1}, \ldots, y_l$, and $-\gamma$ is the additional score due to introducing a gap above y_j. Similarly, in the second alternative, the resulting score would be $P^*(i+1, j) - \gamma$. Finally, in the third alternative the resulting score would be $P^*(i+1, j+1) + F(x_i, y_j)$. Now the principle of optimality tells us that the best thing to do would be to maximize among these three alternatives. It is of course possible that there is a "tie" between two alternatives, in which case we arbitrarily choose one of them; this does not affect the discussion to follow. To apply the above formula, we begin at the ends of the two strings $\mathbf{x}$ and $\mathbf{y}$ with the optimal score $P^*(k, l) = 0$, and work backwards.

Example 8.1 The application of (8.1) is illustrated through a "toy" example. It must be emphasized that in reality the smallest values of l, k for which we would wish to carry out optimal gapped alignment would be of the order of a few hundred.

Suppose the scoring function is given by

$$F = \begin{array}{c} y/x \\ A \\ C \\ G \\ T \end{array} \begin{array}{c} \begin{array}{cccc} A & C & G & T \end{array} \\ \left[\begin{array}{rrrr} 10 & -3 & -2 & 1 \\ -2 & 8 & 1 & -2 \\ -3 & 1 & 9 & -3 \\ 0 & -3 & -2 & 6 \end{array} \right] \end{array}, \gamma = 1.$$

Suppose the two strings to be aligned are

$$\mathbf{x} = CACGAAT, \mathbf{y} = AGTTCAA.$$

Then we can construct the table of optimal payoff functions as follows:

	C	A	C	G	A	A	T	
A	$32 \leftarrow$	$33 \nwarrow$	$22 \nwarrow$	$24 \uparrow$	$17 \nwarrow$	$11 \nwarrow$	$0 \uparrow$	$-7 \uparrow$
G	$26 \nwarrow$	$24 \uparrow$	$23 \uparrow$	$25 \nwarrow$	$15 \uparrow$	$7 \uparrow$	$1 \uparrow$	$-6 \uparrow$
T	$24 \uparrow$	$25 \uparrow$	$24 \uparrow$	$18 \uparrow$	$16 \uparrow$	$8 \uparrow$	$2 \nwarrow$	$-5 \uparrow$
T	$25 \leftarrow$	$26 \nwarrow$	$25 \uparrow$	$19 \uparrow$	$17 \uparrow$	$9 \uparrow$	$3 \nwarrow$	$-4 \uparrow$
C	$24 \nwarrow$	$25 \leftarrow$	$26 \nwarrow$	$20 \nwarrow$	$18 \uparrow$	$10 \uparrow$	$-1 \uparrow$	$-3 \uparrow$
A	$15 \leftarrow$	$16 \leftarrow$	$17 \leftarrow$	$18 \leftarrow$	$19 \nwarrow$	$11 \nwarrow$	$0 \uparrow$	$-2 \uparrow$
A	$4 \leftarrow$	$5 \nwarrow$	$6 \leftarrow$	$7 \leftarrow$	$8 \leftarrow$	$9 \nwarrow$	$1 \nwarrow$	$-1 \uparrow$
	$-7 \leftarrow$	$-6 \leftarrow$	$-5 \leftarrow$	$-4 \leftarrow$	$-3 \leftarrow$	$-2 \leftarrow$	$-1 \leftarrow$	0

In constructing this table, we need to keep track of the optimal choice that we made at each square of the matrix. Thus a left arrow $\leftarrow$ indicates that the optimal choice was to insert a gap above y_j, while a vertical arrow $\uparrow$ indicates that the optimal choice was to insert a gap below x_i. The diagonal arrow $\nwarrow$ indicates that the optimal choice was to match x_i against y_j. In case of ties, one of the possibilities was chosen arbitrarily.

The above procedure gives the optimal gapped alignment of the two sequences, namely

$$\begin{matrix} C & A & - & - & - & C & G & A & A & T \\ - & A & G & T & T & C & - & A & A & - \end{matrix}, P^* = 32.$$

However, the same matrix can also be used to determine the optimal gapped alignment from the ends of the two strings to any pair of intermediate points. For instance, the optimal alignment from the end of string $\mathbf{x}$ until the symbol G, and from the end of string $\mathbf{y}$ until the symbol T is shown below.

$$\begin{matrix} - & - & G & A & A & T \\ T & T & C & A & A & - \end{matrix}, P^* = 18.$$

Similarly, the optimal alignment from the end of string $\mathbf{x}$ until the symbol C, and from the end of string $\mathbf{y}$ until the symbol A is shown below.

$$\begin{matrix} C & G & - & - & - & A & A & T \\ A & G & T & T & C & A & A & - \end{matrix}, P^* = 22.$$

The solution using dynamic programming was first introduced into the biology community by Needleman and Wunsch [98] to solve problems of optimal *global* alignment of two strings. It is possible to make simple modifications to the algorithm to solve the problem of optimal *local* alignment. This was done by Smith and Waterman; see [119].

From the above discussion, it is easy to see that the complexity of optimal gapped alignment using the above procedure $O(kl)$, where k, l are the lengths of $\mathbf{x}, \mathbf{y}$ respectively. If both strings are of comparable length, then the complexity is quadratic in the length. There are several improvements available that trade off storage for time or vice versa, but these need not concern us here. It is also possible to modify (8.1) to incorporate more sophisticated scoring functions. For example, one can have different penalties for gap *creation* versus gap *extension*; it is believed by biologists that the gap extension penalty should be much smaller than the gap creation penalty. Similarly, it possible to make the scoring function depend not only on the two symbols being matched, but also on their positions within the two strings. The modifications required are relatively straightforward, and the reader is referred to [63] for more detailed discussion. Finally, it is worth noting that the complexity of aligning n strings of length k is $O(2^n k^2)$. In other words, while the complexity of aligning two strings is proportional to k^2, the complexity increases *exponentially* with respect to the number of strings to be aligned. We shall see in Section 8.4 that a first step in classifying proteins into one of a small number of families is gapped alignment of all the protein sequences from each family. It is not possible to carry out such an alignment in an optimal fashion, due to the exponential complexity with respect to the number of sequences to be aligned. Instead one uses a great many heuristics to achieve near-optimal alignment. The interested reader is referred to [97].

8.3 GENE FINDING

8.3.1 Genes and the Gene-Finding Problem

Roughly speaking, a "gene" is a stretch of DNA that gets converted into a protein. Within the continuous stretch of a single gene, some regions are called the **coding regions** while others are called **noncoding regions**. The conversion of a gene to a protein can happen in one of two ways. In so-called prokaryotes or lower-level organisms, each gene consists of one continuous stretch of DNA. In so-called eukaryotes or higher organisms, the gene can actually consist of several "exons" separated by "introns." When the gene produces the corresponding protein, the noncoding regions all get cut out, and all the coding regions come together. When this happens, the concatenation of all the coding regions gets converted to a protein according to the genetic code shown in Figure 8.5. Figure 8.6 depicts the process whereby a gene codes for a protein.

It goes without saying that the total length of all the coding regions put together is an exact multiple of three (so that there is an integer number of codons). Moreover, the last codon is one of the three stop codons TAA, TAG, TGA, while the first codon is one of the start codons ATG, GTG. However, as mentioned above, a start codon can also occur as an intermediate codon. There are other features that characterize a gene. For example, roughly 10 to 15 positions upstream (that is, towards the 5' end) of the start codon, the tetramer[3] $TATA$ usually occurs. This tetramer is called

[3] A tetramer is a quadruplet of nucleotides. Other commonly used expressions such as

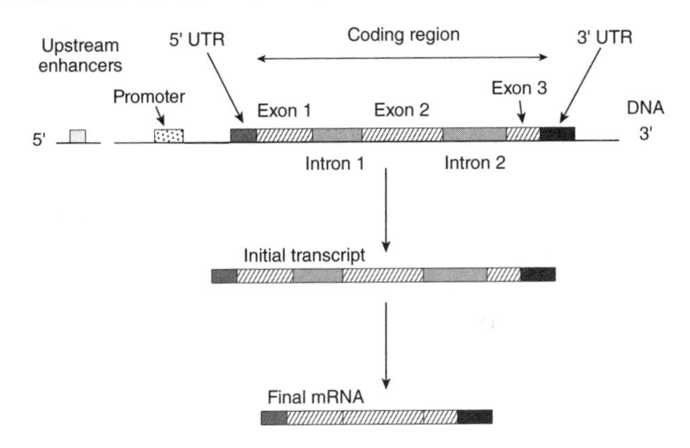

Figure 8.6 Coding by Genes for Proteins.

the $TATA$-box. In eukaryotes the sequence is sometimes taken as the hexamer $TATAAA$ and is also referred to as the Goldberg-Hogness box, while in prokaryotes the sequence is sometimes taken as $TATAAT$ and is also referred to as the Pribnow box or the Pribnow-Schaller box. Taking all of these features into account, a stretch of DNA that possesses the following features is a possible gene, and is referred to as an ORF (Open Reading Frame):

1. The sequence begins with a start codon ATG or GTG.

2. There is a $TATA$-box 10 to 15 positions upstream of the sequence.

3. The sequence ends with one of the three stop codons TAA, TAG, TGA.

4. The total length of the sequence is "reasonable," not less than 300 nucleotides, and not more than 12,000 nucleotides.

The last convention merely indicates that the shortest known protein is 57 amino acids long, corresponding to 161 nucleotides, while the longest known proteins contain about 3,500 amino acids, corresponding to 10,500 nucleotides. Hence the upper limit is set at 12,000 nucleotides.

In the case of eukaryotes (higher organisms) an additional complication arises: A gene need not consist of one continuous stretch of DNA; instead, in general it can consist of several "exons" interspersed by "introns." When a gene produces the corresponding protein, first the introns get "cut out" and all the exons come together. Then the concatenation of the exons gets converted into a sequence of amino acids via the genetic code discussed earlier. This process is also illustrated in Figure 8.6. The boundary between an exon and intron is called a "donor site" and is the dimer AG, while the boundary between and intron and an exon is called an "acceptor site" and

trimer, hexamer, etc. are self-explanatory.

consists of the dimer GT. Of course, the main difficulty is that while every donor site is the dimer AG, the converse is not at all true: Just because we see the dimer AG at some point in the genome, we cannot automatically conclude that it is a donor site. In fact it is fairly easy to see that *only a very small fraction* of the occurrences of AG correspond to donor sites. Similar remarks apply to acceptor sites. Together the donor sites and acceptor sites are referred to as "splice sites." In general, introns tend to be considerably longer than exons. In eukaryotes, the genes are separated by intergenic regions, sometimes called "junk DNA." Thus, in order to determine where a gene begins and ends, we need first to weed out the intergenic regions, and then, within each gene, weed out the introns, leaving only the exons.

It is clear from the above that determining the ORFs from the genome is relatively straightforward. However, the difficulty is that not all ORFs are genes. (If they were, life would be very simple indeed.) Thus one of the fundamental challenges in genomics is to determine which ORFs are actually genes. One can think of ORFs as a "candidate genes," or "putative genes" as some persons prefer to call them.

An obvious first step in determining whether an ORF is actually a gene is to compare it to known genes from other organisms, at a symbol by symbol level. In case there is a sufficiently good match, then one can be confident that the ORF under study is also a gene, and moreover, its function is broadly similar to that of the known gene. This is precisely the motivation behind the gapped alignment problem studied in Section 8.2. Therefore the focus in this section is on the case where there is no symbol by symbol similarity between the ORF under study and any known gene. In such a case, one is obliged to adopt statistical methods of analysis.

There are two distinct kinds of algorithms used in the literature to solve the gene-finding problem. These can be described as *ab initio* methods and "bootstrapping" methods. In *ab initio* methods, one begins with the "raw" genome and does not assume anything at all about which sections of the genome actually correspond to genes. The original GLIMMER algorithm [113] for identifying genes in prokaryotes, which is described in greater detail in Section 8.3.2, begins with the assumption that *all* ORFs that are longer than 500 base pairs are genes, and uses them to train an appropriate statistical model. In reality, the vast majority (more than 80%) of ORFs in bacteria are *not* genes. See [134, Table IV] for a summary of the situation for several bacterial genomes. And yet algorithms based on this assumption seem to work remarkably well. Moreover, such an assumption allows one to analyze a given genome without *any* prior knowledge, which is the meaning of the expression *ab initio*.

In bootstrapping algorithms, one needs at least a few ORFs that are "known" to be genes. By "known" genes, we mean either that the ORF has been experimentally verified to be a gene, or else that the ORF sequence is sufficiently close to an experimentally verified gene in some other organism, that we can be very confident that the ORF really is a gene without bothering with experimental verification of this particular ORF. In either case, this kind of "known" gene is commonly referred to as an "annotated" gene in the literature. When there are a few "known" (or annotated) genes, these are used as the starting point to analyze other ORFs and to make predictions as to whether those ORFs are genes or not. If the algorithm predicts some ORFs (whose status is previously unknown) to be genes, then the experimenters would go to work to validate these predictions. If the predictions are accurate and the predicted gene is indeed a gene, then the predicted (and now validated) gene is added to the database of annotated genes. If the prediction is not borne out by experiment, presumably the originators of the prediction algorithm would introspect on how to improve the accuracy of their algorithms.

8.3.2 The GLIMMER Family of Algorithms

GLIMMER is not just a single algorithm; rather, the name refers to a series of algorithms with a broadly similar approach. The original algorithm [113] was aimed at discovering genes in prokaryotes. Subsequent versions of the algorithms [114, 104, 122, 94, 35] incorporate several improvements, such as extension of the core idea from prokaryotes to eukaryotes, detecting splice sites, detecting start codons, and so on. In the interests of simplicity, only the most basic version of GLIMMER is described here, and the reader is referred to the above papers for more details.

The three letters IMM in GLIMMER stand for "interpolated Markov model." To put this approach in perspective, let us begin with conventional multistep Markov models for genic regions in genomes. These multistep Markov models are then "interpolated" in GLIMMER.

Let us suppose that several exemplars of genic regions as well as nongenic regions are available. To be specific, suppose that strings $\mathbf{a}_1, \ldots, \mathbf{a}_k$ are examples of genic regions, while $\mathbf{b}_1, \ldots, \mathbf{b}_m$ are examples of nongenic regions, where each string is from the four-symbol nucleotide alphabet $\{A, C, G, T\}$. The lengths of the various strings need not be the same. The premise underlying the use of Markov models for finding genes is that the statistics of the nucleotide frequencies in genes are quite distinct from the statistics in nongenic regions. Accordingly, two different Markov models are constructed: one using $\mathbf{a}_1, \ldots, \mathbf{a}_k$ as sample paths and another using $\mathbf{b}_1, \ldots, \mathbf{b}_m$ as sample paths. Let us denote these models as $\mathcal{G}$ (for genic region model) and $\mathcal{NG}$ (for nongenic region model) respectively. Then, given an ORF $\mathbf{c}$ whose status is not known, one would compute the likelihood of the sample path $\mathbf{c}$ using the two Markov models, say $L_{\mathcal{G}}(\mathbf{c})$ and $L_{\mathcal{NG}}(\mathbf{c})$, and choose the model for which the likelihood is higher. For good measure, the log likelihood ratio provides a confidence figure that can be attached with the classification.

The previous paragraph describes the general philosophy. Now let us provide the quantitative details. A crude "zeroth-order" Markovian model would be simply to compute the distribution of each of the four nucleotides in the two sets of strings $\{\mathbf{a}_1, \ldots, \mathbf{a}_k\}$ and $\{\mathbf{b}_1, \ldots, \mathbf{b}_l\}$, and thereby arrive at two different four-dimensional probability vectors. Call them ϕ for $\mathcal{G}$ and

ψ for $\mathcal{NG}$. Then one would compute the likelihood of the path $\mathbf{c}$ by

$$\log L_{\mathcal{G}}(\mathbf{c}) = \sum_{i=1}^{4} c_i \log \phi_i,$$

where c_i is the number of times that the i-th nucleotide symbol occurs in the string $\mathbf{c}$, and similarly for $\log L_{\mathcal{NG}}(\mathbf{c})$. In this approach, it is obvious that the string $\mathbf{c}$ is being viewed as a sample path of an i.i.d. process over the nucleotide alphabet, whose distribution is either $\boldsymbol{\phi}$ or $\boldsymbol{\psi}$.

A more refined approach would be to consider the nucleotide sequence corresponding to a gene to be a sample path of a Markov process. In Section 4.1.2, we discuss the problem of finding the maximum likelihood estimate of the transition matrix of a Markov process, given its sample path. Specifically the maximum likelihood estimate can be computed using (4.6) for conventional "one-step" Markov models, and (4.7) for multistep Markov models. There are however a couple of details that need to be addressed. First, the discussion in Section 4.1.2 addresses the situation where there is just one sample path of the Markov process that is available. But in the present situation, there are multiple sample paths. Therefore the question arises as to how they should be treated. One possibility is to derive a separate transition matrix corresponding to each string, and then average them. To illustrate, suppose l_i is the length of the string $\mathbf{a}_i$, and let $A_i \in [0,1]^{4\times 4}$ be the transition matrix corresponding to the string $\mathbf{a}_i$. Then one can define

$$A = \sum_{i=1}^{k} \lambda_i A_i, \text{ where } \lambda_i = \frac{l_i}{l}, \text{ and } l = \sum_{i=1}^{k} l_i.$$

The second issue is known as "three-periodicity" and is specific to genic regions but not to nongenic regions. To explain this issue, suppose $\mathbf{a}_1, \ldots \mathbf{a}_k$ are all ORFs that are known to be genes. Therefore the length of each $\mathbf{a}_i$ is a multiple of three. Now suppose we wish to construct an s-step Markov model for this collection of sample paths, where s is a prespecified integer. To keep notation simple, we drop the subscript and suppose $\mathbf{a} = a_1 \ldots a_l$ is an ORF to which it is desired to fit an s-step Markov model. Thus it is desired to compute the conditional probability $\Pr\{X_t = a_t | X_{t-s}^{t-1} = a_{t-s}^{t-1}\}$. Now it has been observed that *this conditional probability depends on the value of* t. Specifically, the starting point of the ORF is designated as frame 0, the next position as frame 1, and the next position as frame 2. In other words, the t-th nucleotide is in the frame $r = t \bmod 3$.

The effect of three-periodicity for bacterial genomes is analyzed in [134]. In that paper, sixth order Markov models are constructed for both coding regions and noncoding regions of several bacteria, for all the three frames (frames 0, 1, and 2). Then the relative entropy rate between each pair of the three Markov models is computed using Definition 5.11, specifically (5.24). From [134, Table I], it can be seen that in the coding regions there is quite a pronounced three-periodicity effect, whereas [134, Table II] shows that in the noncoding regions the effect is essentially absent. Therefore, in the discussion below, it can be assumed that three-periodicity is taken into account for the coding regions, by deriving three different Markov models; in contrast, only one Markov model is used for the noncoding regions.

Now the basic idea behind the original GLIMMER algorithm is described. As shown in Section 4.1.2, in order to fit an s-step Markov model to a given sample path, it is necessary to compute the frequency of occurrence of $(s+1)$-tuples. Since there are four nucleotides, it is necessary to compute the frequency of occurrence of 4^{s+1} different $(s+1)$-tuples within the given string of length l. Clearly, if $l \ll 4^{s+1}$, it is quite possible that many of the $(s+1)$-tuples simply do not occur in the sample path. As a result the estimates of the entries in the transition matrix would become quite unreliable. On the other hand, *some* of the $(s+1)$-tuples might occur sufficiently frequently to permit an accurate estimation of *their* frequency of occurrence. So in the GLIMMER algorithm, one begins by setting s to be as high as 8. If the number of occurrences of a particular (s_1)-tuple is larger than 400, it is deemed that its frequency can be estimated reliably. If not, then one looks at shorter tuples, and "interpolates" the frequency estimates from shorter tuples to arrive at an overall value. Specifically, let M denote a collection of Markov models with memories ranging from zero (i.e., an i.i.d. model) to eight, and let S denote the string whose likelihood $P(S|M)$ is to be computed. Let l denote the length of the string S, and for each x between 1 and l, let S_x denote the substring of length 8 (referred to as an "oligomer" in [113]) ending at location x. Then

$$P(S|M) = \sum_{x=1}^{l} \mathrm{IMM}_8(S_x),$$

where $\mathrm{IMM}_8(S_x)$ denotes the local interpolated likelihood associated with the substring S_x. In other words, the overall likelihood score is just a sum of the local positional scores associated with the substrings S_x as x sweeps over the length of the string S. The above formula becomes explicit once the method for computing the local positional score $\mathrm{IMM}_8(S_x)$ is given. This is done next. Let s denote the step size of a Markov model, ranging from 8 down to 0. Then

$$\mathrm{IMM}_s(S_x) = \lambda_s(S_{x-1})P_s(S_x) + [1 - \lambda_s(S_{x-1})]\mathrm{IMM}_{s-1}(S_x).$$

In this expression, the quantity $\lambda_s(S_{x-1})$ is a kind of confidence figure associated with the oligomer S_x. If the oligomer S_x occurs more than 400 times in the overall string S, then $\lambda_s(S_{x-1}) = 1$. In other words, we have full confidence in our estimate, and it is not necessary to examine shorter substrings. If this is the case, then the likelihood associated with the substring S_x is given by (4.7), that is

$$P_s(S_x) = \frac{f(S_{x,i})}{\sum_{b \in \{A,C,G,T\}} f(S_{x,b})}.$$

On the other hand, if the oligomer S_x occurs fewer than 400 times in the overall string S, then the confidence $\lambda_s(S_{x-1})$ is determined through a fairly complicated χ^2-test. The reader is referred to [113] for full details. It is pointed out in this reference that the IMM (interpolated Markov model) outperforms a fixed fifth-order Markov model.

8.3.3 The GENSCAN Algorithm

In this subsection we discuss the GENSCAN algorithm, introduced in [25, 26], which is an elaboration of the Genemark algorithm introduced in [93]. Though Genemark continues to exist as a separate algorithm, the focus here is on GENSCAN, because it employs a hidden Markov model with a larger state space than Genemark (fourteen states as opposed to nine on each of the forward and reverse strands).

In a eukaryotic genome, genes can occur on either the forward strand, or the reverse strand, or both. Accordingly, the hidden Markov model used by the GENSCAN algorithm consists of two models that are mirror images of each other. To save space, we display only the part of the HMM corresponding to the forward strand. To indicate this, all states in this HMM have a superscript $+$. The mirror image HMM is identical except that all quantities have a subscript of $-$. In [25, Figure 3], the full HMM is shown, including both the forward strand and the reverse strand. However, to save space, in Figure 8.7 only the forward strand is shown. Moreover, the superscript $+$ on all nodes is omitted, in contrast to [25, Figure 3].

Now the various states of the HMM (actually half of a HMM) are described. In a eukaryote, the genes can either consist of a single exon (that is, the gene is one contiguous stretch of DNA), or else there can be multiple exons. A single exon constituting a gene is denoted by E_S. In the case of multi-exon genes, the first and last exons are referred to as the initial and terminal exons respectively, and denoted by E_I and E_T respectively. In the case of multiple exons (even just two), the introns can occur in phase 0, phase 1, or phase 2. The notation is self-explanatory. If an intron occurs after the end of a codon, then it is in phase 0, as is the succeeding exon. If an intron occurs after the first symbol of a codon, then it is in phase 1 as is the succeeding exon. If an intron occurs after the second symbol of a codon, then it is in phase 2 as is the succeeding exon. Note that, in all cases, an intron in phase k can only be succeeded by an exon in phase k. However, depending on the length of the exon, an exon in any phase can be succeeded by an intron in any other phase. The part of the genome preceding the first exon (which could in fact be the only exon) is called the $5'$ UTR (Untranslated Region) and is denoted by the symbol F. Similarly the part of the genome succeeding the last exon (which could in fact be the only exon) is called the $3'$ UTR and is denoted by T. The promoter region is denoted by P while the polyadenation region is denoted by A. The fundamental functional units of the eukaryotic genome can occur in any order that is consistent with biology. Figure 8.7 shows the (forward strand half of the) HMM associated with the GENSCAN algorithm.

In [25, 26], the GENSCAN HMM is trained using a part of the data

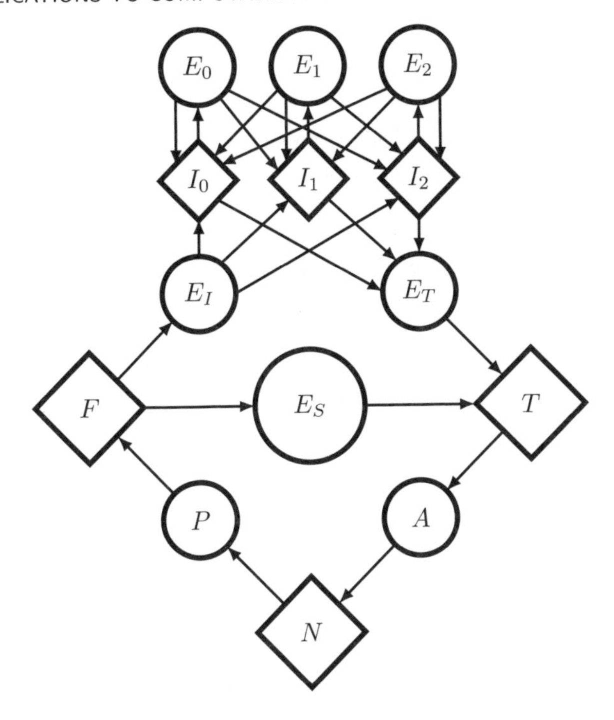

Figure 8.7 Hidden Markov Model Used by the GENSCAN Algorithm. Patterned after Burge and Karlin 1997, [Figure 3].

available at that time, and tested on the remainder. The performance is quite good. Since then a great many improvements have been made in the algorithm. The reader is referred to the current literature for the latest developments.

8.4 PROTEIN CLASSIFICATION

8.4.1 Proteins and the Protein Classification Problem

Proteins are at the next level of complexity after genes. As stated above, the central dogma of biology describes how genes get converted into proteins. The original and rather simplistic recipe of one gene, one protein, one function has long since been revised in favor of far more subtle models. For instance, it is now known that the same gene can, in different kinds of cells, code for different proteins. These subtleties of biochemistry are beyond the scope of this book. For present purposes, we will stick with the simple model whereby the relevant parts of a gene come together during translation, codons get converted into amino acids as per the genetic code, and the resulting sequence of amino acids forms the protein. Thus, once we know that a particular stretch of DNA represents a gene, and we know the functional parts of the gene (coding regions and/or exons), we can unambiguously determine the sequence of amino acids produced by that gene.

The sequence of amino acids is known as the primary structure of a protein. It is the lowest level description of a protein, just as the genome is the lowest level description of the DNA of an organism. As befits a low level description, knowing the primary structure of a protein does not get us very far in terms of knowing how a protein works. In order to understand how a protein performs its assigned function, it is highly desirable to know how the protein folds, that is, the three-dimensional structure of the protein, which is known as the tertiary structure of the protein, and also its so-called active sites. In between the primary and the tertiary structure is the so-called secondary structure, which is a simplified representation of the tertiary structure. Finally, when two or more proteins bond, the result is a molecule that is still more complex. The 3-D structure of the protein-protein complex is referred to as the quaternary structure.

The tertiary structure of a protein corresponds to the 3-D conformation that minimizes the potential energy of the conformation. While this simple-sounding statement is consistent with the laws of physics, in reality the potential energy function of a protein is a highly complex function of its conformation. The conformation itself consists of $2(n-1)$ angles where n is the number of amino acids, representing the two degrees of freedom at each joint between two amino acid molecules. Each amino acid molecule can be thought as being essentially rigid; however, the orientation at each joint represents two degrees of freedom. The potential energy term must also include the interaction of each amino acid with the surrounding medium, usually water. In principle it is possible to determine the structure of a protein by minimizing its potential energy. In practice however, the minimization problem is all but intractable unless the protein has a very short amino acid sequence.

If a protein can be crystallized, then its tertiary structure can be determined on the basis of experimental methods such as NMR or X-ray diffraction. However, many proteins of interest cannot be crystallized. In such a case one is forced to resort to other methods to predict the 3-D structure of the protein. Even if a protein can be crystallized, the procedure for determining the structure is both time-consuming and expensive. Thus there is a definite need for computational methods for predicting the tertiary structure of a protein. This has led to a number of methods for predicting protein structure, which are discussed at length in [11]. These methods include some *ab initio* methods, as well as methods based on neural networks. Among the most popular are homology-based methods, whereby several proteins whose structures are known are grouped into a small number of families, typically three or four families. Then the protein of interest is classified as being most similar to one of the protein families. If the similarity is sufficiently high, then one makes a guess that the 3-D structure of the new protein is similar to those of the known proteins. This approach has the advantage (from the standpoint of the present book) of being based on hidden Markov models.

8.4.2 Protein Classification Using Profile Hidden Markov Models

In this section we study the problem of protein classification using a special type of hidden Markov model, referred to as a profile HMM. In this class of HMMs, all the states of the underlying Markov process are inessential, except for a so-called end state. Such models are well-suited for a specific type of protein classification problem, described next. Suppose there are several proteins that have been clustered into a small number of families. The clustering reflects the insights of the biology community as to which proteins belong within the same family. The outcome of the clustering process is a set of distinct protein families. Note that the number of proteins within a particular family can vary from one family to another. Initially, the primary sequences of the various proteins within the same family need not all have the same lengths. However, once a set of proteins has been assigned to a common family, the next step is to carry out a gapped alignment of all the proteins, so as to make the length of the gapped sequences exactly the same for all proteins within a family. As pointed out earlier, the complexity of aligning n sequences optimally is $O(2^n)$. Hence the gapped alignment of all the sequences within the same family has to be carried out in some suboptimal fashion, for example as described in [97].

In any case, the starting point of the protein classification is a collection of protein families $\mathcal{F}_1, \ldots, \mathcal{F}_k$, where all proteins within a family have the same length (after assigning gaps as applicable). Now suppose a new protein is specified in terms of its primary, or amino acid, sequence. The objective is to assign the new protein to one of the k families. This problem is addressed as follows: First, the new protein to be classified is also gap-aligned to the proteins within each family. Note that, because the gap-aligned proteins within family $\mathcal{F}_i$ might have a different length from the proteins within family $\mathcal{F}_j$, the gap-aligned version of the new protein to be classified also varies from one family to another. Next, for each family $\mathcal{F}_i$, a corresponding profile HMM is constructed as described below. Then the likelihood of the gap-aligned new protein is computed for each of the k profile HMMs. The new protein is then assigned to the family for which the likelihood is maximum. Log likelihood ratios are used to quantify one's confidence in the assignment.

The above description makes it clear that there are two key steps in this approach. First, given a set of gapped sequences, all of the same length, how can one construct a corresponding profile HMM? Second, given a new gapped sequence of the same length, how can one compute the likelihood of the sequence for the given profile HMM? Each of these steps is now described. As one would expect, constructing a profile HMM is a far more complicated affair than computing the likelihood of a sequence given the profile HMM. In this section, a possible way of constructing a profile HMM is described. Unfortunately, in the computational biology literature, there does not seem to be a tradition of describing the procedures and algorithms completely. Instead the finer details are often left out of the publications and embedded within the code. Thus it is not possible to know, merely from reading published descriptions of various algorithms, what steps precisely comprise the algorithmic procedure. Therefore the discussion below represents the author's best guess as to what the currently used algorithms might be.

Accordingly, suppose one is given several sequences of equal length over a finite alphabet $\mathbb{A} \cup \{-\}$, where $-$ is the gap symbol. In the case of protein families, the set $\mathbb{A}$ is the alphabet consisting of the 20 amino acid symbols; it does not matter whether one uses the three-letter code or the single-letter code described in Table 8.1. However, for ease of exposition and ease of depiction, throughout this section it is assumed that $\mathbb{A} = \{A, C, G, T\}$, the alphabet of nucleotides. To repeat, this is just to make the various displayed vectors have dimension 4×1 thus making them easy to read, as opposed to 20×1. However, in the overwhelming majority of cases profile HMMs are used to classify proteins, not nucleotide sequences. The modifications to the procedures described below to convert from the four-symbol nucleotide alphabet to the 20-symbol amino acid alphabet are straightforward and are left to the reader.

The first step is to determine the length of the profile HMM, which is bounded above by the length of the gap-aligned sequences, and is often less. The length is computed as follows: At each location of the aligned sequences, determine whether the majority of symbols belongs to the alphabet $\mathbb{A}$, or is the gap symbol. If the majority of symbols belongs to the alphabet, call that a "match" location, and if the majority of symbols is the gap symbol, then call that an "insert" location. Let l denote the number of match locations. Then the associated profile HMM has $3l + 3$ states in the underlying Markov process, namely: A "start" state S, an "end" state E, l match states $M_1, \ldots, M_l$, l "delete" states $D_1, \ldots, D_l$, and $l + 1$ "insert" states $I_0 \ldots, I_l$. Note that the index for the insertion state starts with 0, whereas for the match and delete states the index starts with 1. The Markov process starts in the start state S and proceeds through the various intermediate states, and terminates in the end state E, which is the only essential state, the rest being inessential.

The following rules would assist in determining the various transition probabilities.

- Only the insert states I_0 through I_l may have self-transitions, but not the rest.
- From a delete state, say D_t, transitions are possible only to D_{t+1} and to M_{t+1}, but not to I_t.
- From a match state, say M_t, transitions are possible to M_{t+1}, D_{t+1}, and I_t.
- From an insert state, say I_t, transitions are possible only to D_{t+1}, M_{t+1}, and back to itself, but not to I_{t+1}.

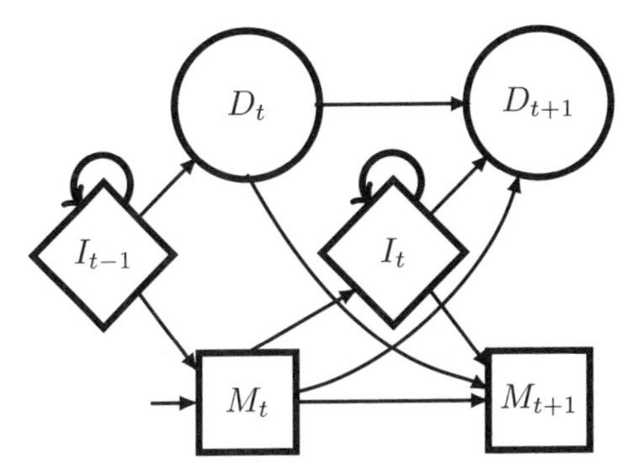

Figure 8.8 Possible Transitions in a Profile Hidden Markov Model

Figure 8.8 illustrates the possible transitions in a profile HMM.

Once the length of the profile HMM is determined and the $3l + 3$ states are labeled, the next step is to determine the various transition probabilities. Note that, since every location in the gap-aligned sequence is labeled as either a match location or an insert location, there are only four possible pairs of adjacent locations, namely: match-match, match-insert, insert-match, and insert-insert. The last possibility is not entertained, as any number of insert locations in an unbroken sequence are grouped into one insert symbol. Therefore the possible transitions are the following: $M_t \to M_{t+1}$, $M_t \to I_t$, $I_t \to M_{t+1}$, and $I_t \to I_t$. In what follows, rules are given for computing the transition probabilities for each of the four cases.

Example 8.2 We begin with the case match to match. In this case there are four transitions: M_t to M_{t+1}, M_t to D_{t+1}, D_t to M_{t+1}, and D_t to D_{t+1}. These correspond precisely to the four cases symbol to symbol, symbol to gap, gap to symbol, and gap to gap. To illustrate, suppose two successive locations are as shown below where to save space we write the sequences horizontally rather than vertically.

$$\begin{array}{c|cccccccccc} M_t & G & C & - & - & A & - & T & - & G & A \\ M_{t+1} & - & G & C & - & C & G & C & G & T & G \end{array}$$

There are six symbols in the current row, out of which one is followed by a gap and five are followed by symbols. So the "raw" transition probabilities are given by

$$\Pr\{M_{t+1}|M_t\} = 5/6, \Pr\{D_{t+1}|M_t\} = 1/6, \Pr\{I_t|M_t\} = 0/6.$$

But if we apply the so-called Laplacian correction, the probabilities now become

$$\Pr\{M_{t+1}|M_t\} = 6/9, \Pr\{D_{t+1}|M_t\} = 2/9, \Pr\{I_t|M_t\} = 1/9.$$

Similarly, there are four gap symbols in the current row, out of which three are followed by symbols in the next row and one is followed by a gap. Therefore the raw transition probabilities from D_t are given by

$$\Pr\{M_{t+1}|D_t\} = 3/4, \Pr\{D_{t+1}|D_t\} = 1/4, \Pr\{I_t|D_t\} = 0/4.$$

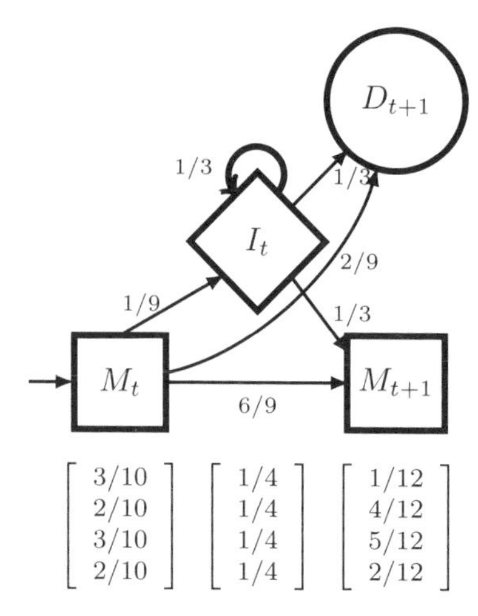

Figure 8.9 One Step in the Hidden Markov Model of Example 8.2

The Laplacian-corrected probabilities are

$$\Pr\{M_{t+1}|D_t\} = 4/6, \Pr\{D_{t+1}|D_t\} = 2/6, \Pr\{I_t|D_t\} = 0/6.$$

Note that *the correction is not added to the* $D_t \to I_t$ *transition*, as that is forbidden under the rules of the profile HMM.

Thus far we have computed the transition probabilities. We also need to compute the output or emission probabilities. Note that the delete state does not emit any output, and only the match state does. At time t, there are six symbols, so the raw and Laplace-corrected emission probability vectors are

$$M_t : [2/6 \;\; 1/6 \;\; 2/6 \;\; 1/6], \text{ and } [3/10 \;\; 2/10 \;\; 3/10 \;\; 2/10].$$

Similarly, at time $t+1$ the raw and Laplace-corrected emission probability vectors are

$$M_{t+1} : [0/8 \;\; 3/8 \;\; 4/8 \;\; 1/8], \text{ and } [1/12 \;\; 4/14 \;\; 5/12 \;\; 2/12].$$

To conclude the example, we need also to determine the transmission and emission probabilities for the insert state I_t. To compute the transition probability *out* of the state I_t, and the emission probabilities, we begin with the observation that all frequencies are zero, because there is no data! So we assign the default outward transition probability of 1/3 to each of the transitions $I_t \to I_t$, $I_t \to M_{t+1}$, and $I_t \to D_{t+1}$. This can be thought as the Laplace-corrected answer for the all zero vector. Similarly, we assign the emission probability of 1/4 to each nucleotide, which is again the Laplace-corrected version of the all zero vector.

The complete set of transition and emission probabilities is depicted in Figure 8.9.

Example 8.3 Next we consider the situation of two match location interspersed by one or more insert locations. Consider again the same two match locations as in Example 8.2, with three insert locations, as shown below.

$$\begin{array}{l|cccccccccc} M_t & G & C & - & - & A & - & T & - & G & A \\ I_t & C & - & - & - & A & - & T & - & - & A \\ I_t & - & G & - & G & C & - & - & C & - & - \\ I_t & A & - & T & - & A & - & - & A & - & - \\ M_{t+1} & - & G & C & - & C & G & C & G & T & G \end{array}$$

To analyze this situation, we first write down the ten "sample paths," using the following conventions: In a match location, use the symbol M for a symbol and D for a gap. In an insert location, use the symbol I for a symbol, and D for a gap. This leads to the following ten sample paths, corresponding to the ten columns:

$$MIDID, MDIDM, DDDIM, DDIDD, MIIIM,$$

$$DDDDM, MIDDM, DDIIM, MDDDM, MIDDM.$$

Next, these sample paths are trimmed using the following rules:

- All D symbols in insert locations are removed. This causes the ten sample paths to become as follows:

$$MIID, MIM, DIM, DID, MIIIM,$$

$$DM, MIM, DIIM, MM, MIM.$$

- Any remaining transitions of the form DI are replace by just I. This causes the ten sample paths to become as follows:

$$MIID, MIM, IM, ID, MIIIM, DM, MIM, IIM, MM, MIM.$$

- Next, the pairwise transitions are computed, as shown below, with the row corresponding to the current location and the column corresponding to the next location.

	M	D	I
M	1	0	5
D	1	0	0
I	6	2	4

- Next the Laplacian correction of adding one to each transition is applied, except for the $D \to I$ transition. This leads to

	M	D	I
M	2	1	6
D	2	1	0
I	7	3	5

Now normalizing the rows gives the transition matrix at this location.

$$A = \begin{bmatrix} 2/9 & 1/9 & 6/9 \\ 2/3 & 1/3 & 0/3 \\ 7/15 & 3/15 & 5/15 \end{bmatrix}.$$

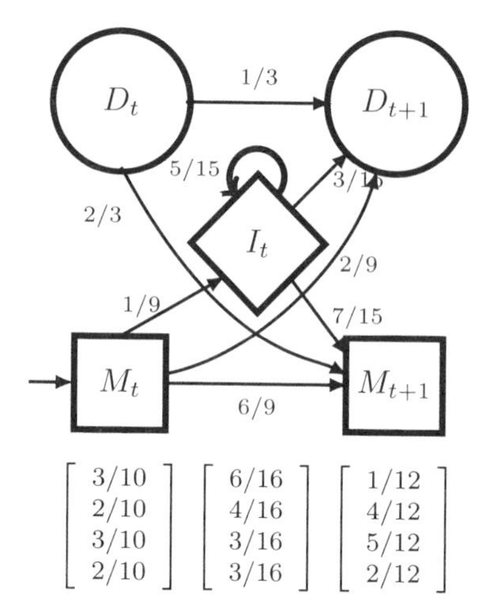

Figure 8.10 One Step in the Hidden Markov Model of Example 8.3

Thus far we have computed the transition probabilities. The output or emission probabilities are computed exactly in analogy with Example 8.2. For states M_t and M_{t+1} the same approach as in that example can be applied, resulting in the following raw and Laplace-corrected probability vectors:

$$M_t : [2/6 \;\; 1/6 \;\; 2/6 \;\; 1/6], \text{ and } [3/10 \;\; 2/10 \;\; 3/10 \;\; 2/10],$$

$$M_{t+1} : [0/8 \;\; 3/8 \;\; 4/8 \;\; 1/8], \text{ and } [1/12 \;\; 4/14 \;\; 5/12 \;\; 2/12].$$

For the state I_t, the same approach is adopted, except that the symbols in *all three rows* labeled as insert locations are added. This leads to the following raw and Laplace-corrected probability vectors:

$$I_t : [5/12 \;\; 3/12 \;\; 2/12 \;\; 2/12], \text{ and } [6/16 \;\; 4/16 \;\; 3/16 \;\; 3/16].$$

To conclude the example, if we have a chain $S \to I_0 \to M_1$ or $M_l \to I_l \to E$ where S and E denote the start and end states respectively, then we can use the same procedure as above, with S replacing M_t or E replacing M_{t+1}.

Chapter Nine

BLAST Theory

BLAST (Basic Local Alignment Search Tool) is a widely used statistical method for finding similarities between sequences of symbols from finite alphabets. While the theory is completely general, the most widely used applications are to comparing sequences of nucleotides and sequences of amino acids. Though the letter B in BLAST stands for "basic," in fact the theory itself is anything but basic. The objective of this chapter therefore is to present an accessible treatment of the theory.

The theory of BLAST was developed through a series of papers coauthored by Samuel Karlin; see [77, 74, 75, 76, 39, 40]. The notation and problem formulations across these papers are not always consistent, making it very difficult for the nonexpert reader to navigate through these papers. It is hoped that the present chapter will assist somewhat in this process. The treatment here follows [39, 40]. The reader is cautioned that there are several modifications of the theory presented here; these modifications do not always have a theoretical justification. In the interests of brevity and clarity, we treat here only the most "basic" version of BLAST theory.

In Section 8.2, we discussed the problem of optimal gapped alignment between two sequences. Though an "exact" solution to this problem can be found using dynamic programming, it turns out that an approximate solution is often good enough in many situations. This was one of the motivations for the development of BLAST theory. In Section 9.1, we present the problem that BLAST theory addresses, state the main results, and show how these main results can be applied in practice. In Section 9.2, we present the proofs of all the main results. A reader who is not interested in knowing how the theorems that underlie BLAST are proved can skip this section.

9.1 BLAST THEORY: STATEMENTS OF MAIN RESULTS

9.1.1 Problem Formulations

The fundamental objective of BLAST theory is to align sequences as well as possible, and then make a determination as to the level of statistical significance of the alignment. Thus one computes a "maximal segmental score" of the alignment between the two sequences, and tests to see whether the maximal segmental score could have been obtained purely as a matter of chance. If the match is better than could be explained by chance, then one would be able to conclude that the two sequences do indeed show some similarity. Thus, in order to apply the theory, one needs to be able to compute two things: The expected maximal segmental score for sequences of a given length, and the "tail probability distribution" of the likelihood that the maximal segmental score will exceed this expected value. In the remainder of the chapter, we derive answers to these and other related questions. Thus we begin by surveying the various problem formulations.

Suppose $\mathbb{A}$ and $\mathbb{B}$ are finite sets, and that X, Y are independent random variables assuming values in $\mathbb{A}$ and $\mathbb{B}$ respectively, with probability distributions $\boldsymbol{\phi}$ and $\boldsymbol{\psi}$ respectively. We can also consider the "product" random variable $Z = (X, Y)$ that assumes values in the product set $\mathbb{A} \times \mathbb{B}$ and has the product distribution $\boldsymbol{\mu} := \boldsymbol{\phi} \times \boldsymbol{\psi}$. Note that the marginal distributions of $\boldsymbol{\mu}$ are given by $\boldsymbol{\mu}_{\mathbb{A}} = \boldsymbol{\phi}$ and $\boldsymbol{\mu}_{\mathbb{B}} = \boldsymbol{\psi}$.

Now suppose we draw i.i.d. samples of X and Y of length l according to their respective laws; call the sample paths $x_1, \ldots, x_l$ and $y_1, \ldots, y_l$ respectively. As discussed in Section 8.2, let us define a scoring function $F : \mathbb{A} \times \mathbb{B} \to \mathbb{R}$. Then we can define the cumulative score of the sample paths as

$$\sum_{i=1}^{l} F(x_i, y_i).$$

However, this cumulative score is not of interest to us. Rather, we are interested in the "maximal segmental score." This can be defined in one of two ways. If we insist that the starting points of the two segments must coincide, then we examine the quantity

$$R_l := \max_{L \geq 0, 0 \leq i \leq l-L} \sum_{k=1}^{L} F(x_{i+k}, y_{i+k}). \tag{9.1}$$

The quantity R_l examines all subsequences of length L within the two sample paths, and then computes only the *segmental score* over this segment of length L; then the maximum of all these segmental scores over all possible segment length L becomes R_l. If we *don't* insist that the starting points of the two segments must coincide, or in other words, if we allow the two segments to be shifted with respect to each other, then we look at the quantity

$$M_l := \max_{L \geq 0, 0 \leq i,j \leq l-L} \sum_{k=1}^{L} F(x_{i+k}, y_{j+k}). \tag{9.2}$$

To repeat, the main difference between the quantities R_l and M_l is that in defining R_l, we insist that the starting points of the two segments being aligned must coincide, whereas in defining M_l, we permit the starting points of the two segments to be shifted with respect to each other. Note that in [74, 76], the quantity R_l is referred to as M_l; thus the notation changes from [74, 76] to [39, 40].

When we examine the issue of maximal scores, we can ask four distinct questions:

1. Given sample paths of length l, what is the expected value of M_l or R_l?

2. Let L_l denote the length of a maximal scoring segment. What is the expected value of L_l? That is, how long is a maximally scoring segment on average, from a sample path of length l?

3. What is the empirical distribution of the symbols x_{i+k}, y_{j+k} in a maximally scoring segment?

4. What is the tail probability distribution of the quantities M_l and R_l beyond their expected values? In other words, suppose M_l exceeds its expected value by some ϵ. Can we compute the likelihood that this has happened purely due to chance? This would give us the *significance* of the high-scoring segment.

In the sequel, we will answer all of these questions.

9.1.2 The Moment Generating Function

In this subsection we revisit the moment generating function and the logarithmic moment generating function, which were very briefly introduced in Section 1.3.5. Suppose $\mathbb{A} = \{x_1, \ldots, x_n\}$ is a subset of the real numbers $\mathbb{R}$ (and not an abstract set of labels, as in other places in this book). Suppose X is a random variable assuming values in the set $\mathbb{A}$ with the distribution $\boldsymbol{\mu}$. Thus μ_i denotes $\Pr\{X = x_i\}$ for all i. It can be assumed without loss of generality that $\mu_i > 0$ for all i, because if $\mu_i = 0$ for some i, then the corresponding element x_i can simply be deleted from the set $\mathbb{A}$. For each positive integer k, the quantity

$$M_k(X) := \sum_{i=1}^{n} x_i^k \mu_i = E[X^k, \boldsymbol{\mu}]$$

is called the **k-th moment** of the random variable X. In particular, the quantity $M_1(X)$ is just the mean of the random variable X, while $M_2(X) - [M_1(X)]^2$ is the variance of X. Note that, since $\mathbb{A}$ is a finite set, all of the summations above are also finite; as a result, $M_k(X)$ is well-defined for every integer $k \geq 1$. Next, the function

$$\text{mgf}(\lambda; X) := E[\exp(\lambda X), \boldsymbol{\mu}] = \sum_{i=1}^{n} \mu_i \exp(\lambda x_i)$$

is called the **moment-generating function (mgf)** of the random variable X. Note that

$$\left[\frac{d^k \text{mgf}(\lambda; X)}{d\lambda^k}\right]_{\lambda=0} = \left[\sum_{i=1}^{n} \mu_i x_i^k \exp(\lambda x_i)\right]_{\lambda=0} = \sum_{i=1}^{n} \mu_i x_i^k = M_k(X)$$

for every integer $k \geq 1$. This explains the nomenclature. Note that $\text{mgf}(0; X) = 1$ for every random variable X.

Next we define the so-called **logarithmic moment generating function** $\Lambda(\lambda; X)$ by

$$\Lambda(\lambda; X) := \log \text{mgf}(\lambda; X) = \log E[\exp(\lambda X), \boldsymbol{\mu}].$$

Since the function log is concave, it follows from Jensen's inequality that

$$\log E[\exp(\lambda X), \boldsymbol{\mu}] \geq E[\log \exp(\lambda X), \boldsymbol{\mu}] = \lambda M_1(X), \ \forall \lambda.$$

A very useful property of the mgf and its logarithm are brought out next.

Lemma 9.1 *For a fixed nontrivial random variable X, both* $\text{mgf}(\lambda; \mathbf{x})$ *and* $\Lambda(\lambda; X)$ *are strictly convex functions of λ.*

Here by a "nontrivial" random variable, we mean a random variable that assumes at least two distinct values. (Otherwise the "random variable" would be just a constant!)

Proof. For a fixed random variable X, we have

$$\text{mgf}(\lambda; X) = \sum_{i=1}^{n} \mu_i \exp(\lambda x_i)$$

is a linear combination of strictly convex functions $\lambda \mapsto \exp(\lambda x_i)$. Hence $\text{mgf}(\lambda; X)$ is also a strictly convex function of λ. To show that $\Lambda(\lambda; X)$ is also strictly convex in λ, let us, for the purposes of this proof alone, use η to denote the moment generating function and η' to denote $d\eta/d\lambda$. Then we have

$$\Lambda = \log \eta, \Lambda' = \frac{\eta'}{\eta}, \Lambda'' = \frac{\eta\eta'' - (\eta')^2}{\eta^2}.$$

Hence the strict convexity of Λ follows if it can be established that

$$\eta\eta'' > (\eta')^2 \ \forall \lambda.$$

For this purpose, note that

$$\eta = \sum_{i=1}^{n} \mu_i \exp(\lambda x_i), \eta' = \sum_{i=1}^{n} \mu_i x_i \exp(\lambda x_i) \eta'' = \sum_{i=1}^{n} \mu_i x_i^2 \exp(\lambda x_i).$$

So the inequality that we desire to establish can be written as

$$(E[X e^{\lambda X}, \boldsymbol{\mu}])^2 < E[e^{\lambda X}, \boldsymbol{\mu}] \cdot E[X^2 e^{\lambda X}, \boldsymbol{\mu}].$$

Now we make use of Schwarz's inequality, which says in this setting that

$$(E[fg, \boldsymbol{\mu}])^2 \leq E[f^2, \boldsymbol{\mu}] \cdot E[g^2, \boldsymbol{\mu}],$$

with equality if and only if f and g are multiples of each other. Apply Schwarz's inequality with the choices

$$f = \exp(\lambda X/2), g = X \exp(\lambda X/2),$$

and observe that if X is a nontrivial random variable, then f and g are *not* multiples of each other. The desired inequality follows. □

Now we state a very useful fact about the moment generating function.

Lemma 9.2 *Suppose X is a random variable assuming values in a finite set $\mathbb{A} = \{x_1, \ldots, x_n\} \subseteq \mathbb{R}$, with the probability distribution $\boldsymbol{\mu}$, where $\mu_i > 0$ for all i. Suppose in addition that*

(i) $\mathbb{A}$ contains both positive and negative numbers. In other words, there exist indices i and j such that $x_i > 0$ and $x_j < 0$ (and by assumption $\mu_i, \mu_j > 0$).

(ii) $E[X, \boldsymbol{\mu}] \neq 0$.

Under these conditions, there exists a unique $\lambda^* \neq 0$ such that $\text{mgf}(\lambda^*; X) = 1$ or equivalently $\Lambda(\lambda^*; X) = 0$. Moreover, λ^* has sign opposite to that of $E[X, \boldsymbol{\mu}]$.

Proof. Note that $\text{mgf}(0; X)$ always equals 1. Now

$$\text{mgf}'(0; X) = \left[\frac{d\text{mgf}(\lambda; X)}{d\lambda} \right]_{\lambda=0} = E[X, \boldsymbol{\mu}] \neq 0$$

by assumption. We have already seen from Lemma 9.1 that the mgf is a strictly convex function. Finally, it follows from Condition (i) again that

$$\text{mgf}(\lambda; X) \to \infty \text{ as } \lambda \to \pm\infty.$$

This is because at least one of the x_i is positive and at least one is negative. From this information we conclude that the equation $\text{mgf}(\lambda; X) = 1$ has precisely two solutions, one of which is $\lambda = 0$, and the other one, denoted by λ^*, has sign opposite to that of $E[X, \boldsymbol{\mu}]$. □

Note that if the set X consists of only nonnegative or only nonpositive numbers, then the above lemma is false. In particular, if $x_i \geq 0$ for all i, then $\text{mgf}(\lambda; X) \to 0$ as $\lambda \to -\infty$, and the only solution of $\text{mgf}(\lambda; X) = 1$ is $\lambda = 0$. The situation when every element of X is nonpositive is similar.

The vector $\boldsymbol{\theta}$ defined by

$$\theta_i = \exp(\lambda^* x_i)\mu_i$$

belongs to $\mathbb{S}_n$ because clearly $\theta_i > 0$ for all i, and in addition,

$$\sum_{i=1}^{n} \theta_i = \text{mgf}(\lambda^*; X) = 1.$$

The distribution $\boldsymbol{\theta}$ is referred to as the **conjugate distribution** of $\boldsymbol{\mu}$ with respect to the random variable X. Note that $\boldsymbol{\theta}$ depends on both the distribution $\boldsymbol{\mu}$ and the corresponding values of the random variable X.

9.1.3 Statement of Main Results

In this subsection, we state the main results of BLAST theory. Specifically, we answer the four questions raised earlier. Recall that we are given a probability distribution $\boldsymbol{\mu} = \boldsymbol{\phi} \times \boldsymbol{\psi}$ on the product set $\mathbb{A} \times \mathbb{B}$ and a scoring function $F : \mathbb{A} \times \mathbb{B} \to \mathbb{R}$. So we can think of F as a real-valued random variable that assumes the value $F(x_i, y_j)$ with probability $\mu_{ij} = \phi_i \psi_j$. To simplify notation, let us denote $F(x_i, y_j)$ by F_{ij}.

In order to state these results, we introduce one "standing assumption."

$$E[F, \boldsymbol{\mu}] < 0, \text{ and } \exists i, j \text{ s.t. } F(x_i, y_j) > 0. \tag{9.3}$$

Thus the standing assumption states that there is at least one pair (x_i, y_j) for which the score F_{ij} is positive, but the expected value of the score over all pairs is negative. With this standing assumption, it follows from Lemma 9.2 that there exists a unique number $\lambda^* > 0$ such that $E[\exp(\lambda^* F), \boldsymbol{\mu}] = 1$. As before, let $\boldsymbol{\theta}$ denote the conjugate distribution of $\boldsymbol{\mu}$ with respect to F, that is

$$\theta_{ij} = \exp(\lambda^* F_{ij})\mu_{ij}, \ \forall i, j. \tag{9.4}$$

The results are stated next. Note that all results are asymptotic; that is, they apply as $l \to \infty$. In the theorem, the notation $a_l \sim b_l$ as $l \to \infty$ means that the ratio $a_l / b_l \to 1$ as $l \to \infty$.

Theorem 9.3 *Let all symbols be as defined above. Then*

1. $R_l \sim (\ln l)/\lambda^*$ *as* $l \to \infty$.
2. *Let* $\boldsymbol{\theta}$ *denote the conjugate distribution of* $\boldsymbol{\mu}$*, as defined in (9.4). Then the length of a maximal scoring segment* L_l *is asymptotically equal to* $l/H(\boldsymbol{\theta}\|\boldsymbol{\mu})$.
3. *On any maximal scoring segment, the empirical distribution of* (X, Y) *is asymptotically equal to* $\boldsymbol{\theta}$.

Now we come to the quantity M_l, the maximum segmental score when we *don't* insist that the starting points of the two segments must match.

Theorem 9.4 *Let all symbols be as defined above. Suppose in addition that the two sets* $\mathbb{A}, \mathbb{B}$ *are the same, that the marginal distributions* $\boldsymbol{\phi}, \boldsymbol{\psi}$ *are the same, and that the scoring function* F *is symmetric in the sense that* $F(x_i, y_j) = F(y_j, x_i)$. *Then*

1. $M_l \sim (2 \ln l)/\lambda^*$ *as* $l \to \infty$.
2. *Let* $\boldsymbol{\theta}$ *denote the conjugate distribution of* $\boldsymbol{\mu}$*, as defined in (9.4). Then the length of a maximal scoring segment* L_l *is asymptotically equal to* $2l/H(\boldsymbol{\theta}\|\boldsymbol{\mu})$.
3. *On any maximal scoring segment, the empirical distribution of* (X, Y) *is asymptotically equal to* $\boldsymbol{\theta}$.

While it is possible to analyze the asymptotic behavior of R_l and M_l without making any additional assumptions about the two sets $\mathbb{A}, \mathbb{B}$, the two marginal distributions, or the scoring function, adding some assumptions allows us to get nice results. A perusal of the detailed proofs in the next section would reveal that these assumptions are made only to enable us to give "closed-form" formulas; but if one is willing to forgo these, there is no need to make these additional assumptions.

Taken together, these theorems establish the following facts:

- On a sample of length l, the expected maximal segmental score if we allow different starting points for the two segments is twice the expected maximal segmental score if we insist that the starting points must match.
- The expected length of a maximal scoring segment is asymptotically twice as long in the case of M_l compared to R_l.
- In either case, the empirical distribution of the symbols (x_i, y_j) on a maximal scoring segment is given by the conjugate distribution $\boldsymbol{\theta}$.

In order to state the next theorem, we need to introduce the notion of a "lattice" random variable. We say that a random variable assuming values in a finite (or even a countably infinite) subset S of the real numbers $\mathbb{R}$ is a "lattice" random variable if S is contained in an arithmetic progression. In other words, a random variable is a lattice random variable if all its possible values are of the form $a + md$ for real numbers a, d and integers m. In this case, the set $\{a + kd, k = 0, \pm 1, \pm 2, ...\}$ is the associated lattice. In the present instance, the random variable of interest is $F(X,Y)$, the score as X, Y vary over their respective sets. We shall first state the theorems and then discuss their implications.

Theorem 9.5 *Suppose the score $F(X,Y)$ is not a lattice random variable. Then there exists a constant K^*, which can be estimated and in some cases computed explicitly, such that for all $x > 0$, the following inequality holds:*

$$\lim_{l \to \infty} \Pr\{R_l - \frac{\log l}{\lambda^*} \leq x\} = \exp(-K^* \exp(-\lambda^* x)). \tag{9.5}$$

In case the score $F(X,Y)$ is a lattice random variable, the following statement is true:

$$\lim_{l \to \infty} \Pr\{R_l - \frac{\log l}{\lambda^*} \leq x_l\} \cdot \exp(K^* \exp(-\lambda^* x_l)) = 1 \tag{9.6}$$

whenever $\{x_l\}$ is a bounded sequence such that $x_l - \log l / \lambda^*$ belongs to the lattice for each value of l.

Theorem 9.6 *Suppose the score $F(X,Y)$ is not a lattice random variable. Then there exists a constant K^*, which can be estimated and in some cases computed explicitly, such that for all $x > 0$, the following inequality holds:*

$$\lim_{l \to \infty} \Pr\{M_l - \frac{2 \log l}{\lambda^*} \leq x\} = \exp(-K^* \exp(-\lambda^* x)). \tag{9.7}$$

In case the score $F(X,Y)$ is a lattice random variable, the following statement is true:

$$\lim_{l \to \infty} \Pr\{M_l - \frac{2 \log l}{\lambda^*} \leq x_l\} \cdot \exp(K^* \exp(-\lambda^* x_l)) = 1 \tag{9.8}$$

whenever $\{x_l\}$ *is a bounded sequence such that* $x_l - \log l/\lambda^*$ *belongs to the lattice for each value of* l.

Note that (9.5) is the same as (9.6), just written differently. Thus the difference is that if $F(X,Y)$ is a non-lattice variable, then we have a tail probability estimate for *every* value of x, whereas if $F(X,Y)$ is a lattice random variable, we have a tail probability estimate only for *some* values of x. Similar remarks apply to (9.7) and (9.8). This topic is discussed further in the next few paragraphs.

Equation (9.5) gives an extremely precise estimate of the rate at which the tail probability that R_l exceeds its expected value $\log l/\lambda^*$ by an amount x decays as x increases. The distribution on the right side of (9.5) is called a **Gumbel distribution (of Type I)**. Note that as $x \to \infty$, the exponential term $\exp(-\lambda^* x)$ approaches zero, as result of which the right side of (9.5) approaches one. More precisely, suppose x is sufficiently large that

$$K^* \exp(-l^* x) \ll 1, \text{ or equivalently } x \gg \frac{\log K^*}{\lambda^*}.$$

Then, using the approximation $\exp(-\alpha) \approx 1 - \alpha$ when α is small, we can rewrite (9.5) as

$$\lim_{l\to\infty} \Pr\{R_l - \frac{\log l}{\lambda^*} > x\} \approx K^* \exp(-\lambda^* x) \text{ whenever } x \gg \frac{\log K^*}{\lambda^*}. \quad (9.9)$$

Thus, while the formula (9.5) is extremely precise, in practice the tail probability decays exponentially with respect to x. Similar remarks apply to the tail estimate of M_l as well.

Note that all of the above equations from (9.5) through (9.8) give estimates of the *absolute excess* of the score from a maximal segment beyond its expected value. However, it is obvious that if we replace the scoring function F by some multiple cF, then we still have the same problem, and the maximal segments would still be the same. However, the expected value of the maximal segmental score would be scaled by the same factor c, and λ^* gets replaced by λ^*/c. Thus, in order to make the estimates more meaningful and "scale-free," we should perhaps look at the excess *as a fraction of the expected value*, and not as an absolute excess. Accordingly, suppose

$$x = \alpha \frac{\log l}{\lambda^*}.$$

Then, after some routine algebra, the counterpart of (9.5) is

$$\lim_{l\to\infty} \Pr\{R_l - \frac{\log l}{\lambda^*} \le \alpha \frac{\log l}{\lambda^*}\} = \exp(-K^* l^{-\alpha}). \quad (9.10)$$

Moreover, if

$$l \gg \frac{1}{\alpha} \log \frac{1}{K^*},$$

then

$$\Pr\{R_l - \frac{\log l}{\lambda^*} > \alpha \frac{\log l}{\lambda^*}\} \approx K^* l^{-\alpha}.$$

Similar modifications of (9.6) through (9.8) are routine and are left to the reader.

Now let us discuss the implications of the lattice vs. non-lattice variable. This discussion unfortunately borders on the pedantic, and the reader would miss very little by assuming that the relationship (9.9) always holds. However, since we have attempted to make mathematically precise statements in this text, we discuss this issue.

Quite often one would assign integer values to the scoring function; in other words, $F(x_i, y_j)$ is always assigned an integer value, as in Example 8.1 for instance. This would make $F(X, Y)$ a lattice random variable with the spacing d equal to one. More generally, suppose all entries of $F(X, Y)$ are specified to k significant decimal places. Then clearly every element of $F(X, Y)$ is of the form $k_{ij} \cdot 10^{-k}$ for suitable integers k_{ij}, which would again make $F(X, Y)$ a lattice random variable with the spacing $d = 10^{-k}$. The only way to ensure that $F(X, Y)$ is a non-lattice random variable is to specify the values of F to infinite precision, which makes no sense in practice. In this case, one cannot use the exact formulas (9.5) and (9.7). However, given any x, it is clear that we can always find a bounded sequence $\{x_n\}$ such that $x_n - d \leq x \leq x_n$ (where d is the lattice spacing) and $x_n + \log l/\lambda^*$ is a lattice point. Note that

$$\Pr\{R_l - \frac{\log l}{\lambda^*} > x_l - d\} \leq \Pr\{R_l - \frac{\log l}{\lambda^*} > x\} \leq \Pr\{R_l - \frac{\log l}{\lambda^*} > x_n\}.$$

Moreover, the two extreme probabilities do indeed satisfy the Gumbel type of tail probability estimate of the form (9.5). Hence, for all practical purposes, we need not worry about the distinction between lattice and non-lattice random variables.

9.1.4 Application of Main Results

A very nice discussion of the application of the above theorems to detecting similarity of sequences can be found in [3]. There are two distinct ways in which the theorems can be used. First, suppose the scoring function F is specified, and we are given two sample paths $\mathbf{x}_1^l, \mathbf{y}_1^l$ of known statistics. Suppose we compute the maximal segmental score M_l. Now we wish to know whether the two sample paths are similar or not. If M_l significantly exceeds the expected score $2 \log l/\lambda^*$ by some quantity x, then we can compute the likelihood that this has happened purely by chance, by using the tail probability estimates in Theorem 9.6. This would allow us to say, with appropriate confidence, whether or not the segmental score M_l connotes sequence similarity.

The second application is to "reverse engineer" the scoring function F itself. In a given problem, suppose we do not know what the "right" scoring function is, but we *do have* at hand several pairs of similar sequences. Often these pairs of similar sequences would be generated by biologists, who know that, under any reasonable alignment procedure, these pairs of sequences should have very high similarity scores. So the problem then becomes one of choosing a scoring function F in such a way that these known answers are "automatically" generated by the theory. In this context, Theorems 9.3 and 9.4 are useful. These theorems tell us that, on maximal scoring segments, the empirical distribution looks like $\boldsymbol{\theta}$. So we can proceed as follows: Suppose we are given several pairs of high-scoring segments. By definition, the two segments are of equal length. So we can just concatenate the various segments to generate one really huge segment of the form $(\mathbf{x}_1^l, \mathbf{y}_1^l)$. And then:

- Construct the probability distribution $\boldsymbol{\phi}$ as the empirical distribution of the symbols in $\mathbb{A}$ among $\mathbf{x}_1^l$. Similarly, construct the probability distribution $\boldsymbol{\psi}$ as the empirical distribution of the symbols in $\mathbb{B}$ among $\mathbf{y}_1^l$.

- Similarly, compute the probability distribution $\boldsymbol{\theta}$ on $\mathbb{A} \times \mathbb{B}$ as the *joint empirical distribution* of the pair (a_i, b_j) on the sequence $(\mathbf{x}_1^l, \mathbf{y}_1^l)$.

- Using $\boldsymbol{\mu} = \boldsymbol{\phi} \times \boldsymbol{\psi}$ and $\boldsymbol{\theta}$ as defined above, define the scoring function

$$F_{ij} := \frac{1}{c} \log \frac{\theta_{ij}}{\mu_{ij}} = \frac{1}{c} \log \frac{\theta_{ij}}{\phi_i \cdot \psi_j}. \tag{9.11}$$

  Here c is any constant that we choose. As discussed earlier, if we scale the scoring function uniformly by any constant, the problem remains unchanged.

It might be mentioned that the above approach to constructing scoring functions has been used extensively in the computational biology community.

9.2 BLAST THEORY: PROOFS OF MAIN RESULTS

In this section, we present the proofs of Theorems 9.3 and 9.4. Those who wish only to *use* BLAST theory can perhaps skip reading this section, but those who aspire to understand the basis of BLAST theory and to generalize it to other contexts may perhaps benefit from reading this section. Unfortunately the proofs of Theorems 9.5 and 9.6 are beyond the scope of the book. The interested reader is referred to [37] for the proof of Theorem 9.5 and [40] for the proof of Theorem 9.6. Even the proofs of Theorems 9.3 and 9.4 are quite complicated, as can be seen from the contents of this section. However, on the basis of these proofs, it may perhaps be possible to relax some of the assumptions underlying these two theorems, such as that the two sample paths $\mathbf{x}_1^l$ an $\mathbf{y}_1^l$ come from i.i.d. processes, and instead replace them by, say, sample paths of Markov processes.

The principal source for this section is the seminal paper by Dembo, Karlin and Zeitouni [39]. To assist the reader in following this paper, throughout this section we mention the corresponding theorem or lemma number from [39], and also highlight departures from their notation if any.

Let us recall the framework of the problem under study. We are given sets $\mathbb{A}, \mathbb{B}$ and probability distributions $\boldsymbol{\phi}$ on $\mathbb{A}$ and $\boldsymbol{\psi}$ on $\mathbb{B}$; for convenience let $\boldsymbol{\mu} := \boldsymbol{\phi} \times \boldsymbol{\psi}$ denote the corresponding product distribution on $\mathbb{A} \times \mathbb{B}$. We define the quantities

$$R_l := \max_{L \geq 0, 0 \leq i \leq l-L} \sum_{k=1}^{L} F(x_{i+k}, y_{i+k}),$$

and

$$M_l := \max_{L \geq 0, 0 \leq i,j \leq l-L} \sum_{k=1}^{L} F(x_{i+k}, y_{j+k}).$$

The objective is to answer the following questions:

1. Given sample paths of length l, what is the expected value of M_l or R_l?
2. Let L_l denote the length of a maximal scoring segment. What is the expected value of L_l? That is, how long is a maximally scoring segment on average, from a sample path of length l?
3. What is the empirical distribution of the symbols x_{i+k}, y_{j+k} in a maximally scoring segment?

Answering the last question raised earlier, namely to describe the tail probability distribution of the maximal segmental score beyond its expected value, is beyond the scope of this book. The interested reader is referred to the papers by Dembo and Karlin [37, 38, 76].

Given the set $\mathbb{C} = \mathbb{A} \times \mathbb{B}$, let $\mathcal{M}(\mathbb{C})$ denote the set of all probability distributions on this set. Thus $\mathcal{M}(\mathbb{C})$ can be identified with the nm-dimensional simplex $\mathbb{S}_{nm}$. If $\boldsymbol{\nu} \in \mathcal{M}(\mathbb{C})$, we use the symbols $\boldsymbol{\nu}_{\mathbb{A}}, \boldsymbol{\nu}_{\mathbb{B}}$ to denote its marginal distributions on $\mathbb{A}, \mathbb{B}$ respectively.

To state and prove the theorems, we introduce a great deal of preliminary notation. First, for any $\boldsymbol{\nu} \in \mathcal{M}(\mathbb{C})$, we define

$$H^*(\boldsymbol{\nu} \| \boldsymbol{\mu}) := \max\{0.5H(\boldsymbol{\nu} \| \boldsymbol{\mu}), H(\boldsymbol{\nu}_{\mathbb{A}} \| \boldsymbol{\mu}_{\mathbb{A}}), H(\boldsymbol{\nu}_{\mathbb{B}} \| \boldsymbol{\mu}_{\mathbb{B}})\}.$$

It is clear that $H^*(\boldsymbol{\nu} \| \boldsymbol{\mu}) > 0$ unless $\boldsymbol{\nu} = \boldsymbol{\mu}$. Next, for any given set $U \subseteq \mathcal{M}(\mathbb{C})$, define

$$J(\boldsymbol{\nu}) := \frac{E[F, \boldsymbol{\nu}]}{H^*(\boldsymbol{\nu} \| \boldsymbol{\mu})}, J(U) := \sup_{\boldsymbol{\nu} \in U} \max\{J(\boldsymbol{\nu}), 0\}.$$

Now by the standing hypothesis we have that $E[F, \boldsymbol{\mu}] < 0$. Hence, as $\boldsymbol{\nu} \to \boldsymbol{\mu}$, the quantity $E[F, \boldsymbol{\nu}]$ approaches some negative number, while $H^*(\boldsymbol{\nu} \| \boldsymbol{\mu})$ approaches zero. As a result $J(\boldsymbol{\nu}) \to -\infty$ as $\boldsymbol{\nu} \to \boldsymbol{\mu}$. If the set J does not contain any $\boldsymbol{\nu}$ such that $J(\boldsymbol{\nu}) > 0$, then by definition $J(U)$ is taken to equal zero.

The third piece of notation we need is the following: Suppose we are given the sample paths $\mathbf{x}_1^l, \mathbf{y}_1^l$, and that $L \geq 0, i, j \leq l - L$. Then we define

$\hat{\boldsymbol{\mu}}(i,j,L)$ to be the pairwise empirical distribution on $\mathbb{C}$ defined by the sample $(\mathbf{x}_{i+1}^{i+L}, \mathbf{y}_{i+1}^{i+L})$. Similarly, given the sample path we define $\hat{\boldsymbol{\mu}}(i,\cdot,L)$ to be the empirical distribution on $\mathbb{A}$ defined by the sample $\mathbf{x}_{i+1}^{i+L}$, and analogously we define $\hat{\boldsymbol{\mu}}(\cdot,j,L)$ to be the empirical distribution on $\mathbb{B}$ defined by the sample $\mathbf{y}_{i+1}^{i+L}$. Since the set $\mathbb{C}$ has nm elements and the sample path has length L, it follows that the empirical distribution $\hat{\boldsymbol{\mu}}(i,j,L)$ belongs to the set $\mathcal{E}(L,nm)$ defined in Chapter 5. In the same way, $\hat{\boldsymbol{\mu}}(i,\cdot,L) \in \mathcal{E}(L,n)$ and $\hat{\boldsymbol{\mu}}(\cdot,j,L) \in \mathcal{E}(L,m)$.

Given a set $U \subseteq \mathcal{M}(\mathbb{C})$, define

$$M_l^U := \max\{\sum_{k=1}^{L} F(x_{i+k}, y_{i+k}) : L \geq 0, 0 \leq i,j \leq l-L, \hat{\boldsymbol{\mu}}(i,j,L) \in U\}. \tag{9.12}$$

Thus M_l^U is exactly the same as M_l, except that the maximum is taken only over those segments of length L such that the corresponding pair empirical distribution $\hat{\boldsymbol{\mu}}(i,j,L)$ belongs to U. Other segments where $\hat{\boldsymbol{\mu}}(i,j,L)$ does not belong to U are not included in the computation of the maximum.

Now we state a "global" theorem from which the desired results follow.

Theorem 9.7 *With the notation as above, we have that*

$$P_{\boldsymbol{\mu}}^l\{\limsup_{l\to\infty} \frac{M_l^U}{\log l} \leq J(U)\} \to 0 \text{ as } l \to \infty, \tag{9.13}$$

$$P_{\boldsymbol{\mu}}^l\{\liminf_{l\to\infty} \frac{M_l^U}{\log l} \geq J(U^o)\} \to 0 \text{ as } l \to \infty, \tag{9.14}$$

where U^o denotes the interior of the set U.

This theorem is a slightly weakened version of [39, Theorem 3]. That theorem states the following:

Theorem 9.8 *With the notation above, we have that*

$$J(U^o) \leq \liminf_{l\to\infty} \frac{M_l^U}{\log l} \leq \limsup_{l\to\infty} \frac{M_l^U}{\log l} \leq J(U), \tag{9.15}$$

where both inequalities hold almost surely.

What Theorem 9.7 claims is called "convergence in probability," and we *can* prove *this* theorem using the methods we have developed thus far. It is good enough to allow us to study the behavior of the BLAST algorithm. We have no machinery to discuss "almost sure convergence" as we have scrupulously "avoided the infinite" in this book, so as to keep the exposition both rigorous as well as not so advanced. For those with the appropriate background, going from Theorem 9.7 to Theorem 9.8 is fairly straightforward using the Borel-Cantelli lemma.

The proof of Theorem 9.7 proceeds via a series of preliminary steps. Throughout the sequel, the symbol λ^* denotes the unique positive solution to the equation $E[\exp(\lambda^* F), \boldsymbol{\mu}] = 1$, and $\boldsymbol{\theta}$ denotes the conjugate distribution of $\boldsymbol{\mu}$ as defined in (9.4). Also, to keep the notation simple, we will simply use the symbol P to denote the probability measure $P_{\boldsymbol{\mu}}^l$.

Lemma 9.9 *([39, Lemma 1]) Choose any* $\lambda_0 \in (0, \lambda^*)$. *Then, whenever*

$$L \geq -\frac{5 \log l}{\Lambda(\lambda_0)} =: L_0(l),$$

we have that

$$P\{\sup_{L \geq L_0(l)} \sup_{0 \leq i,j \leq L} \sum_{k=1}^{L} F(x_{i+k}, y_{i+k}) \geq 0\} \leq 1/l^2.$$

Remark: This lemma allows us to focus our attention to segments of length $L \leq L_0(l)$, because the likelihood of having a positive segmental score on very long segments becomes vanishingly small as $l \to \infty$. Note that moment generating function $E[\exp(\lambda F), \boldsymbol{\mu}] \in (0,1)$ for $0 < \lambda < \lambda^*$. Hence the logarithmic moment generating function $\Lambda(\lambda) < 0$ for $0 < \lambda < \lambda^*$, and $L_0(l) > 0$. Note that hereafter we suppress the dependence of L_0 on l.

Proof. Since there are at most l^3 possible choices of i, j, L, the conclusion follows if it can be shown that

$$P\{\sum_{k=1}^{L} F(x_{i+k}, y_{i+k}) \geq 0\} \leq 1/l^5 \; \forall L \geq L_0.$$

To establish this inequality, apply Markov's inequality as in Corollary 1.26 with $\epsilon = 0$. This leads to

$$\begin{aligned} P\{\sum_{k=1}^{L} F(x_{i+k}, y_{i+k}) \geq 0\} &\leq E\left[\exp\left(\lambda_0 \sum_{k=1}^{L} F(x_{i+k}, y_{i+k})\right), P_{\boldsymbol{\mu}}^L\right] \\ &\leq E\left[\prod_{k=1}^{L} \exp(\lambda_0 F(x_{i+k}, y_{i+k})), P_{\boldsymbol{\mu}}^L\right] \\ &= (E[\exp(\lambda_0 F(x,y)), P_{\boldsymbol{\mu}}])^L \\ &= (\exp(\Lambda(\lambda_0)))^L = \exp(L\Lambda(\lambda_0)) \\ &\leq \exp(-5 \log l) = 1/l^5. \end{aligned}$$

□

Lemma 9.10 *([39, Lemma 2]) Let* $\bar{M}_l^U$ *be the same as* M_l^U, *except that* $L \leq L_0(l)$. *Suppose* $J_U > 0$. *Then for all* $t > 1$, *we have that*

$$P\{\bar{M}_l^U \geq t J_U \log l\} \leq \frac{(L_0 + 1)^{nm}}{l^{t-1}}.$$

Remark: This lemma is a counterpart to Lemma 9.9. In that lemma we could say that the maximal segmental length on very long segment (defined as L exceeding L_0) will not even exceed zero, with high probability. In the present lemma we examine the complementary situation where put an upper bound of L_0 on the length of the segments, and call the resulting maximal segmental length as $\bar{M}_l^U$.

Proof. For each $\boldsymbol{\nu} \in \mathcal{E}(L, nm)$, let $A(\boldsymbol{\nu}, L)$ denote the event

$$A(\boldsymbol{\nu}, L) := \{\exists i, j, 0 \leq i, j \leq l - L, \hat{\boldsymbol{\mu}}(i, j, L) = \boldsymbol{\nu}\}.$$

We recognize that $A(\boldsymbol{\nu}, L)$ is the event that some segment of length L generates the empirical distribution $\boldsymbol{\nu}$. Now an upper bound for the likelihood of the event $A(\boldsymbol{\nu}, L)$ is readily available from the earlier discussion in Chapter 5, specifically from the method of types upper bound in Lemma 5.5. If $\hat{\boldsymbol{\mu}}(i, j, L) = \boldsymbol{\nu}$, then definitely the marginals also match, so that $\hat{\boldsymbol{\mu}}_{\mathbb{A}}(i, j, L) = \boldsymbol{\nu}_{\mathbb{A}}$ and $\hat{\boldsymbol{\mu}}_{\mathbb{B}}(i, j, L) = \boldsymbol{\nu}_{\mathbb{B}}$. For any fixed i, j, we have the method of types bounds

$$P\{\hat{\boldsymbol{\mu}}(i, j, L) = \boldsymbol{\nu}\} \leq \exp(-LH(\boldsymbol{\nu} \| \boldsymbol{\mu})),$$

$$P\{\hat{\boldsymbol{\mu}}_{\mathbb{A}}(i, j, L) = \boldsymbol{\nu}_{\mathbb{A}}\} \leq \exp(-LH(\boldsymbol{\nu}_{\mathbb{A}} \| \boldsymbol{\mu}_{\mathbb{A}})),$$

$$P\{\hat{\boldsymbol{\mu}}_{\mathbb{B}}(i, j, L) = \boldsymbol{\nu}_{\mathbb{B}}\} \leq \exp(-LH(\boldsymbol{\nu}_{\mathbb{B}} \| \boldsymbol{\mu}_{\mathbb{B}})).$$

Now there are l^2 possible choices for the pair (i, j), and l choices for i and j respectively. So we can write

$$\begin{aligned} P\{A(\boldsymbol{\nu}, L)\} \leq \min\{ & l^2 \exp(-LH(\boldsymbol{\nu} \| \boldsymbol{\mu})), \\ & l \exp(-LH(\boldsymbol{\nu}_{\mathbb{A}} \| \boldsymbol{\mu}_{\mathbb{A}})), l \exp(-LH(\boldsymbol{\nu}_{\mathbb{B}} \| \boldsymbol{\mu}_{\mathbb{B}})), 1\}. \end{aligned}$$

The last term 1 reflects the obvious fact that every probability is bounded above by one. Now let us note that

$$l^2 \exp(-LH(\boldsymbol{\nu} \| \boldsymbol{\mu})) = [l \exp(-0.5H(\boldsymbol{\nu} \| \boldsymbol{\mu}))]^2,$$

and use the inequality $\min\{a^2, 1\} \leq a$ for all $a > 0$. This leads to the final estimate

$$P\{A(\boldsymbol{\nu}, L)\} \leq l \exp(-LH^*(\boldsymbol{\nu} \| \boldsymbol{\mu})). \tag{9.16}$$

We shall make use of this inequality many times in the remainder of the proof. Returning to the above, we conclude that, as a consequence,

$$LH^*(\boldsymbol{\nu} \| \boldsymbol{\mu}) \geq t \log l \Rightarrow P\{A(\boldsymbol{\nu}, L)\} \leq l^{-(t-1)}. \tag{9.17}$$

Next, observe that along any segment of length L, we have

$$\sum_{k=1}^{L} F(x_{i+k}, y_{i+k}) = LE[F, \hat{\boldsymbol{\mu}}(i, j, L)].$$

So the event $\{\bar{M}_l^U \geq tJ_U \log l\}$ is contained in the union of the events

$$\hat{\boldsymbol{\mu}}(i, j, L) = \boldsymbol{\nu} \in U \cap \mathcal{E}(L, nm), LE[F, \boldsymbol{\nu}] \geq tJ(\boldsymbol{\nu}) \log l. \tag{9.18}$$

But the latter inequality implies that

$$LH(\boldsymbol{\nu} \| \boldsymbol{\mu}) = L \frac{E[F, \boldsymbol{\nu}]}{J(\boldsymbol{\nu})} \geq t \log l.$$

In turn, from the definition of the function H^*, the above inequality implies that

$$LH^*(\boldsymbol{\nu} \| \boldsymbol{\mu}) \geq t \log l.$$

So whenever the events in (9.18) hold, we can infer from (9.17) that

$$P\{\sum_{k=1}^{L} F(x_{i+k}, y_{i+k}) \geq tJ_U \log l\} \leq l^{-(t-1)} \; \forall \boldsymbol{\nu} \in U \cap \mathcal{E}(L, nm).$$

Since $|\mathcal{E}(L, nm)| \leq (L+1)^{nm} \leq (L_0+1)^{nm}$, the desired conclusion follows.
□

Lemma 9.11 *(Lemma 3 of [39]) For any $\boldsymbol{\nu} \in \mathcal{E}(L, nm)$ and any $l \geq L$, we have*

$$\begin{aligned} 1 - P\{A(\boldsymbol{\nu}, L)\} \leq & \, 4(L+1)^{nm+1} l_*^{-2} \exp(LH(\boldsymbol{\nu}\boldsymbol{\mu})) \\ & + (L+1)^{nm} l_*^{-1} \exp(LH(\boldsymbol{\nu}_{\mathbb{A}} \| \boldsymbol{\mu}_{\mathbb{A}})) \\ & + (L+1)^{nm} l_*^{-1} \exp(LH(\boldsymbol{\nu}_{\mathbb{B}} \| \boldsymbol{\mu}_{\mathbb{B}})), \end{aligned} \tag{9.19}$$

where $l_* = L\lfloor l/L \rfloor$ is the largest integer multiple of L not exceeding l.

Proof. Clearly, for a fixed $\boldsymbol{\nu}, L$, the quantity $P\{A(\boldsymbol{\nu}, l)\}$ is a nondecreasing function of l. So we can just replace l by l_*; that is, we can assume that $l = ML$ for some integer M.

Divide the sample path $x_1^l = x_1^{ML}$ into M blocks of length L each, and do the same for y_1^l. Given $\boldsymbol{\nu} \in \mathcal{E}(L, nm)$, let N_x denote the number of times that the empirical distribution of a block of length L precisely equals $\boldsymbol{\nu}_{\mathbb{A}}$. In other words, let

$$N_x := \sum_{i=1}^{M} I_{\{\hat{\boldsymbol{\mu}}((i-1)L+1, \cdot, L) = \boldsymbol{\nu}_{\mathbb{A}}\}}.$$

Define the integer N_y analogously. Let p_x denote the probability that the empirical distribution of a block of length L is equal to $\boldsymbol{\nu}_{\mathbb{A}}$. We can even write down a formula for p_x, but it is not necessary. Now N_x is the sum of M independent binary variables, each of which assumes a value of 1 with the probability p_x. Hence N_x is a random variable assuming values in the range $\{0, 1, \ldots, M\}$ with a binomial distribution corresponding to p_x. Similar remarks apply to N_y and p_y.

Now let B_{ij} be the event that, on the i-th X-block with empirical distribution $\boldsymbol{\nu}_{\mathbb{A}}$ and the j-th Y-block with empirical distribution $\boldsymbol{\nu}_{\mathbb{B}}$, the joint empirical distribution of (X, Y) equals $\boldsymbol{\nu}$. Let p denote the probability that the event B_{ij} occurs. Thus

$$p = \{\hat{\boldsymbol{\mu}}(i, j, L) = \boldsymbol{\nu} | \hat{\boldsymbol{\mu}}_{\mathbb{A}} = \boldsymbol{\nu}_{\mathbb{A}} \wedge \hat{\boldsymbol{\mu}}_{\mathbb{B}} = \boldsymbol{\nu}_{\mathbb{B}}\}.$$

Let us define

$$W = \sum_{i=1}^{N_x} \sum_{j=1}^{N_y} I_{B_{ij}}.$$

Then the above definitions imply that

$$E[W | N_x, N_y] = p N_x N_y.$$

Also, the variance of W is bounded by

$$\mathrm{var}[W|N_x, N_y] = N_x N_y \mathrm{var}[B_{ij}|N_x, N_y] = N_x N_y p(1-p) \leq p N_x N_y.$$

So by Chebycheff's inequality, it follows that

$$P\{W = 0|N_x, N_y\} \leq \frac{\mathrm{var}[W|N_x, N_y]}{(E[W|N_x, N_y])^2} \leq \frac{1}{pN_x N_y}.$$

We can also write

$$P\{W = 0\} = E[P\{W = 0|N_x \geq 1 \wedge N_y \geq 1\}] + P\{N_x = 0\} + P\{N_y = 0\}.$$

Now it is clear that

$$P\{N_x = 0\} = (1 - p_x)^M, P\{N_y = 0\} = (1 - p_y)^M.$$

As for the first term, we have

$$E[P\{W = 0 \mid N_x \geq 1 \wedge N_y \geq 1\}] \leq E\left[\frac{1}{pN_x N_y}\middle| N_x \geq 1 \wedge N_y \geq 1\right]$$
$$= \frac{1}{p} E[1/N_x|N_x \geq 1] \cdot E[1/N_y|N_y \geq 1],$$

where we take advantage of the fact that N_x and N_y are independent random variables. Now it can be readily verified from the binomial distribution of N_x that

$$E[1/N_x|N_x \geq 1] = \sum_{i=1}^{M} \frac{1}{i} \binom{M}{i} p_x^i (1-p_x)^{M-i} \leq \frac{2}{Mp_x}.$$

Similarly

$$E[1/N_y|N_y \geq 1] \leq \frac{2}{Mp_y}.$$

This leads to

$$P\{W = 0\} \leq \frac{4}{M^2 p p_x p_y} + (1-p_x)^M + (1-p_y)^M.$$

A cruder estimate is

$$P\{W = 0\} \leq \frac{4}{M^2 p p_x p_y} + \frac{1}{Mp_x} + \frac{1}{Mp_y}. \tag{9.20}$$

Now observe from the method of types lower bounds in Lemma 5.5 that

$$pp_x p_y = P\{\hat{\boldsymbol{\mu}}(1,1,L) = \boldsymbol{\nu}\} \geq (L+1)^{-nm} \exp(-LH(\boldsymbol{\nu}\|\boldsymbol{\mu})), \tag{9.21}$$

$$p_x = P\{\hat{\boldsymbol{\mu}}(1,\cdot,L) = \boldsymbol{\nu}_{\mathbb{A}}\} \geq (L+1)^{-nm} \exp(-LH(\boldsymbol{\nu}_{\mathbb{A}}\|\boldsymbol{\mu}_{\mathbb{A}})), \tag{9.22}$$

$$p_y = P\{\hat{\boldsymbol{\mu}}(1,\cdot,L) = \boldsymbol{\nu}_{\mathbb{B}}\} \geq (L+1)^{-nm} \exp(-LH(\boldsymbol{\nu}_{\mathbb{B}}\|\boldsymbol{\mu}_{\mathbb{B}})). \tag{9.23}$$

Now observe that the event $\{W > 0\}$ implies the event $\{A(\boldsymbol{\nu}, L)\}$. Hence

$$P\{A(\boldsymbol{\nu}, L)\} \geq P\{W > 0\} = 1 - P\{W = 0\},$$

which in turn implies that

$$1 - P\{A(\boldsymbol{\nu}, L)\} \leq P\{W = 0\}.$$

The desired conclusion now follows from substituting the bounds from (9.21) through (9.23) into (9.20). □

Lemma 9.12 *([39, Lemma 4]) Suppose $J_{U^o} > 0$. Then for each $t < 1$, there exists an integer $l_0 = l_0(t)$*[1] *such that*

$$P\{M_l^U \leq t \log l J_{U^o}\} \leq \frac{1}{l^{(1-t)/2}}, \ \forall l \geq l_0.$$

Proof. Suppose $t < 1$, and choose a number $\tau \in (t, (1+t)/2)$. Because $\tau > t$, it follows that

$$\frac{\tau + t}{2\tau} = 0.5 + 0.5 \frac{t}{\tau} < 1.$$

Given the set U, choose $\tilde{\boldsymbol{\nu}} \in U^o$ such that

$$J(\tilde{\boldsymbol{\nu}}) > \frac{\tau + t}{2\tau} J_{U^o}.$$

Define

$$k_l = \lceil \tau \log l / H^*(\tilde{\boldsymbol{\nu}} \| \boldsymbol{\mu}) \rceil,$$

and note that eventually k_l eventually less than any constant multiple of l, because $k_l \sim \text{const} \cdot \log l$. Hence, for large enough l, we can always choose $\tilde{\boldsymbol{\nu}}_l \in \mathcal{E}(k_l, nm)$ such that $\rho(\boldsymbol{\nu}, \tilde{\boldsymbol{\nu}}_l) \leq nm/k_l$, where ρ denotes the total variation metric. Now let c denote an upper bound for the function $F(x, y), x \in \mathbb{A}, y \in \mathbb{B}$. Then

$$\begin{aligned} k_l E[F, \tilde{\boldsymbol{\nu}}_l] &\geq k_l (E[F, \boldsymbol{\nu}] - c\rho(\boldsymbol{\nu}, \tilde{\boldsymbol{\nu}}_l)) \\ &\geq k_l E[F, \boldsymbol{\nu}] - cnm. \end{aligned}$$

Note that cnm is just some constant independent of l. Hence it follows from the definition of the constant k_l that

$$\begin{aligned} k_l E[F, \tilde{\boldsymbol{\nu}}_l] &\geq k_l E[F, \boldsymbol{\nu}] - cnm \\ &\geq \tau \log l \frac{E[F, \boldsymbol{\nu}]}{H^*(\tilde{\boldsymbol{\nu}} \| \boldsymbol{\mu})} - cnm \\ &= \tau \log l J(\tilde{\boldsymbol{\nu}}) - cnm \\ &\geq \frac{\tau + t}{2} \log l J_{U^o} - cnm. \end{aligned}$$

It is clear that $\tilde{\boldsymbol{\nu}}_l \to \tilde{\boldsymbol{\nu}}$ as $l \to \infty$, and $\tilde{\boldsymbol{\nu}} \in U^o$. Hence $\tilde{\boldsymbol{\nu}}_l \in U^o$ for all large enough l. Moreover, since $(\tau + t)/2 > t$, the constant term cnm can be neglected in comparison with the $\log l$ term. Hence for large enough l it can be said that

$$k_l E[F, \tilde{\boldsymbol{\nu}}_l] > t \log l J_{U*o}.$$

Since $\tilde{\boldsymbol{\nu}}_l \in \mathcal{E}(k_l, nm)$ for each l, this means that the event $\{M_l^U \leq t \log l J_{U*o}\}$ is contained in the complement of the event $\{A(\tilde{\boldsymbol{\nu}}_l, k_l)\}$, i.e., that there exists a sequence of length k_l whose empirical distribution is $\tilde{\boldsymbol{\nu}}_l$. Thus

$$P\{M_l^U \leq t \log l J_{U*o}\} \leq 1 - P\{A(\tilde{\boldsymbol{\nu}}_l, k_l)\}. \tag{9.24}$$

Now an upper bound for the right side is given by Lemma 9.11, specifically (9.19). There are three terms in this bound, out of which one contains l^{-2}

[1] Not to be confused with L_0.

and thus decays faster than the other two terms which contain l^{-1}. Next, observe that, with l replaced by k_l, we have

$$\exp(k_l H^*(\boldsymbol{\nu}\|\boldsymbol{\mu})) \sim \exp(\tau \log l) = l^{\tau}.$$

Hence

$$1 - P\{A(\tilde{\boldsymbol{\nu}}_l, k_l)\} \leq \text{const} \cdot l^{-(1-\tau)}.$$

But since $\tau < (1+t)/2$, it follows that $1 - \tau > (1-t)/2$. So we conclude that

$$[1 - P\{A(\tilde{\boldsymbol{\nu}}_l, k_l)\}] l^{(1-t)/2} \to 0 \text{ as } l \to \infty.$$

In view of the bound (9.24), this in turn implies that

$$P\{M_l^U \leq t \log l J_{U*o}\} l^{(1-t)/2} \to 0 \text{ as } l \to \infty.$$

This is the desired conclusion. □

Now we are in a position to prove Theorem 9.7. What has been proven thus far is that:

$$P\{\bar{M}_l^U \geq t J_U \log l\} \to 0 \text{ as } l \to \infty, \ \forall t > 1,$$

$$P\{M_l^U \leq t \log l J_{U^o}\} \to 0 \text{ as } l \to \infty, \ \forall t < 1.$$

Taken together, these two assertions are precisely the desired conclusions.

Bibliography

[1] N. Alon, M. Krivelevich, I. Newman, and M. Szegedy, "Regular languages are testable with a constant number of queries," *Foundations of Computer Sciences*, IEEE, New York, 645–655, 1999.

[2] N. Alon, M. Krivelevich, I. Newman, and M. Szegedy, "Regular languages are testable with a constant number of queries," *SIAM Journal on Computing*, 30, 1842–1862, 2001.

[3] S. F. Altschul, "The statistics of similarity sequence scores," http://www.ncbi.nlm.nih.gov/BLAST/tutorial/Altschul-1.html.

[4] S. F. Altschul, W. Gish, W. Q. Miller, E. W. Myers, and D. J. Lipman, "Basic local alignment search tool," *Journal of Molecular Biology*, 215, 403–410, 1990.

[5] S. F. Altschul, T. L. Madden, A. A. Schaffer, J. Zhang, Z. Zhang, W. Q. Miller, and D. J. Lipman, "Gapped BLAST and PSI-BLAST: A new generation of protein database search programs," *Nucleic Acids Research*, 25, 3389–3401, 1997.

[6] B. D. O. Anderson, "The realization problem for hidden Markov models," *Mathematics of Control, Signals, and Systems*, 12(1), 80–120, 1999.

[7] B. D. O. Anderson, "From Wiener to hidden Markov models," *IEEE Control Systems Magazine*, 19(6), 41–51, June 1999.

[8] B. D. O. Anderson, M. Deistler, L. Farina, and L. Benvenuti, "Nonnegative realization of a system with a nonnegative impulse response," *IEEE Transactions on Circuits and Systems-I: Fundamental Theory and Applications*, 43, 134–142, 1996.

[9] L. R. Bahl, J. Cocke, F. Jelinek, and J. Raviv, "Optimal decoding of linear codes for minimizing symbol error rate," *IEEE Transactions on Information Theory*, 20(2), 284–287, March 1974.

[10] L. R. Bahl and F. Jelinek, "Decoding for channels with insertions, deletions and substitutions with applications to speech recognition," *IEEE Transactions on Information Theory*, 21(4), 404–411, July 1975.

[11] P. Baldi and S. Brunak, *Bioinformatics: A Machine Learning Approach* (Second Edition), MIT Press, Cambridge, MA, 2001.

[12] L. E. Baum and T. Petrie, "Statistical inference for probabilistic functions of finite state Markov chains," *Annals of Mathematical Statistics*, 37(2), 1554–1563, 1966.

[13] L. E. Baum, T. Petrie, G. Soules, and N. Weiss, "A maximization technique occurring in the statistical analysis of probabilistic functions of Markov chains," *Annals of Mathematical Statistics*, 41(1), 164–171, 1970.

[14] R. Begleiter, R. El-Yaniv, and G. Yona, "On prediction using variable order Markov models," *Journal of Artificial Intelligence Research*, 22, 385–421, 2004.

[15] L. Benvenuti and L. Farina, "A tutorial on the positive realization problem," *IEEE Transactions on Automatic Control*, 49, 651–664, 2004.

[16] A. Berman and R. J. Plemmons, *Nonnegative Matrices*, Academic Press, New York, 1979.

[17] E. Bianconi, A. Provesan, F. Facchin, et al., "An estimation of the number of cells in the human body," *Annals of Human Biology*, 40(6), 463–471, November-December 2013.

[18] P. Billingsley, *Probability and Measure*, Wiley, New York, 1986.

[19] P. Billingsley, *Probability and Measure* (Third Edition), Wiley, New York, 1995.

[20] D. Blackwell and L. Koopmans, "On the identifiability problem for functions of finite Markov chains," *Annals of Mathematical Statistics*, 28, 1011–1015, 1957.

[21] V. Blondel and V. Catarini, "Undecidable problems for probabilistic automata of fixed dimension," *Theory of Computing Systems*, 36, 231–245, 2003.

[22] L. Breiman, "The individual ergodic theorem of information theory," *Annals of Mathematical Statistics*, 28, 809–811, 1957.

[23] L. Breiman, *Probability*, SIAM Classics in Applied Mathematics, No. 7, SIAM, Philadelphia, 1992.

[24] T. Brown Jr., *Nucleic Acids Book*, available free at http://www.atdbio.com/nucleic-acids-book.

[25] C. Burge and S. Karlin, "Prediction of complete gene structure in human genomic DNA," *Journal of Molecular Biology*, 268, 78–94, 1997.

[26] C. Burge and S. Karlin, "Finding the genes in genomic DNA," *Current Opinion in Structural Biology*, 8(3), 346–354, 1998.

[27] J. W. Carlyle, "Identification of state-calculable functions of finite Markov chains," *Annals of Mathematical Statistics*, 38, 201–205, 1967.

[28] J. W. Carlyle, "Stochastic finite-state system theory," Chapter 10, in *System Theory*, L. Zadeh and E. Polak (Eds.), McGraw-Hill, New York, 1969.

[29] J.-R. Chazottes and D. Gabrielli, "Large deviations for empirical entropies of g-measures," *Nonlinearity*, 18(6), 2545–2563, November 2005.

[30] T. M. Cover and J. A. Thomas, *Elements of Information Theory* (Second Edition), Wiley Interscience, New York, 2006.

[31] F. Crick, *What Mad Pursuit*, Basic Books, New York, 1988.

[32] I. Csiszár, "The method of types," *IEEE Transactions on Information Theory*, 4(6), 2505–2523, 1998.

[33] I. Csiszár and J. Körner, *Information Theory: Coding Theorems for Discrete Memoryless Systems*, Academic Press, New York, 1981.

[34] L. D. Davisson, G. Longo, and A. Sgarro, "The error exponent for the noiseless encoding of finite ergodic Markov sources," *IEEE Transactions on Information Theory*, 27(4), 431–438, July 1981.

[35] A. L. Delcher, K. A. Bratke, E. C. Powers, and S. L. Salzberg, "Identifying bacterial genes and endosymbiont DNA with Glimmer," *Bioinformatics*, 23(6), 673–679, 2007.

[36] A. L. Delcher, D. Harmon, S. Kasif, O. White, and S. L. Salzberg, "Improved microbial gene identification with GLIMMER," *Nucleic Acids Research*, 27(23), 4636–4641, 1999.

[37] A. Dembo and S. Karlin, "Strong limit theorems of empirical functionals for large exceedances of partial sums of i.i.d. variables," *Annals of Probability*, 19(4), 1737–1755, 1991.

[38] A. Dembo and S. Karlin, "Strong limit theorems of empirical functionals for large exceedances of partial sums of Markov variables," *Annals of Probability*, 19(4), 1756–1767, 1991.

[39] A. Dembo, S. Karlin, and O. Zeitouni, "Critical phenomena for sequence matching with scoring," *Annals of Probability*, 22(4), 1993–2021, 1994.

[40] A. Dembo, S. Karlin, and O. Zeitouni, "Limit distribution of maximal non-aligned two-sequence segmental score," *Annals of Probability*, 22(4), 2022–2039, 1994.

[41] A. Dembo and O. Zeitouni, *Large Deviation Techniques and Applications*, Springer-Verlag, Berlin, 1998.

[42] L. Devroye, L. Gyorfi, and G. Lugosi, *A Probabilistic Theory of Pattern Recognition*, Springer-Verlag, Berlin, 1996.

[43] S. W. Dharmadhikari, "Functions of finite Markov chains," *Annals of Mathematical Statistics*, 34, 1022–1031, 1963.

[44] S. W. Dharmadhikari, "Sufficient conditions for a stationary process to be a function of a Markov chain," *Annals of Mathematical Statistics*, 34, 1033–1041, 1963.

[45] S. W. Dharmadhikari, "A characterization of a class of functions of finite Markov chains," *Annals of Mathematical Statistics*, 36, 524–528, 1965.

[46] S. W. Dharmadhikari, "A note on exchangeable processes with states of finite rank," *Annals of Mathematical Statistics*, 40(6), 2207–2208, 1969.

[47] S. W. Dharmadhikari and M. G. Nadkarni, "Some regular and non-regular functions of finite Markov chains," *Annals of Mathematical Statistics*, 41(1), 207–213, 1970.

[48] J. Dixmier, "Proof of a conjecture by Erdös and Graham concerning the problem of Frobenius," *Journal of Number Theory*, 34, 198–209, 1990.

[49] S. R. Eddy, "Profile hidden Markov models," *Bioinformatics*, 14(9), 755–763, 1998.

[50] Y. Ephraim and N. Merhav, "Hidden Markov processes," *IEEE Transactions on Information Theory*, 48(6), 1518–1569, June 2002.

[51] R. V. Erickson, "Functions of Markov chains," *Annals of Mathematical Statistics*, 41, 843–850, 1970.

[52] W. J. Ewens and G. R. Grant, *Statistical Methods in Bioinformatics*, Springer-Verlag, New York, 2001.

[53] W. J. Ewens and G. R. Grant, *Statistical Methods in Bioinformatics* (Second Edition), Springer-Verlag, New York, 2006.

[54] J. Feng and T. G. Kurtz, *Large Deviations for Stochastic Processes*, American Mathematical Society, Providence, RI, 2006.

[55] M. Fliess, "Series rationelles positives et processus stochastique," *Annales de l'Institut Henri Poincaré, Section B*, XI, 1–21, 1975.

[56] M. Fox and H. Rubin, "Functions of processes with Markovian states," *Annals of Mathematical Statistics*, 39, 938–946, 1968.

[57] G. Frobenius, "Über Matrizen aus positiven Elementen," *S.-B. Preuss. Akad. Wiss. (Berlin)*, 471–476, 1908.

[58] G. Frobenius, "Über Matrizen aus positiven Elementen," *S.-B. Preuss. Akad. Wiss. (Berlin)*, 514–518, 1909.

[59] G. Frobenius, "Über Matrizen aus nicht negativen Elementen," *S.-B. Preuss. Akad. Wiss. (Berlin)*, 456–477, 1912.

[60] E. J. Gilbert, "The identifiability problem for functions of Markov chains," *Annals of Mathematical Statistics*, 30, 688–697, 1959.

[61] B. V. Gnedenko, *The Theory of Probability* (Fourth Edition), Chelsea, New York, 1968.

[62] R. M. Gray, *Probability, Random Processes, and Ergodic Properties*, Springer-Verlag, New York, 1988.

[63] D. Gusfield, *Algorithms on Strings, Trees and Sequences*, Cambridge University Press, Cambridge, UK, 1997.

[64] A. Heller, "On stochastic processes derived from Markov chains," *Annals of Mathematics*, 36, 1286–1291, 1965.

[65] W. Hoeffding, "Probability inequalities for sums of bounded random variables," *Journal of the American Statistical Association*, 58, 13–30, 1963.

[66] J. M. van den Hof, "Realization of continuous-time positive linear systems," *Systems and Control Letters*, 31, 243–253, 1997.

[67] J. M. van den Hof and J. H. van Schuppen, "Realization of positive linear systems using polyhedral cones," *Proceedings of the 33rd IEEE Conference on Decision and Control*, 3889–3893, 1994.

[68] F. Den Hollander, *Large Deviations*, American Mathematical Society, Providence, RI, 2000.

[69] International Human Genome Research Consortium, "Initial sequencing and analysis of the human genome," *Nature*, 409, 860–921, February 2001.

[70] H. Ito, S. Amari, and K. Kobayashi, "Identifiability of hidden Markov information sources and their minimum degrees of freedom," *IEEE Transactions on Information Theory*, 38, 324–333, 1992.

[71] F. Jelinek, *Statistical Methods for Speech Recognition*, MIT Press, Cambridge, MA, 1997.

[72] B-H. Juang and L. R. Rabiner, "A probabilistic distance measure for hidden Markov models," *AT&T Technical Journal*, 64(2), 391–408, February 1985.

[73] S. Kalikow, "Random Markov processes and uniform martingales," *Israel Journal of Mathematics*, 71(1), 33–54, 1990.

[74] S. Karlin and S. F. Altschul, "Methods for assessing the statistical significance of molecular sequence features by using general scoring schemes," *Proceedings of the National Academy of Sciences*, 87, 2264–2268, 1990.

[75] S. Karlin and S. F. Altschul, "Applications and statistics for multiple high-scoring segments in molecular sequences," *Proceedings of the National Academy of Sciences*, 90, 5873–5877, 1993.

[76] S. Karlin and A. Dembo, "Limit distributions of maximal segmental score among Markov-dependent partial sums," *Advances in Applied Probability*, 24(1), 113–140, 1992.

[77] S. Karlin and F. Ost, "Maximal length of common words among random letter sequences," *Annals of Probability*, 16, 535–563, 1988.

[78] S. Karlin and H. Taylor, *A First Course in Stochastic Processes*, Academic Press, New York, 1975.

[79] A. I. Khinchin, *Mathematical Foundations of Information Theory*, Dover, New York, 1957.

[80] H. G. Khorana, H. Büchi, H. Ghosh, et al., "Polynucleotide synthesis and the genetic code," *Cold Spring Harbor Symposia on Quantitative Biology*, 31, 39–49, 1966.

[81] A. N. Kolmogorov, *Foundations of the Theory of Probability*, Chelsea, New York, 1950.

[82] T. Koski, *Hidden Markov Models for Bioinformatics*, Kluwer, Dordrecht, The Netherlands, 2001.

[83] A. Krogh, "An introduction to hidden Markov models for biological sequences," in *Computational Methods in Molecular Biology*, S. Salzberg, D. Searls, and S. Kasif (Eds.), Elsevier, New York, 45–63, 1998.

[84] A. Krogh, M. Brown, I. S. Mian, K. Sjölander, and D. Haussler, "Hidden Markov models in computational biology: Applications to protein modeling," *Journal of Molecular Biology*, 235, 1501–1531, 1994.

[85] A. Krogh, I. S. Mian, and D. Haussler, "A hidden Markov model that finds genes in *E. coli* DNA," *Nucleic Acids Research*, 22(22), 4768–4778, 1994.

[86] L. Kronecker, "Zur Theorie der Elimination einer Variablen aus zwei algebraischen Gleichungen," *Monatsber. Königl. Preuss. Akad. Wiss. Berlin*, 535–600, 1881.

[87] S. Kullback and R. A. Leibler, "On information and sufficiency," *Annals of Mathematical Statistics*, 22, 79–86, 1951.

[88] G. Langdon, "A note on the Ziv-Lempel model for compressing individual data sequences," *IEEE Transactions on Information Theory*, 29, 284–287, 1983.

[89] M. Lewin, "A bound for a solution of a linear Diophantine problem," *Journal of the London Mathematical Society (2)*, 6, 61–69, 1972.

[90] F. Liese and I. Vajda, "On divergences and informations in statistics and information theory," *IEEE Transactions on Information Theory*, 52(10), 4394–4412, October 2006.

[91] R. Lohrmann, D. Söll, H. Hayatsu, E. Ohtsuka, and H. G. Khorana, "Studies on polynucleotides. LI. Syntheses of the 64 possible ribotrinucleotides derived from the four major ribomononucleotides," *Journal of the American Chemical Society*, 88(4), 819–829, 1966.

[92] A. R. Morgan, R. D. Wells, and H. G. Khorana, "Studies on polynucleotides, lix. Further codon assignments from amino acid incorporations directed by ribopolynucleotides containing trinucleotide sequences," *Proceedings of the National Academy of Sciences*, 56(6) , 1899–1906, 1966.

[93] A. V. Lukashin and M. Borodvsky, "GeneMark.hmm: New solutions for gene finding," *Nucleic Acids Research*, 26(4), 1107–1115, 1998.

[94] W. H. Majoros and S. L. Salzberg, "An empirical analysis of training protocols for probabilistic gene finders," *BMC Bioinformatics*, available at http://www.biomedcentral.com/1471-2105/5/206, 21 December 2004.

[95] P. Massart, "The tight constant in the Dvoretzky-Kiefer-Wolfowitz inequality," *Annals of Probability*, 18(3), 1269–1283, 1990.

[96] B. McMillan, "The basic theorems of information theory," *Annals of Mathematical Statistics*, 24(2), 196–219, 1953.

[97] D. W. Mount, *Bioinformatics: Sequence and Genome Analysis*, Cold Spring Harbor Press, Cold Spring Harbor, NY, 2001.

[98] S. B. Needleman and C. D. Wunsch, "A general method applicable to the search for similarities in the amino acid sequences of two proteins," *Journal of Molecular Biology*, 48(3), 443–453, 1970.

[99] M. Nirenberg and P. Leder, "RNA codewords and protein synthesis," *Science*, 145(3639), 1399–1407, 25 September 1964.

[100] J. R. Norris, *Markov Chains*, Cambridge University Press, Cambridge, UK, 1997.

[101] G. Nuel, "LD-SPatt: Large deviations statistics for patterns on Markov chains," *Journal of Computational Biology*, 11(6), 1023–1033, 2004.

[102] D. S. Ornstein and B. Weiss, "How sampling reveals a process," *Annals of Probability*, 18(3), 905–930, 1990.

[103] O. Perron, "Zur Theorie der Matrizen," *Mathematische Annalen*, 64, 248–263, 1907.

[104] M. Pertea, X. Lin, and S. L. Salzberg, "GeneSplicer: A new computational method for splice site detection," *Nucleic Acids Research*, 29(5), 1185–1190, 2001.

[105] P. A. Pevzner, *Computational Molecular Biology*, MIT Press, Cambridge, MA, 2000.

[106] G. Picci, "On the internal structure of finite-state stochastic processes," in *Recent Developments in Variable Structure Systems*, R. Mohler and A. Ruberti (Eds.), Lecture Notes in Economics and Mathematical Systems, No. 162, Springer-Verlag, Heidelberg, 1978.

[107] D. Pollard, *Convergence of Stochastic Processes*, Springer-Verlag, Berlin, 1984.

[108] L. W. Rabiner, "A tutorial on hidden Markov models and selected applications in speech recognition," *Proceedings of the IEEE*, 77(2), 257–285, February 1989.

[109] Z. Rached, F. Alalaji, and L. L. Campbell, "The Kullback-Leibler divergence rate between Markov sources," *IEEE Transactions on Information Theory*, 50(5), 917–921, May 2004.

[110] R. T. Rockafellar, *Convexity*, Princeton University Press, Princeton, NJ, 1970.

[111] G. Rozenberg and A. Salomaa, *Cornerstones in Undecidability*, Prentice Hall, Englewood Cliffs, NJ, 1994.

[112] W. Rudin, *Real and Complex Analysis*, McGraw-Hill, New York, 1959.

[113] S. L. Salzberg, A. L. Delcher, S. Kasif, and O. White, "Microbial gene identification using interpolated Markov models," *Nucleic Acids Research*, 26(2), 544–548, 1998.

[114] S. L. Salzberg, M. Pertea, A. L. Delcher, M. J. Gardner, and H. Tettelin, "Interpolated Markov models for eukaryotic gene finding," *Genomics*, 59, 24–31, 1999.

[115] A. Sayre, *Rosalind Franklin and DNA*, Norton, New York, 2000 (reissue of 1975 edition).

[116] E. Seneta, *Non-negative Matrices and Markov Chains* (Second Edition), Springer-Verlag, Berlin, 1981.

[117] C. E. Shannon, "A mathematical theory of communication," *Bell System Technical Journal*, 27, 379–423, 623–656, July, October 1948.

[118] P. C. Shields, *The Ergodic Theory of Discrete Sample Paths*, Amer. Math. Soc., Providence, RI, 1996.

[119] T. F. Smith and M. S. Waterman, "Identification of common molecular subsequences," *Journal of Molecular Biology*, 147, 195–197, 1981.

[120] E. D. Sontag, "On certain questions of rationality and decidability," *Journal of Computer and System Science*, 11, 375–381, 1975.

[121] D. W. Stroock, *An Introduction to Markov Processes*, Springer-Verlag, Berlin, 2005.

[122] B. E. Suzek, M. D. Ermolaeva, M. Schreiber, and S. L. Salzberg, "A probabilistic method for identifying start codons in bacterial genomes," *Bioinformatics*, 17(12), 1123–1130, 2001.

[123] F. Topsoe, "Bounds for entropy and divergence over a two-element set," *Journal of Inequalities in Pure and Appllied Mathematics*, 2(2), 2001.

[124] V. N. Vapnik, *Statistical Learning Theory*, John Wiley, New York, 1998.

[125] J. C. Venter, M. D. Adams, E. W. Myers, et al., "The sequence of the human genome," *Science*, 291, 1304–1351, 16 February 2001.

[126] M. Vidyasagar, *A Theory of Learning and Generalization*, Springer-Verlag, London, 1997.

[127] M. Vidyasagar, *Learning and Generalization with Applications to Neural Networks*, Springer-Verlag, London, 2003.

[128] M. Vidyasagar, *Nonlinear Systems Analysis*, SIAM Publications, Philadelphia, PA, 2003.

[129] M. Vidyasagar, "Convergence of empirical means with alpha-mixing input sequences, and an application to PAC learning," *Proceedings of the IEEE Conference on Decision and Control*, Seville, Spain, 2005.

[130] M. Vidyasagar, "The realization problem for hidden Markov models: The complete realization problem," *Proceedings of the IEEE Conference on Decision and Control*, Seville, Spain, 2005.

[131] M. Vidyasagar, "The 4M (Mixed Memory Markov Model) algorithm for stochastic modelling over a finite alphabet with applications to gene-finding algorithms," *Proceedings of the Symposium on the Mathematical Theory of Networks and Systems*, Kyoto, Japan, 443–459, July 2006.

[132] M. Vidyasagar, "A realization theory for hidden Markov models: The partial realization problem," *Proceedings of the Symposium on the Mathematical Theory of Networks and Systems*, Kyoto, Japan, 2145–2150, July 2006.

[133] M. Vidyasagar, "The complete realization problem for hidden Markov models: A survey and some new results," *Mathematics of Control, Signals and Systems*, 23(1), 1–65, 2011.

[134] M. Vidyasagar, S. S. Mande, Ch. V. S. K. Reddy, and V. R. Raja Rao, "The 4M (mixed memory Markov model) algorithm for finding genes in prokaryotic genomes," *IEEE Transactions on Automatic Control*, 53(1), 26–27, January 2008.

[135] M. Vidyasagar, "An elementary derivation of the large deviation rate function for finite state Markov processes," *Asian Journal of Control*, 16(1), 1–19, January 2014.

[136] J. D. Watson, *The Double Helix: A Personal Account of the Discovery of the Structure of DNA*, Touchstone, New York, 2001 (reprint of 1968 edition).

[137] L. R. Welch, "Hidden Markov models and the Baum-Welch algorithm," *IEEE Information Theory Society Newsletter*, 53(4), 1–24, December 2003.

[138] T. A. Welch, "A technique for high performance data compression, *IEEE Computer*, 17(6), 8–19, 1984.

[139] H. Wielandt, "Unzerlegbare, nicht negativen Matrizen," *Mathematische Zeitschrift*, 52, 642–648, 1950.

[140] M. Wilkins, *The Third Man of the Double Helix*, Oxford University Press, Oxford, UK, 2003.

[141] L. Xie, A. Ugrinovskii, and I. R. Petersen, "Probabilistic distances between finite-state finite-alphabet hidden Markov models," *IEEE Transactions on Automatic Control*, 50(4), 505–511, April 2005.

[142] B-J. Yoon, "Hidden Markov models and their applications in biological sequence analysis," *Current Genomics*, 10, 402–415, 2009.

[143] M. Q. Zhang, "Computational prediction of eukaryotic protein-coding genes," *Nature Reviews*, 3, 698–710, September 2002.

[144] J. Ziv and A. Lempel, "A universal algorithm for sequential data compression," *IEEE Transactions on Information Theory*, 23(3), 337–343, 1977.

[145] J. Ziv and A. Lempel, "Compression of sequences via variable rate encoding," *IEEE Transactions on Information Theory*, 24(5), 530–536, 1978.

Index

PRINCETON SERIES IN APPLIED MATHEMATICS

Chaotic Transitions in Deterministic and Stochastic Dynamical Systems: Applications of Melnikov Processes in Engineering, Physics, and Neuroscience, Emil Simiu

Selfsimilar Processes, Paul Embrechts and Makoto Maejima

Self-Regularity: A New Paradigm for Primal-Dual Interior-Point Algorithms, Jiming Peng, Cornelis Roos, and Tamás Terlaky

Analytic Theory of Global Bifurcation: An Introduction, Boris Buffoni and John Toland

Entropy, Andreas Greven, Gerhard Keller, and Gerald Warnecke, editors

Auxiliary Signal Design for Failure Detection, Stephen L. Campbell and Ramine Nikoukhah

Thermodynamics: A Dynamical Systems Approach, Wassim M. Haddad, VijaySekhar Chellaboina, and Sergey G. Nersesov

Optimization: Insights and Applications, Jan Brinkhuis and Vladimir Tikhomirov

Max Plus at Work, Modeling and Analysis of Synchronized Systems: A Course on Max-Plus Algebra and Its Applications, Bernd Heidergott, Geert Jan Olsder, and Jacob van der Woude

Impulsive and Hybrid Dynamical Systems: Stability, Dissipativity, and Control, Wassim M. Haddad, VijaySekhar Chellaboina, and Sergey G. Nersesov

The Traveling Salesman Problem: A Computational Study, David L. Applegate, Robert E. Bixby, Vasek Chvátal, and William J. Cook

Positive Definite Matrices, Rajendra Bhatia

Genomic Signal Processing, Ilya Shmulevich and Edward R. Dougherty

Wave Scattering by Time-Dependent Perturbations: An Introduction, G. F. Roach

Algebraic Curves over a Finite Field, J.W.P. Hirschfeld, G. Korchmáros, and F. Torres

Distributed Control of Robotic Networks: A Mathematical Approach to Motion Coordination Algorithms, Francesco Bullo, Jorge Cortés, and Sonia Mart´inez

Robust Optimization, Aharon Ben-Tal, Laurent El Ghaoui, and Arkadi Nemirovski

Control Theoretic Splines: Optimal Control, Statistics, and Path Planning, Magnus Egerstedt and Clyde Martin

Matrices, Moments, and Quadrature with Applications, Gene H. Golub and Gérard Meurant

Totally Nonnegative Matrices, Shaun M. Fallat and Charles R. Johnson

Matrix Completions, Moments, and Sums of Hermitian Squares, Mihály Bakonyi and Hugo J. Woerdeman

Modern Anti-windup Synthesis: Control Augmentation for Actuator Saturation, Luca Zaccarian and Andrew W. Teel

Graph Theoretic Methods in Multiagent Networks, Mehran Mesbahi and Magnus Egerstedt

Stability and Control of Large-Scale Dynamical Systems: A Vector Dissipative Systems Approach, Wassim M. Haddad and Sergey G. Nersesov

Mathematical Analysis of Deterministic and Stochastic Problems in Complex Media Electromagnetics, G. F. Roach, I. G. Stratis, and A. N. Yannacopoulos

Topics in Quaternion Linear Algebra, Leiba Rodman

Hidden Markov Processes: Theory and Applications to Biology, M. Vidyasagar